Understanding the Scientific Bases of Human Movement

SECOND EDITION

Barbara A. Gowitzke, Ph.D.

F.A.A.H.P.E.R.
Assistant Professor of Physical Education
School of Physical Education and Athletics
McMaster University, Hamilton, Ontario, Canada

Morris Milner, Ph.D., P.Eng.

F.I.E.E.
Director, Rehabilitation Engineering Department
Ontario Crippled Children's Centre
Adjunct Associate Professor
Departments of Mechanical Engineering, and Surgery
Institute of Biomedical Engineering
University of Toronto, Toronto, Ontario, Canada

D1441616

WILLIAMS & WILKINS
Baltimore/London

SANS TACHE

Made in the United States of America

Library of Congress Cataloging in Publication Data

Gowitzke, Barbara A
    Understanding the scientific bases of human movement.

    First ed. (1972) by A. L. O'Connell and E. B. Gardner.
    Includes index.
    1. Human mechanics. 2. Kinesiology. I. Milner, Morris, joint author. II. O'Connell, Alice Louise, 1907– Understanding the scientific bases of human movement. III. Title. [DNLM: 1. Movement. WE103.3 G723u]
QP303.026 1979   612'.76   78-14723
ISBN 0-683-03592-4

First Edition, 1972
    Reprinted, 1973
    Reprinted, 1976

Composed and printed at the
Waverly Press, Inc.
Mt. Royal and Guilford Aves.
Baltimore, Md. 21202, U.S.A.

Cover Illustration: The display on the cover was obtained with Lise Arsenault-Goertz fitted with illuminating sources at the wrist, shoulder, hip, knee and ankle, then executing a back walkover while being tracked optoelectronically in one plane by use of a Selspot system coupled to a PDP 11/10 computer. A maximum of 711 frames at a sampling frequency of 161 Hz were collected, the data were processed using a 5 Hz low-pass digital filter and the display was generated by taking every seventh sample set, showing the pertinent stick diagram and generating the trajectories of each light source. The walkover commenced on the right side of the display with the subject upright. (J. A. Wallace provided technical assistance and Dr. H. de Bruin attended to the computational aspects of the display. We are grateful to both of them.)

Understanding the Scientific Bases of Human Movement

SECOND EDITION

Title Page for the First Edition

Understanding
The Scientific Bases of
Human Movement

ALICE L. O'CONNELL, Ph.D.

F.A.C.S.M.; F.A.A.H.P.E.R.
Associate Professor of Biomechanics
Boston University
Sargent College of Allied Health Professions

ELIZABETH B. GARDNER, Ph.D.

F.A.C.S.M.
Professor of Biology and Physiology, Emerita
Boston University
Sargent College of Allied Health Professions

WILLIAMS & WILKINS
Baltimore/London

Dedications

Former authors:

This book is dedicated to Ruth B. Glassow who has stimulated and inspired so many, including the authors, to pursue the study of human movement.

A. L. O'Connell
E. B. Gardner

Present authors:

To my husband, David B. Waddell, who initially encouraged me to write the second edition, and who subsequently provided the moral support, patience, and dedication to do so.

B. A. Gowitzke

For Maureen, Joanne, and Alan Sean, my family, who made the task so much lighter by their understanding.

M. Milner

Foreword to the Second Edition

Setting aside the flattering implications of being asked to write a foreword, I awaited arrival of the proofs of this book with special interest because it was written by two close colleagues. Ready to say nice things about it, I soon found that it is a book over which it is easy to wax enthusiastic. This is the first modern comprehensive book on human movement based on current scientific thought. Yet it is comprehensible for both students and old-timers alike.

The first edition—also by dear friends Alice O'Connell and Elizabeth Gardner became an instant classic; but like all classics in science, it required updating in order to remain alive as a student's text-book. Doctors Gowitzke and Milner have succeeded in a way that I had hoped for; but they have gone beyond. The book is now essential reading for all who deal with human performance. This includes both physical education professionals and medical personnel who treat musculoskeletal disorders.

A word or two about the authors of the new edition: Barbara Gowitzke is a professor of physical education whose hallmark is thoroughness and dedication to her profession of scientific teaching; "Mickey" Milner, with whose career my own efforts have intertwined for more than a decade, is an energetic and imaginative biomedical engineer. The melding of the talents and special skills each brought to this book demonstrates the high levels that can be achieved by interdisciplinary work. While neither person in isolation could have produced the special nature of this volume in a reasonable length of time, yet in a few short years they have succeeded quite dramatically. It is a book of substance and quality!

John V. Basmajian, MD, FACA, FRCP(C)
Director, Chedoke Rehabilitation &
Continuing Care Centres and
Professor of Medicine, McMaster University
Hamilton, Ontario, Canada

Foreword to
the First Edition

The term kinesiology, long used in referring to the science or study of human motion, has been giving way to the term biomechanics. Certainly this latter term is more self-explanatory and hence its rising popularity. In the past half century the change in the content of textbooks in this field reflects the changing concepts of kinesiology; the early texts were basically applications of the principles of anatomy and mechanics to human movement. Some physiology of muscle was included, and a little material on the central nervous system was added as that field advanced. The neuromuscular physiology which appeared in these early kinesiological texts was greatly simplified, or perhaps it seems that way now in the light of our present knowledge.

Early kinesiologists were concerned primarily with the anatomy of motion: with action occurring at the different joints, with the identity of muscles that could produce these actions and, in many cases, with the use of this knowledge to develop those particular muscles to a high peak of efficiency. That this is still the case with a majority of the kinesiology instructors of undergraduate students today is illustrated indirectly by a survey made during the middle 1960's by Neuberger of Eastern Michigan University which was concerned with the visual aids used by these instructors. Of the replies reported in detail, 60% leaned very heavily to almost entirely on anatomical materials. Also, of the seven best known kinesiology texts on the market as this is written, all but two devote 50% or more of their content to musculoskeletal anatomy, and some laboratory manuals do likewise. However, we seem to be moving away from the applied anatomy concept; the trend in some instances has swung so far afield that there appeared a tendency to include the effects of emotions, physiological condition, progress in motor learning, etc. under a kinesiological umbrella. This text has no such ambitions: it is written for those kinesiologists who are concerned with the rapidly expanding ways and means of studying, and so arriving at a better understanding of, human motor control.

Today the modern kinesiologist asks many questions about human movement and eagerly seeks the answers. He not only wants to know what movement pattern(s) or sequence of movements is involved in a given skill, but how those movements can be made more efficient. This in turn involves him in seeking answers to a number of other questions such as:

What is the mass of the body, body part(s), or extremity?

What is the angular or linear acceleration of a body, body segment, or extremity?

What force does a body, or body part, exert as a result of its mass and acceleration?

How, and in what direction, is this force applied to an external object, (ball, club, discus, etc.) or to the body itself (as in jumping or running)?

Where is the center of gravity of the body, not only when it is motionless but during performance of a motor skill?

Where is the center of gravity of any combination of body parts: e.g., an entire or partial extremity, the entire or upper trunk plus one or both upper limbs, the entire or lower trunk plus one or both of the lower limbs, etc.?

What is the angular velocity of a body segment at any given instant?

What linear velocity becomes available at the distal end of a kinematic chain as a result of the angular velocity of a particular joint action, and which can be imparted to a ball or other throwing or striking implement?

Finding these answers is made possible by our modern technology. The motion picture camera with speeds ranging up to thousands of frames per second provides the means for capturing a single motor performance for an indefinite period of study. Repeated viewing of the film with the aid of a time-motion study projector, a film editor-viewer or a microfilm reader gives ample opportunity for analysis. Finally, as has been illustrated by Plagenhoef* at the University of Michigan and then by Garrett et al.,† the computer can be elegantly used to expedite matters. Many of the available methods for attaining these answers and for using various of the investigative tools‡ available will be found in the text of this book.

Perhaps more difficult to answer are the questions asked by some of today's kinesiologists involving the initiation of movement and control of the body in performing motor skills. They are aware that true understanding of human movement requires knowledge of the means by which the central nervous system integrates proprioceptive input and coordinates the activity of the muscles so that each will contribute properly to the intended movement. However, knowledge of the functioning of the nervous system is now so extensive and is increasing so rapidly that it is difficult for the neurophysiologist, and impossible for the layman, to keep abreast of it. There is therefore a real need for material which will assist the kinesiologist in maintaining a general overview of advances in this area which may be of significance to his field. Up to the present, relatively few writers in the fields of physical education and physical therapy have been able to write in this field with assurance and authority. Hopefully, however, this situation is in the process of change. This text is one of the first to attempt to interpret and apply some of the expanding

* Plagenhoef, S. C., 1962. *An Analysis of the Kinematics and Kinetics of Selected Symmetrical Body Actions.* Doctoral Dissertation, University of Michigan.

† Garrett, R. E., Widule, C. J., and Garrett, G. E., 1968. Computer aided analysis. *Kinesiol. Rev.* pp. 1–4; Garrett, G. E., Widule, C. J., Reed, W. S., and Garrett, R. E., 1969. Human movement via computer graphics. Paper presented at convention of the American Association for Health, Physical Education and Recreation, Boston, 1969.

‡ Methodology of computer usage per se has not been included in this text.

knowledge of neurophysiology to the field of motor performance. It includes four chapters on the neuromuscular bases of movement (Chapters 10 to 13), to supplement the too often scanty coverage of such information in undergraduate curricula. The inclusion of Section III on proprioceptive reflexes is unique with this text and is presented to provide a background for enlarging the scope of kinesiological analysis. The final chapter deals with speculative postulations of reflex involvement in certain skills. It is offered in the hope that it may encourage the kinesiologist to include consideration of this aspect of human movement in his analysis and research, to recognize and investigate reflexes which may be assisting a performance, and to identify those which may be interfering and require voluntary inhibition. With such information available, he should be better equipped to understand the difficulties encountered by the beginner in learning a new skill, and why the use of one method or technique produces better results than another. He can then improve the best of the older techniques and design new and more effective methods based on his expanded knowledge.

A. L. O'C.
E. B. G.

Preface

In revising the original text, we have adhered to a logical sequence: overall body movements were described through the medium of skeletal linkages and pertinent definitions related to them. There followed definitions of fundamental mathematical tools and physical concepts pertinent to descriptors of motions and forces that can be applied in a biomechanical context. Then, an examination was made of the mechanical characteristics of muscle in the human skeletal system. Next, biomechanical calculations of muscle forces for both static and dynamic situations were pursued. Having dealt with the skeleton and its prime movers (muscles) and examined a number of force situations, we were able then to add the control paths for the overall neuromusculoskeletal system. Hence, the muscle and nervous systems and their anatomical arrangements and essential facets of function were described. Effectors and affectors could thus be illuminated and a number of instances examined for overall system performance.

Fortunately, the previous authors provided a substantial base upon which we were able to build. Revising and rearranging the contents of the first edition, adding new material and unifying the whole has been an exciting interdisciplinary venture for us.

Chapter 1 is well illustrated with 52 new photographs; over twice the number used in the first edition. This chapter is augmented by new approaches to describing skeletal system movements. While some of the content of several chapters of the first edition are included, Chapters 2 and 3 constitute primarily new material of an applied mechanics character. The contents of the original Chapters 2 and 10 are interwoven in Chapter 4 which is now augmented principally with general facets of electromyography. A large number of biomechanical examples are encompassed in Chapters 5 and 6, many of which have been taken from the previous edition, but with substantial rearrangements, and the inclusion of many new explanations. Chapter 6, in particular, incorporates details of recent technological instrumentation systems which facilitate kinesiological and biomechanical studies. The original Chapter 10 is replaced by Chapter 7 with little change in content except for the addition of recent information on the sliding filament theory of muscle contraction and updated photographs pertinent to muscle structure. Formerly Chapter 12, Chapter 8 is reorganized with new material. New content constitutes Chapter 9 which has been added because the authors believe that an overview of the central nervous system would enable the reader to assimilate more easily the chapters which follow without immediately having to consult a neuroanatomy textbook. Chapter 10 is essentially Chapter 13 of the first edition with some revisions and clarifications. The content of the original Chapters 14 and 15 are included in Chapter 11, with significant changes in organization; practical aspects of proprioception have been added. The Appendices have been changed to more fully meet the needs of the student by deleting trigonometric and other tables that are readily available elsewhere, or by using modern hand-held calculators, and adding pertinent conversion factors, mathematical aids and functions related to systems of measurement, as well as solutions to some problems posed in the text.

Barbara A. Gowitzke
Morris Milner

Acknowledgments

Former Authors:

The authors wish to record their gratitude to the Boston University Graduate School for a grant-in-aid which contributed to the preparation of this manuscript.

A. L. O'C.
E. B. G.

Present Authors:

We gratefully acknowledge the cooperative efforts of many colleagues and friends, some of whom are acknowledged within the pages of the text. Special thanks are due to our spouses, Maureen Milner and David B. Waddell, who helped us significantly with our illustrations. Maureen produced the artwork for the new illustrations and David contributed his photographic and layout talents. Lise Arsenault-Goertz, past Canadian gymnastic champion, is the subject in many illustrations describing body positions; she gave generously of her time and extraordinary talent. The authors also wish to acknowledge the staff of McMaster University Audio-Visual Services and, in particular, Geoff Brown, for their photographic work. Facets of technical support for various illustrations were kindly provided by Dr. Hubert DeBruin, John Moroz and Tony Wallace. We are especially indebted to Sarah Ann Fick, who took responsibility for typing the manuscript and attending to the many details of preparing final copy; she was assisted by Barbara Harris and Mary Hickey. The assistance and cooperation of all of The Williams & Wilkins Company staff have greatly expedited this work. Finally, we acknowledge the contributions of our students in perhaps unwittingly providing the motivation and impetus for our efforts.

B. A. G.
M. M.

Contents

CHAPTER 1
ON THE SKELETAL SYSTEM

CHAPTER 2
MOTION

CHAPTER 3
FUNDAMENTALS OF APPLIED MECHANICS

CHAPTER 4
MUSCLE—THE BODY'S FORCE GENERATOR

CHAPTER 5
MUSCULOSKELETAL FORCES

CHAPTER 6
ANALYSIS OF HUMAN MOVEMENTS

CHAPTER 7
STRUCTURE AND CHEMISTRY OF SKELETAL MUSCLE

CHAPTER **1**

On the Skeletal System

INTRODUCTION

The skeletal system is the bony framework that supports body organs, protects many of them, and forms the hard core of all body segments. Its many articulations provide mobility, and it is the functions of these mobile articulations that are of concern in facilitating descriptions of human movement.

The driving forces for skeletal components or links derive from muscular actions which in turn are controlled by the nervous system. Overall movements are thus the manifestation of integrated activities within a complex which is often termed the neuro-musculo-skeletal system. For the sake of convenience and to facilitate our understanding of the component parts of the neuromusculoskeletal system we tend to decompose the system, looking at its parts in relative isolation and making simplifications with regard to the descriptions of overall performance or behavior of these parts. Turning to the skeletal system alone we realize that it can be thought of as a system of interconnected links (Fig. 1.1).

LINKS IN THE BODY

Dempster (1955) was the first kinesiologist to adapt the link concept initially used by engineers to those problems involving kinetic and kinematic treatments of movements of the human body. Since engineering links involve overlapping articulating members held together by pins which act as axes of rotation, a link is considered to be a straight line of constant length running from axis to axis. A system of links is essentially a geometric entity for analysis of motion by geometric or kinematic methods. " . . . In engineering mechanisms the links move in relation to a framework, and this framework itself forms a link in the system. Thus, to transmit power, the links of machinery must form a closed system in which the motion of one link has determinate relations to every other link in the system" (Reuleaux, quoted by Dempster (1955)).

Reuleaux (1875) also introduced the term "kinematic chain" to refer to a mechanical system of links. In engineering, the chain forms a closed system where, as quoted earlier, "the motion of one link has

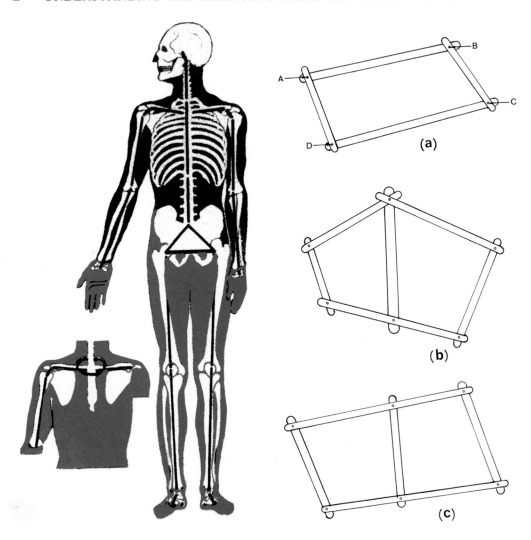

Figure 1.1. The human skeleton as a system of links. Note the simplified link segments, drawn between major joints that constitute anatomical links. (From Dempster, W. T. 1955. *Space Requirements for the Seated Operator.* Wright Air Development Center Technical Report 55159.)

Figure 1.2. Types of closed kinematic chains. All joints are pin-centered and free to move. If for example in (a), link AB (pinned at A and B) is held fixed, links BC, CD, and DA will, when influenced by externally applied forces on one or more of them, be able to move in a predictable pattern which can be determined by constructing circles with A and B as centers and respective radii AD and BC. The initial distance between C and D must be maintained throughout the ranges of movement of C and D to describe the resulting possible movement pattern of the link system. The interested student might care to attempt such constructions and make simple models to obtain appropriate visualizations.

determinate relations to every other link in the system," and "the closed system assures that forces are transmitted in positive predetermined ways." Thus, in engineering a kinematic chain is a closed system of links joined in such a manner that if any one is moved on a fixed link, all of the other links will move in a predictable pattern (Fig. 1.2). With a few exceptions the system of skeletal links in the human body is generally not composed of closed chains, but of open ones, as the peripheral ends of the extremities are free (Fig. 1.1). Forces may be transmitted in positive ways, predetermined by the central nervous system, but the central nervous system is notorious for never accomplishing the same act in exactly the same way from one time to the next, even though the external results may appear similar. Thus, when speaking of a living kinematic chain, we are usually speaking of a series of links arranged in an open system, whose dimensions are determined by the linear distance from joint axis to joint axis, ignoring muscle mass, bone structure, and type of articulation between body segments.

Although most living kinematic chains are open, Brunnstrom (1962) defines two closed kinematic chains in the body. The first is the pelvic girdle which is made up of three bony segments united at the two sacroiliac joints and at the symphysis pubis. This can hardly be classified as a kinematic chain because normally no movement occurs at the joints mentioned. Dempster classes the pelvis as a single triangular link (Fig. 1.1). The second closed kinematic chain in the body according to Brunnstrom is the thorax where the upper 10 ribs are jointed to the vertebral column and sternum. The rib cage, however, does constitute a system of closed kinematic chains because the upper 10 ribs of the left side cannot move without similar movement by the upper 10 ribs of the right side when they lift the sternum on inhalation. Steindler* (1973) considers a closed living chain (which he terms kinetic rather than kinematic) to exist in "all situations in which the peripheral joint of the chain meets with overwhelming resistance" (Fig. 1.3).

Because the bones of the body rarely overlap at joints as at the ankle and, except

Figure 1.3. Example suggested by Steindler of a human closed kinematic chain. The wall and floor provide "overwhelming" resistance at the peripheral joints.

for the atlantoaxial joint, have no pin-centered axes, and because at many joints movement can occur in different directions and planes, the engineering concept of links must be redefined to fit the need of the kinesiologist. Dempster has proposed the use of the term "link" in kinesiology as the distance between joint axes; e.g., the leg link becomes the linear distance between the joint axes passing through the distal end of the femur and the proximal end of the talus (or through the two malleoli), thus spanning both the knee and ankle joints (Fig. 1.4).

In the appendicular skeleton a body segment consists of a hard core made up of one or more bones enclosed in an irregular mass of soft tissue (muscle, connective tissue, and skin). Ligaments and muscle tendons cross the joint between contiguous body segments, anchoring to the adjacent bone or bones and holding the segments together. The instantaneous axis around which one segment moves on an-

---

* While Brunnstrom and Steindler used the term "chain" modified by kinematic or kinetic, respectively, when referring to a series of body segments, they did not use the term "links" in their discussions.

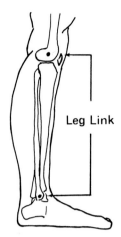

Figure 1.4. Leg link.

other generally passes through one of the bones through an area near the joint.

JOINT AXES AND DEGREES OF FREEDOM

Anatomists and kinesiologists speak of joints as being uniaxial, biaxial, or multiaxial and as having certain degrees of freedom (Steindler (1973); Brunnstrom (1962); Terry and Trotter (1955)). A joint with only one axis (uniaxial) has one degree of freedom; that is to say, the articulating bones can move only in one plane. Examples in the human body include hinge and pivot joints. Hinge joints occur at the elbow, knee, interphalangeal, and ankle joints. The pivot joints are the atlantooccipital in the vertebral column and the radioulnar joints in the forearm. Joints that can move about two axes (biaxial) have two degrees of freedom and so produce movement in two different planes. The wrist, the metacarpophalangeal, and the metatarsophalangeal joints are biaxial. Joints that can permit movement in all three planes have three degrees of freedom but are called multiaxial rather than triaxial. This is because movement can occur in oblique planes as well as in the three major planes which are defined as three mutually perpendicular planes. Examples of multiaxial joints include the ball and socket joints at the hips and shoulders and the numerous plane joints of the axial skeleton. In this instance the term "plane" is an adjective referring to the almost flat articular surfaces which can glide over one another, with movement being limited only by ligaments or by the joint capsule. Examples include those joints between the articular processes of the vertebrae and between the ribs and the vertebrae. These joints have such a limited amount of movement at any one articulation that total movement of the torso occurs only because of the combined action of many or all of the joints and their degrees of freedom.

For the kinesiologist there is a distinct advantage in using the term "degrees of freedom." While no one joint can have more than three degrees of freedom, the degrees at adjacent joints can be summed to express the total amount of freedom of motion of a distal segment relative to a proximal one. For instance, the distal phalanges of a pianist enjoy 17 degrees of freedom relative to his trunk: one degree at each of the distal and proximal phalangeal joints; two degrees at the metacarpophalangeal joints; two degrees at the wrist joint; one degree in the forearm at the radioulnar joints; one degree at the elbow; three degrees at the shoulder; three degrees at the acromioclavicular joint; three degrees at the sternoclavicular joint. Observation of many pianists might, however, lead us to add three more degrees of freedom arising from the motion in the vertebral column. This would express the freedom of the phalanges relative to the pelvis which is resting on the piano bench, rather than relative to his torso, making a total of 20 degrees of freedom available at the fingertips.

Summing up the degrees of freedom of the pianist's fingertips involved listing the joints occurring between the distal phalanges and the pelvis. These joints unite the various body segments which move upon each other in the manner of links in a chain. Thus, we have a further example of an open, living kinematic chain.

LIMITATIONS OF MOVEMENT

The type and range of movement about any given joint depend upon the structure of the joint and the number of its axes, the restraints imposed by ligaments and muscles crossing the joint, and the bulk of adjacent tissue. A joint with three degrees of freedom may, because of its structure, have a very limited range of motion as was indicated earlier in regard to the intervertebral joints, while a joint with only one degree of freedom may have a large range of motion. For example the forearm can move through an average range of 150° from the position in line with the arm to the fully bent position (full flexion). The range may be increased by from 5° to 15° in the individual who has a smaller than average olecranon process or a deeper than average olecranon fossa which permits the forearm to go beyond the position in line with the arm (i.e., hyperextend). Conversely, in an individual with overdeveloped biceps and brachialis muscles or with excessive adipose tissue, flexion may be limited by the very bulk of the soft tissue of the arm. Similar factors can also affect the range of mobility of other joints.

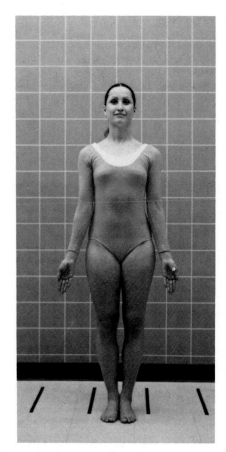

Figure 1.5. Anatomical position.

REFERENCE POSITIONS AND PLANES

For convenience in specifying the positions of anatomical members in space a reference system must be established. This system may be established in various locations depending upon the details of the issue being investigated. For example, while whole body movements might be referred to a reference system whose origin is located at the center of gravity of the body, finger movements might more conveniently be defined by a reference system whose origin is chosen to be at a selected metacarpal joint. Traditionally, kinesiologists have made use of an orientation system which defines three cardinal mutually perpendicular planes of orientation having a common intersection at the *center of gravity*† of the body while it occupies the anatomical position exemplified in Figure 1.5. Three essential, mutually perpendicular planes of concern from a practical standpoint are the sagittal, transverse, and frontal planes. Views in these respective planes are depicted in Figure 1.6, (a), (b),

† This is the point in or about the body through which the resultant body force will act due to the gravitational pull of the earth (or other environment) upon the masses of the various body parts. The location of the center of gravity changes as limbs vary their relative positions. From a practical standpoint it is not the best choice of a reference origin since a great deal of anatomical information must be supplied to specify its location exactly. The location of the center of gravity can be determined using the methods depicted in Chapter 5 under "Locating the Center of Gravity."

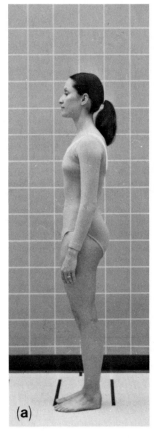

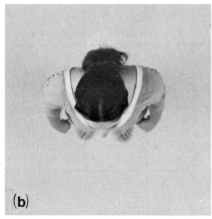

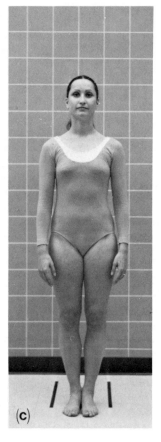

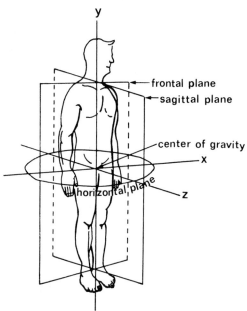

Figure 1.7. Diagrammatic representation of cardinal reference planes.

and (c). The cardinal planes are the mid-sagittal which divides the body into right and left halves, the frontal plane which divides the body anterioposteriad into front and back halves, and the transverse plane which divides the body into upper and lower halves. Figure 1.7 is a diagrammatic representation of the cardinal reference planes. Location of any anatomical members can be specified by noting their distances from the three mutually perpendicular axes which pass through the origin (in this case the center of gravity of the body).

Figures 1.8 through 1.16 show the joint axes for major movements about the shoulder, elbow, forearm and wrist, fingers and thumb, the hip joint, the knee joint, the ankle and foot, and finally the axial skeleton. A knowledge of the movement descriptors discussed in the following paragraphs will be helpful in interpreting the aforementioned figures.

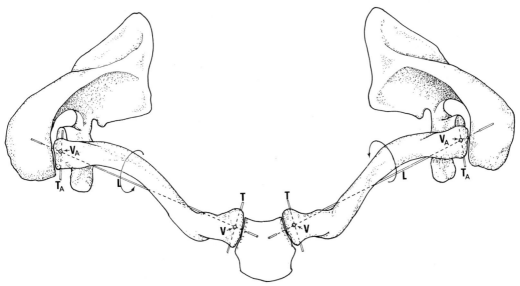

Figure 1.8. Major axes of the shoulder girdle as seen from above. V, vertical axis for protraction and retraction of the shoulder girdle; T, transverse axis for elevation and depression of the shoulder girdle; L, longitudinal axis for the limited rotational movements of the clavicle; V$_A$, vertical axis and T$_A$, transverse axis at the acromial end of clavicle for scapular motion. (Joint axes in Figs. 1.8 to 1.16 drawn after Grant, J. C. B. and Smith, C. G., 1953. In *Morris' Human Anatomy*, edited by J. P. Schaeffer. New York: The Blakiston Company.)

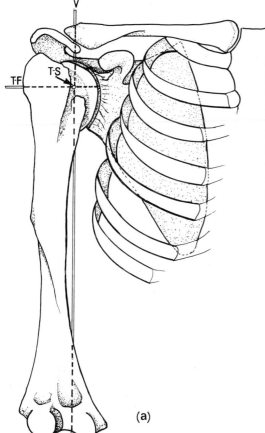

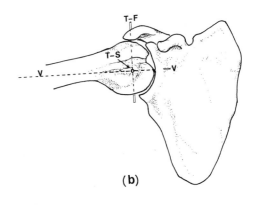

Figure 1.9. Axes for movements at the shoulder joint. (a), from the anatomical position. T-F, transverse axis in the frontal plane for movements of flexion and extension; T-S, transverse axis in the sagittal plane for movements of adbuction and adduction: V, vertical axis running the length of the humerus for movements of inward and outward rotation of the arm. (b), same axes but arm is abducted 90°. Note altered positions of T-F axes.

(a)

(b)

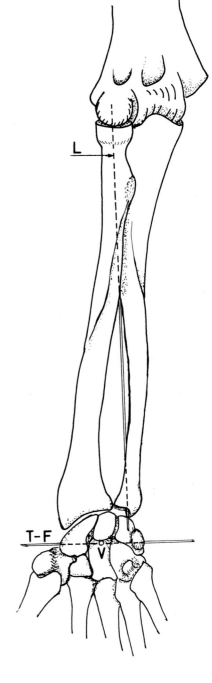

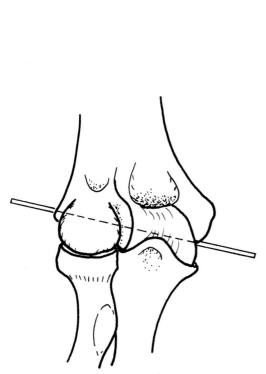

Figure 1.10. Transverse axis through the elbow joint. Movements are flexion and extension.

Figure 1.11. Axes of the forearm and wrists. L, long axis of the forearm for pronation and supination: T-F, compromise transverse axis in the frontal plane for wrist flexion and extension; V, volardorsal axis for radial and ulnar deviation of the hand at the wrist.

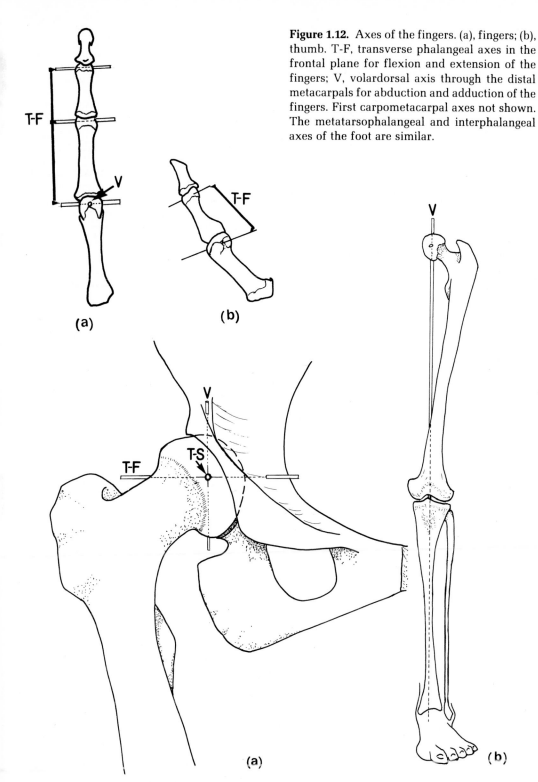

Figure 1.12. Axes of the fingers. (a), fingers; (b), thumb. T-F, transverse phalangeal axes in the frontal plane for flexion and extension of the fingers; V, volardorsal axis through the distal metacarpals for abduction and adduction of the fingers. First carpometacarpal axes not shown. The metatarsophalangeal and interphalangeal axes of the foot are similar.

Figure 1.13. Axes for movements at the hip joint. T-F, transverse axis in the frontal plane for flexion and extension of the thigh; T-S, transverse axis in the sagittal plane for abduction and adduction of the thigh; V, vertical axis for inward and outward (lateral and medial) rotation of the thigh. Note location of axis through the lower limb in (b).

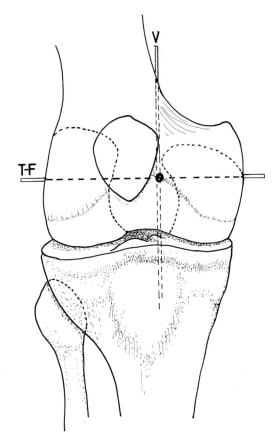

Figure 1.14. Axes through the knee joint. T-F, transverse axis in the frontal plane for flexion and extension; V, vertical axis around which the tibia can rotate when the knee is flexed.

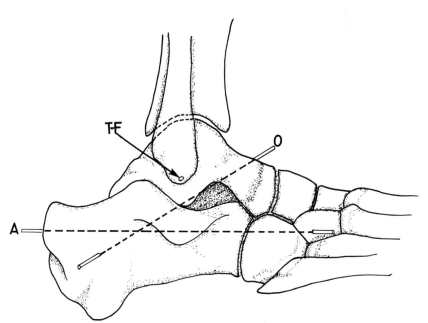

Figure 1.15. Axes of the ankle and foot. T-F, transverse axis in the frontal plane passing through both malleoli and the talar head. The only movements are dorsiflexion and plantar flexion. A and O are compromise axes for inversion and eversion at the intertarsal joints, A through Chopart's joint and O through the talocalcaneal and talonavicular joints.

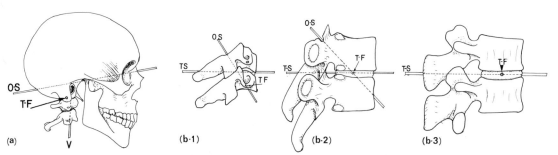

Figure 1.16. (a), joint axes of the axial skeleton axes for movement of the skull on the vertebrae. O-S, oblique axis in the sagittal plane for lateral flexion of the head; T-F, transverse axis in the frontal plane for dorsi- and ventriflexion of the head; V, vertical axis for head rotation, right and left. (b), axes for movement of one vertebra on the adjacent one below: 1, cervical; 2, thoracic; 3, lumbar vertebrae. O-S, oblique axis in the sagittal plane for movements of rotation combined with abduction; T-F, transverse axis in the frontal plane for movements of flexion and extension; T-S, transverse axis in the sagittal plane for movements of lateral flexion (abduction) right and left.

MOVEMENT DESCRIPTORS

To facilitate the explanations allied with the pertinent movement descriptors, photographic views have been utilized in what follows. It should be noted that movements in the frontal plane are best depicted by frontal views, while those in the sagittal plane rely chiefly on a side view. In an attempt to describe movements recourse has been taken to still photographic representations. It should be recognized that the positions depicted are arrived at from the anatomical position. This implies that movement occurs before the particular pose shown was achieved.

Flexion-Extension

Flexion is popularly considered to be a movement which decreases the angle between the moving part and the adjacent segment (as in elbow or finger flexion), and extension is considered to be a movement which increases this angle (Fig. 1.17).

For the purpose of defining joint movements it is convenient to assume that the body is in the anatomical position (Fig. 1.5). Then flexion and extension are movements in which the moving segments travel in a sagittal plane around a horizontal axis defined by anatomical frontal and transverse planes through the axis. While this definition is adequate to a degree, it is not applicable to all joints. A more satisfactory one which can be applied to all except the shoulder joints is based on the anatomical concept that flexion is the approximation of ventral or volar surfaces. This concept is based on the embryological development of the human fetus. Soon after the limb buds first appear in the embryo (Fig. 1.18(a)), they project laterally with the thumbs and great toes uppermost (Fig. 1.18(b)). As the limbs develop, they bend ventrad at the elbows and knees so that the apices of these joints are pointed outward and the palms of the hands and soles of the feet (the volar surfaces) face the torso (Fig. 1.18(c)). Finally, both pairs of limbs rotate 90° but in opposite directions, the rotations taking place about the long axes of the limbs (Fig. 1.18(d)). The upper extremities rotate laterad so that the elbow points backward, the thumbs are outward, and the ventral and volar (palmar) surfaces face forward. The lower extremity rotates mediad so that the knees point forward, the great toes are inward, and the ventral surfaces face backward, as do the soles of the feet (volar surface) when one is standing on the toes. The proximal surface of the limbs retains its embryological orientation in the regions of

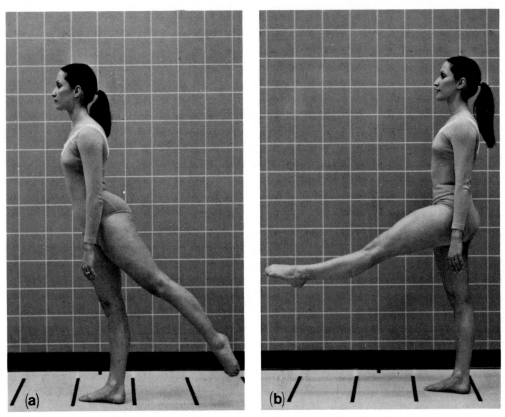

Figure 1.17. Viewing left side in sagittal plane. (a), hip extension; (b), hip flexion.

the axillae and groin. Because of this situation a large portion of the upper part of the thigh still presents some ventral surface on the anterior aspect; therefore, a movement of the lower extremity forward and upward at the hip joint (Fig. 1.17(b)) is an approximation of ventral surfaces and conforms to the latter definition of flexion. Because the rotation is complete at the knee, flexion of this joint also meets the anatomical definition. Shoulder flexion and extension are not easily reconciled to either definition, so these movements are correlated with the direction of the movements at the hip: flexion of the arm at the shoulder is defined as movement forward and upward in the sagittal plane and extension as movement downward and backward in the same plane (Fig. 1.19). Figure 1.19(c) depicts *hyperflexion* of the

shoulder. The prefix "hyper" always describes an exceptional continuation of a movement. In hyperextension this usually means a continuation of extension beyond the anatomical position.

Flexion of the elbow, wrist, fingers, toes, and vertebral column all conform to both concepts, i.e., the anatomical concept of approximating the ventral or volar surfaces and the popular concept of decreasing the angle between the body segments. Extension of these same joints is, of course, movement in the opposite direction. However, this agreement between anatomical and popular definitions breaks down when the concepts on which the definitions are based are applied to the movements occurring at the ankle joint.

Study of Figure 1.20 illustrates the divergence. Decrease of the angle between

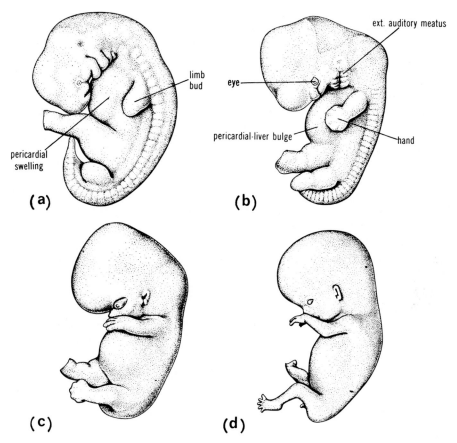

Figure 1.18. The human embryo. (a), at 5 weeks; (b), at 6 weeks; (c), at 7 weeks; (d), at 8 weeks. (From Langman, J., 1969. *Medical Embryology: Human Development, Normal and Abnormal.* Baltimore: The Williams & Wilkins Company.)

the foot and the leg, anatomical extension, is popularly called ankle flexion; "pointing the toes," anatomical flexion, is popularly known as ankle extension. If Figure 1.21(b) is consulted one sees how the anatomical terminology is derived. Because of this paradoxical situation the term *dorsi-flexion* has been adopted for anatomical extension/popular flexion, and *plantar-flexion* is the term applied for anatomical flexion/popular extension.

Abduction-Adduction

This pair of movements takes place in the frontal plane and occurs at biaxial (metacarpophalangeal and metatarsophalangeal) joints and at multiaxial (shoulder, hip, and first carpometacarpal) joints. Abduction of the fingers and toes is movement away from the middle digit, while adduction is movement toward that digit.

Abduction at a ball and socket joint (shoulder or hip) is movement of the limb upward and away from the midline (Fig. 1.22). At the glenohumeral joint the arm can be raised only 90° before the greater tuberosity of the humerus contacts the acromion process. Further abduction is accomplished by upward rotation of the glenoid fossa of the scapula (as depicted in

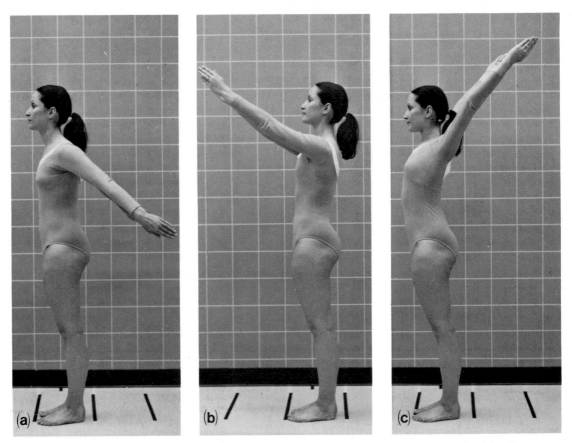

Figure 1.19. Viewing the left side in sagittal plane. (a), shoulder extension; (b), shoulder flexion; (c), shoulder hyperflexion.

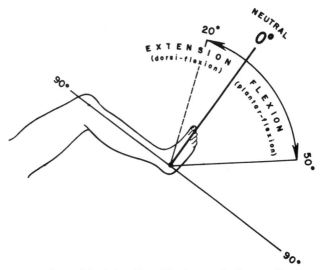

Figure 1.20. Movements at the ankle joint. Dorsiflexion and plantar flexion. (From American Academy of Orthopaedic Surgeons, 1965. *Measuring and Recording Joint Motion.*)

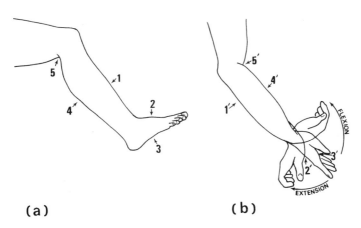

(a)   (b)

Figure 1.21. (a), lateral aspect of right lower extremity; (b), lateral aspect of upper extremity. 1, anterior surface of leg, homologous to back of forearm, 1'. 2, instep or dorsum of the foot, homologous to the back of the hand, 2'. 3, sole of foot, homologous to palm of the hand, 3'. 4, back of leg, homologous to ventral surface of forearm, 4'. 5, popliteal fossa homologous to the cubital fossa of the forearm, 5'.

Figure 1.22. Abduction. Both shoulders and the left hip are abducted.

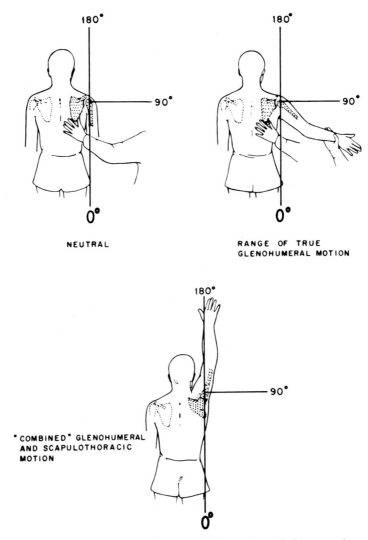

Figure 1.23. Glenohumeral motion. Note the upward rotation of the scapula as abduction is completed. (From American Academy of Orthopaedic Surgeons, 1965. *Measuring and Recording Joint Motion.*)

Fig. 1.23). As a result, the total range of abduction of the upper extremity can be as much as 180°. Further details on movements of the scapula are presented later. In adduction of either the upper or lower extremity the limb may be drawn across the midline of the body (Fig. 1.24).

Horizontal Flexion Extension

In making reference to Figure 1.25 it will be seen that horizontal flexion is performed by the upper extremity from a position of abduction; the motion of the extremity is in a horizontal plane and the

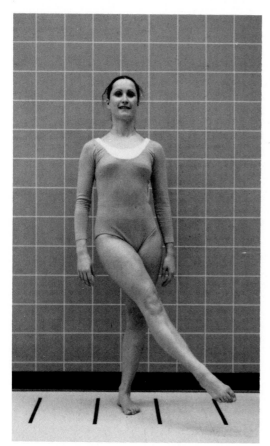

Figure 1.24. Adduction at the right hip.

Medial (Inward) and Lateral (Outward) Rotation

These movements occur at multiaxial joints around a longitudinal axis running the length of the bone of the rotating segment but not necessarily within the shaft of the bone (Figs. 1.9, 1.11, and 1.13). With the exception of the long axis of the clavicle (which rotates slightly with most movements of the upper extremity, see below under "Movements of the Shoulder Girdle"), the rotation axes are vertical or near vertical when the body is in the anatomical position. As vertical axes are defined by the intersection of the frontal and sagittal planes, it follows that the movement of rotation around these axes must occur in the transverse plane. Rotation of an upper extremity in either direction is seen most clearly when the elbow is flexed, as this has the effect of visually magnifying the amount of rotation because of the length of the forearm and hand (Fig. 1.26). Elbow flexion also eliminates the addition of pronation (or supination) to the position of the hand as an indication of the amount of shoulder rotation. Medial or lateral rotation of the humerus (Fig. 1.27) or femur (Fig. 1.28) may take place with the bone in any starting position, but rotation at the hip is more limited than at the shoulder.

Circumduction

This may occur at any biaxial or multiaxial joint and is a combination of flexion-abduction-extension-adduction or the reverse, and it may involve rotation of the limb concerned. The extremity travels in a cone-shaped path with the apex at the fulcrum of the joint at which the movement originates (Fig. 1.29).

The hip and shoulder joints, described as multiaxial joints, differ in the amount of flexibility permitted by their structures. In the hip joint, which requires stability, the socket of bone and cartilage is exceptionally deep. In the shoulder, which requires more flexibility, the bony and soft tissue parts of the socket are shallow. The

limb is carried forward across the front of the body. Horizontal extension is similar movement in the opposite direction. There has been a tendency among therapists to use the term horizontal adduction for horizontal flexion because the arm is moved across the midline of the body. It should be noted, however, that those horizontal movements occur around the same axis through the head of the humerus as do flexion and extension of the shoulder performed from the anatomical position. The abduction of the arm has merely changed the orientation of this axis through the head of the humerus from the horizontal to the vertical (Fig. 1.9).

Figure 1.25. Horizontal flexion/extension at the right shoulder. (a), horizontal flexion or shoulder adduction; (b), horizontal extension or shoulder abduction.

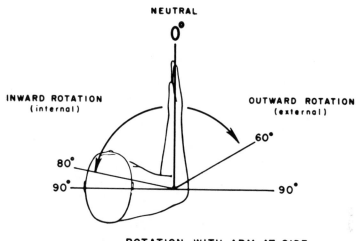

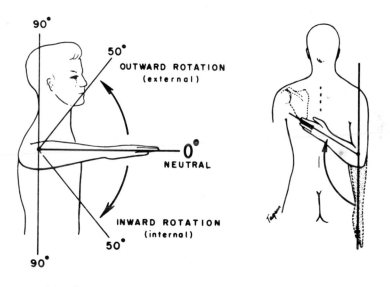

Figure 1.26. Inward and outward (medial and lateral) rotation at the shoulder. (From American Academy of Orthopaedic Surgeons, 1965. *Measuring and Recording Joint Motion.*)

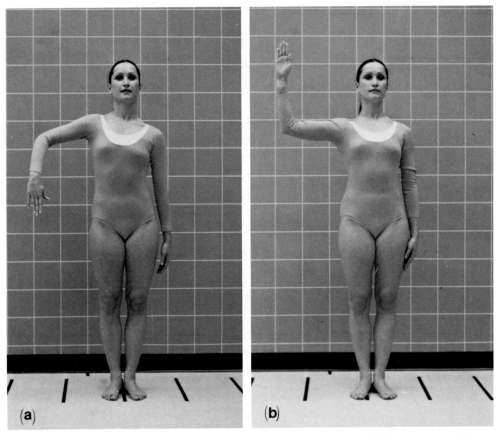

Figure 1.27. Rotation of the humerus at the right shoulder. (a), medial rotation with upper extremity abducted about 90° at the shoulder and flexed 90° at the elbow. (b), lateral rotation with upper extremity abducted 90° at the shoulder and flexed 90° at the elbow.

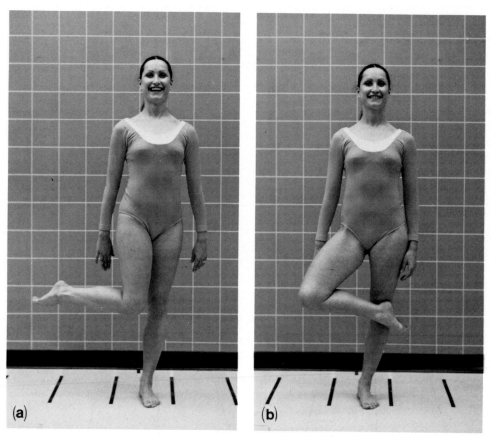

Figure 1.28. Rotation of the right thigh at the hip. (Knee is flexed to exaggerate the movement.) (a), medial (inward) rotation; (b), lateral (outward) rotation.

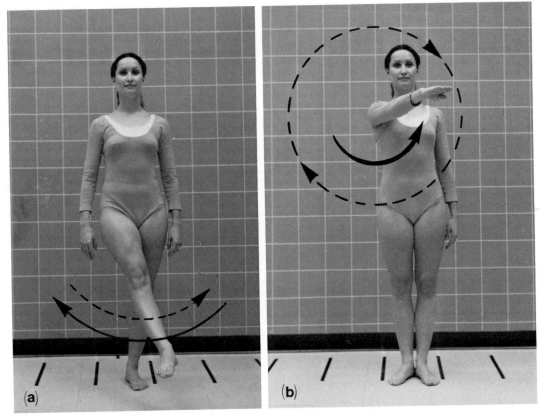

Figure 1.29. Circumduction. (a), of right lower extremity; (b), of right upper extremity.

resulting looseness of the upper extremity makes it possible to move through large ranges of flexion/extension and abduction /adduction.

Codman's Paradox

An interesting paradox results when large ranges of movement are attempted. Concomitant with any of the movements of flexion, extension, abduction, and adduction through large ranges is the rotation of the upper extremity around its long axis. Figure 1.30 illustrates the fact that approximately 180° of lateral rotation has automatically occurred as a direct result of moving the right upper extremity first,

through 180° of shoulder flexion until the upper arm rests beside the ear (Fig. 1.30, (a), (b), and (c)) and second, through 180° of shoulder adduction until the upper extremity rests beside the body (Fig. 1.30, (d), (e), and (f)). Notice the position of the hand at the start and finish of the series (Fig. 1.30, (a) and (f)). The reader is invited to demonstrate for himself/herself the fact that 180° of medial rotation occurs at the shoulder joint when combinations of abduction and extension are used: from anatomical position with the hand in a neutral position (*i.e.*, palm facing the thigh), abduct the upper extremity 180° until the upper arm rests beside the ear; then, move the upper extremity through 180° of extension until the upper extremity is vertical beside the body; the back of the hand will now be facing the thigh. Ranges of movement less than 180° result in smaller amounts of rotation occurring, commensurate with the amount of flexion, extension, abduction or adduction. Clearly, none of the shoulder joint sagittal and frontal plane movements can be functionally classified as purely flexion/extension, abduction/adduction, respectively, because the skeletal framework requires that rotation at the shoulder joint must also occur at the same time.

MOVEMENTS OF THE AXIAL SKELETON (VERTEBRAL COLUMN)

The appearances of major movements of the vertebral column are reflected in Figure 1.31.

The definition of flexion as an approximation of ventral surfaces and of extension as movement in the opposite direction is appropriate here (Fig. 1.31, (a) to (d)). Movement of the trunk to either side is called lateral bending by the orthopaedic surgeons and lateral flexion (or sometimes abduction) by anatomists (Fig. 1.31(e)). Rotation to right or left can occur around the lengthwise axis of the vertebrae (Fig. 1.31(f)). The way that these same movements affect the head when they occur in the cervical vertebrae is illustrated in Figures 1.32 and 1.33. Figure 1.34 shows atlantooccipital movements.

MOVEMENTS AT JOINTS WITHIN THE FOREARM

Pronation and Supination

These movements occur at the pivot joints within the forearm, not at the wrist or elbow. The axis runs from the center of the head of the radius to the center of that

of the ulna (Fig. 1.11). In the anatomical position the palm faces forward because the forearm is supinated. When the elbow is flexed 90° with the palm perpendicular to the floor and the thumb uppermost, the forearm is in the neutral or midposition (Fig. 1.35). Rotating the forearm medially so that the palm turns downward is the movement of pronation; rotating the forearm in the opposite direction, i.e., laterally, turns the palm upward and is the movement of supination. In both of these movements the ulna remains comparatively stationary while the radius is the bone that moves. When the forearm is pronated the radius lies almost parallel to the ulna (Fig. 1.36). Pronation and supination are illustrated in Figure 1.37; both extremities are abducted at the shoulder joints and flexed at the elbow joints.

Radial and Ulnar Deviation of the Hand at the Wrist

These actions at the wrist joint occur in the frontal plane as the hand is moved toward the radial aspect of the forearm (radial deviation) or toward the ulnar side (ulnar deviation) (Fig. 1.38). Radial deviation replaces the older term of radial abduction, as ulnar deviation replaces that of ulnar abduction.‡

MOVEMENTS AT JOINTS WITHIN THE FOOT

Inversion and Eversion

Inversion and eversion (Fig. 1.39) are a result of many small gliding movements between the intertarsal and tarsometatarsal joints. Hicks (1953) located as many as 12 different axes in the foot around which these movements occur, each contributing to an end result of either inversion or

‡ The terms radial and ulnar flexion are sometimes used to describe these movements but, as radial and ulnar movements do not meet the anatomical criteria for flexion (see under "Flexion-Extension," above), the term deviation is preferred.

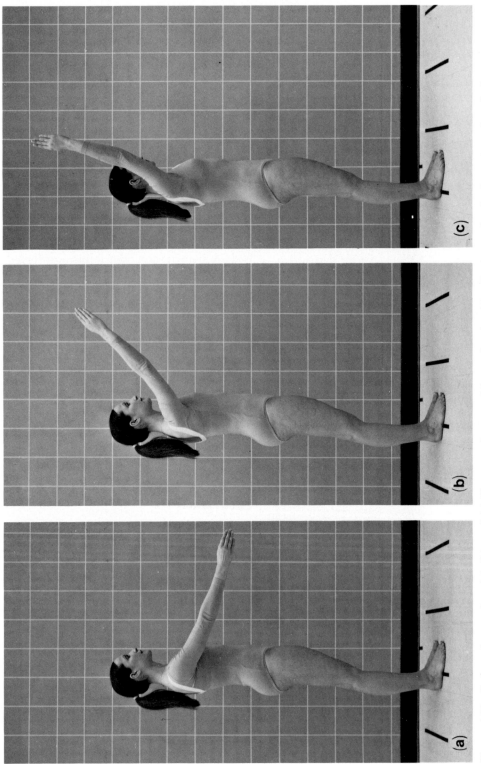

Figure 1.30. Codman's paradox. (a), (b), and (c) illustrate three positions of a shoulder flexion task which originated with the arm resting beside the body. (d), (e), and (f) illustrate three positions of a shoulder abduction task which originated with the arm beside the ear, as in (c). See text for explanation.

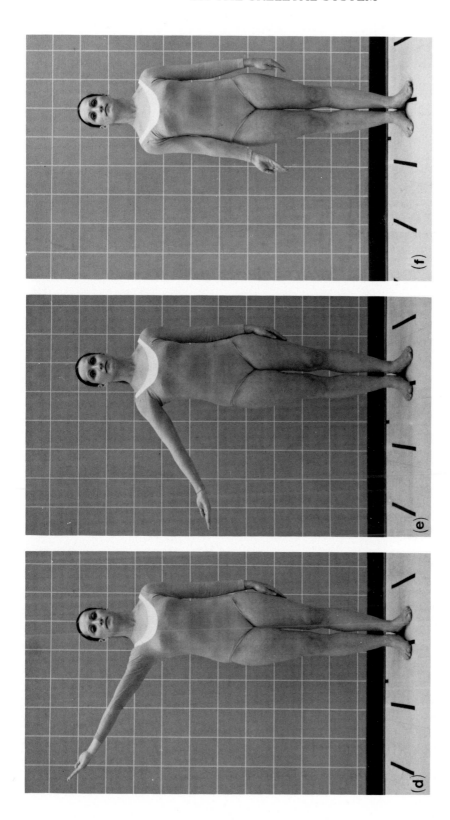

Figure 1.31. Movements of the vertebral column. (a), flexion; (b), hyperflexion; (c), extension; (d), hyperextension; (e), lateral flexion (abduction) to the right; (f), rotation to the left.

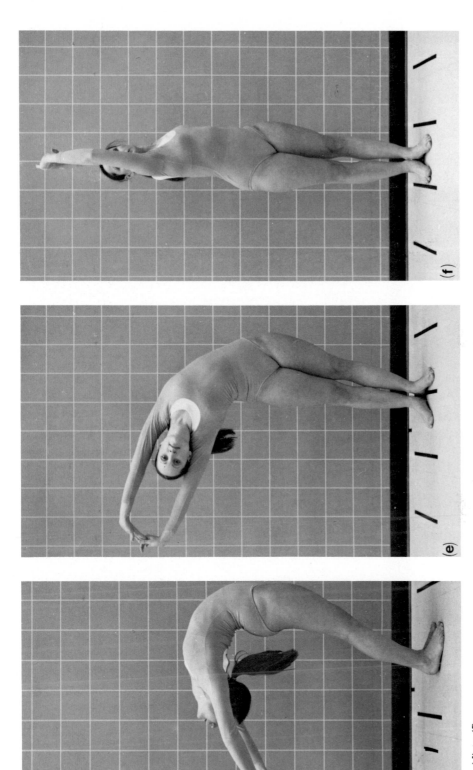

Figure 1.31, (d) to (f)

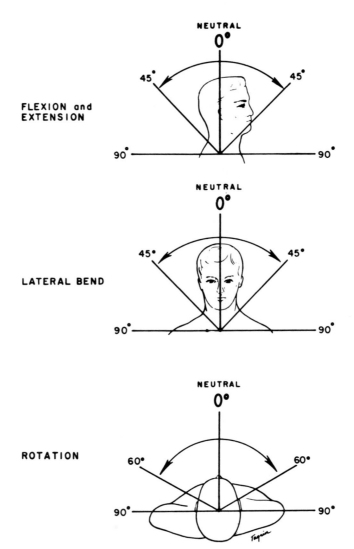

Figure 1.32. Actions at the cervical spine. (From American Academy of Orthopaedic Surgeons, 1965. *Measuring and Recording Joint Motion.*)

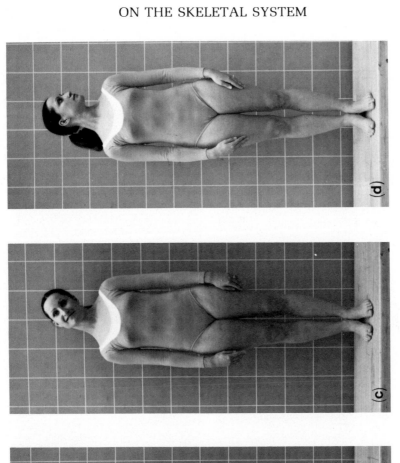
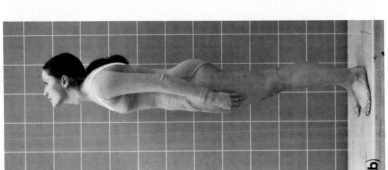
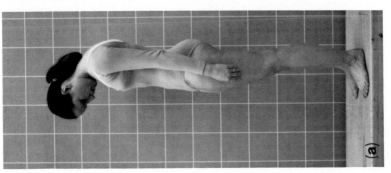

Figure 1.33. Cervical spine movements. (a), ventriflexion; (b), dorsiflexion; (c), lateral flexion; (d), rotation to the left.

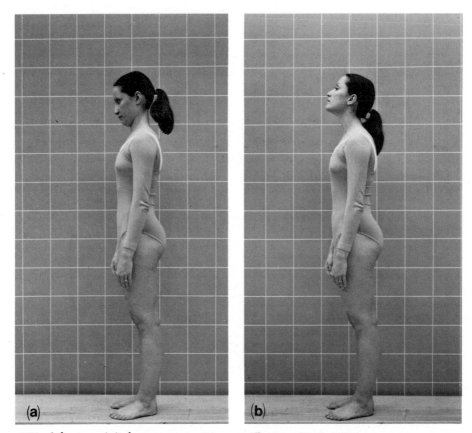

Figure 1.34. Atlantooccipital movements. (a), ventriflexion; (b), dorsiflexion.

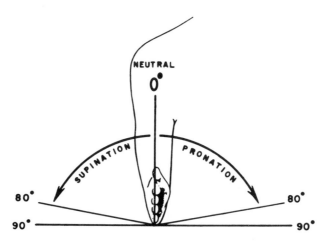

Figure 1.35. Pronation and supination at the radioulnar joints. (From American Academy of Orthopaedic Surgeons, 1965. *Measuring and Recording Joint Motion.*)

proximal end

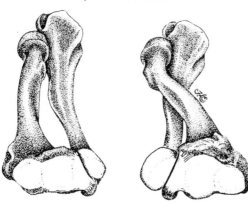

distal end

Figure 1.36. The bones of the right forearm as viewed from the volar surface: left, in supination; right, in pronation (From MacConaill, M. A., and Basmajian, J. V., 1969. *Muscles and Movements—A Basis for Human Kinesiology.* Baltimore: The Williams & Wilkins Company.)

Figure 1.37. Pronation and supination. right hand is pronated; left hand supinated.

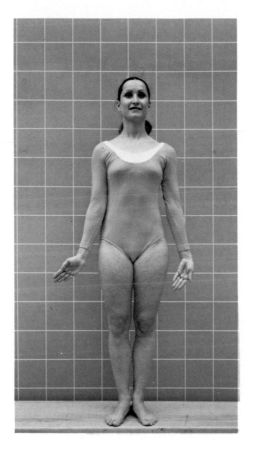

Figure 1.38. Deviation of hand on forearm. Subject exhibits right radial and left ulnar deviations of hand on forearm.

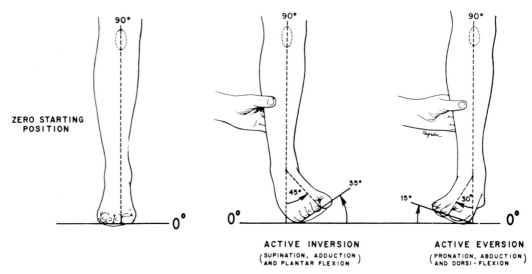

Figure 1.39. Movements of inversion and eversion of the foot. (From American Academy of Orthopaedic Surgeons, 1965. *Measuring and Recording Joint Motion.*)

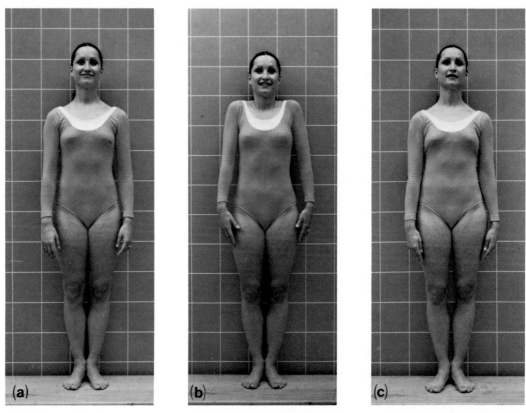

Figure 1.40. Movements of the shoulder girdle at the sternoclavicular joint (frontal views). (a), neutral; (b), elevation; (c), depression.

eversion. Compromise axes for these movements are indicated in Figure 1.15.

The first of these movements, inversion, has been defined as raising the medial border of the foot and/or rotating the sole of the foot inward. Eversion is movement in the opposite direction and involves raising the lateral border of the foot and/or rotating the sole of the foot outward. Walking over rough, uneven, or sloping terrain would be difficult, if not impossible, without the availability of these movements.

MOVEMENTS OF THE SHOULDER GIRDLE

The shoulder and pelvic girdles are comparable in that they both provide support and attachment for the limbs. While the innominate bones of the pelvis are anchored securely to the axial skeleton at the sacrum, the shoulder girdle is tied to the axial skeleton only at the two small sternoclavicular joints. The sacroiliac joint is so firmly bound together by fibrous ligaments that only a negligible amount of movement is possible, while the sternoclavicular joints allow a limited range of movement around three different axes passing through the proximal end of the clavicle (Fig. 1.8). These movements discussed below are elevation-depression, protraction-retraction, inward and outward rotation, and circumduction.

Elevation-Depression

These movements, illustrated in Figure 1.40, occur around a transverse axis slightly oblique to the sagittal plane (T in

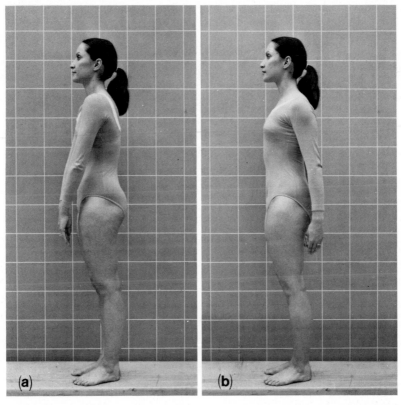

Figure 1.41. Movements of the shoulder girdle at the sternoclavicular joint (sagittal views). (a), protraction (flexion); (b), retraction (extension).

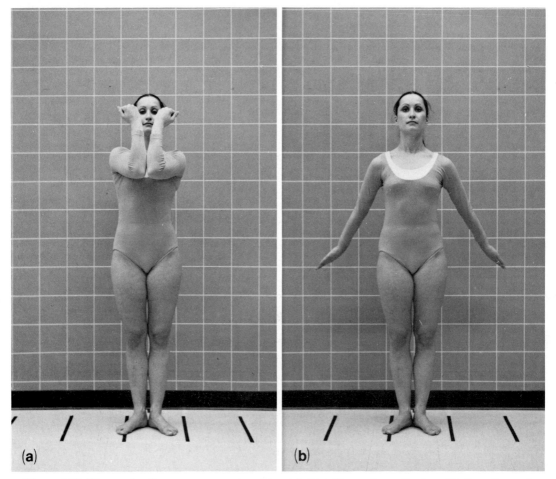

Figure 1.42. Diagonal patterns—upper extremities. (a), first diagonal arm flexion; (b), first diagonal arm extension;

Fig. 1.8). The total angular excursion of the clavicle from maximal depression to maximal elevation has been estimated at 60° (Steindler (1973)).

Protraction-Retraction (Flexion-Extension)

These movements shown in Figure 1.41 take place around a vertical axis through the proximal (sternal) end of the clavicle (V in Fig. 1.8). Anatomists prefer the terms protraction for a forward movement of the shoulders and retraction for a

backward movement, although the flexion-extension terminology selected by the orthopaedic surgeons meets the criteria presented earlier for these movements (see under "Flexion-Extension").

Rotation

Rotation occurs around the long axis L in Figure 1.8. Inward (or upward) rotation of the clavicle around this axis is largely responsible for the movement of the glenoid fossa forward during shoulder flexion, while outward (or downward) ro-

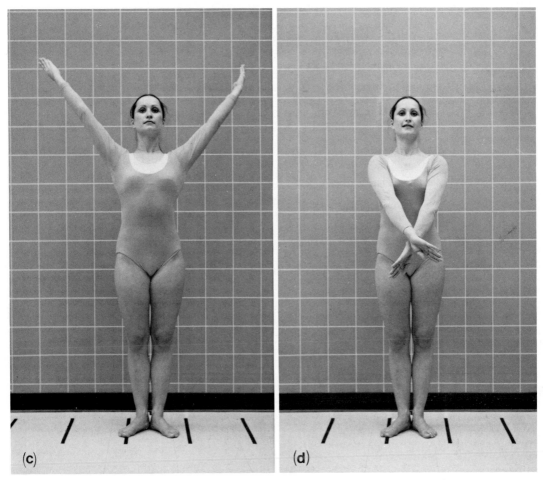

Figure 1.42. Diagonal patterns—upper extremities (c), second diagonal arm flexion; (d), second diagonal arm extension.

tation moves the glenoid downward and backward during extension of the humerus at the shoulder.

Circumduction

When the upper extremity is circumducted the range of the movement is increased by circumduction of the clavicle.

Acromioclavicular Movements

While the range of movement at this joint is limited in comparison to many others, the joint is multiaxial and it does have three degrees of freedom. It is this freedom that enables the scapula to move on the clavicle so that the glenoid fossa can be further rotated upward and sideward, upward and forward, and downward, thus making possible the extreme ranges of motion of the upper extremity.

Movements of the Scapula

The movements described above of the clavicle on the sternum are caused not as much by muscles acting directly on the clavicle as they are by muscles which act

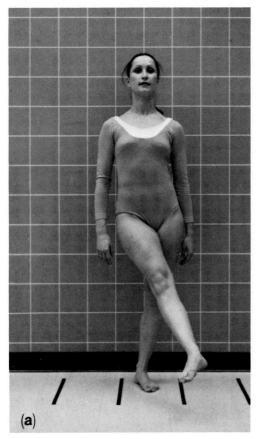 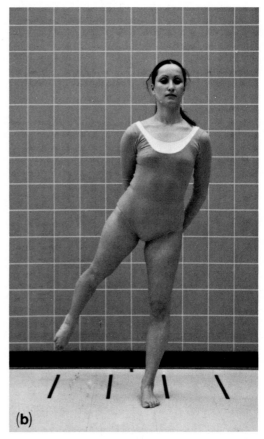

Figure 1.43. Diagonal patterns—lower extremities. (a), first diagonal leg flexion; (b), first diagonal leg extension;

indirectly by holding and moving the scapula on the thorax. For this reason many authorities prefer to omit discussion of sternoclavicular actions and simply to describe the observed behavior of the scapula in anatomical terms as follows.

Elevation moves the scapula higher on the rib cage, while depression moves it downward.

Abduction of the scapula is movement away from the vertebral column, while adduction of the scapula draws it closer to the spine.

Rotation upward: the inferior angle of the scapula moves laterad and upward while the glenoid fossa also turns upward.

Rotation downward: the inferior angle moves mediad as the glenoid fossa shifts slightly laterad and faces downward.

MOVEMENTS OCCURRING AT MORE THAN ONE JOINT

Diagonal Patterns

Seldom, if ever, does the human being isolate movement to one joint unless it be flexion/extension at one of the distal phalangeal joints in the upper or lower extremity. All of us are aware of the variety of multiple joint movement combinations which can occur at any one instant in time. Some of the multiple combinations have captured the interests of therapists because of the predictable patterns which seem to be assured when a particular joint movement is involved. These have been identified by Kabat, Knott and Voss as the diagonal of movement (Knott and Voss

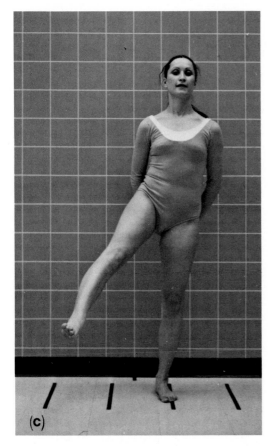
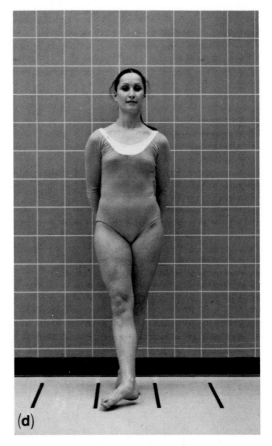

Figure 1.43. Diagonal patterns—lower extremities (c), second diagonal leg flexion; (d), second diagonal leg extension.

(1968)). They are included in this chapter because their existence is directly associated with the anatomical arrangements and functional abilities of the musculoskeletal system.

The alignment of muscle associated with shoulder and hip actions is rotatory and diagonal in direction suggesting that seldom, if ever, are joint movements performed in the pure cardinal planes of the body.

Two diagonal patterns of motion have been identified. The diagonal patterns for the extremities are described according to the three degrees of freedom, hence the three components of motion possible at the proximal joints—the shoulder and the hip. Each pattern is defined by one of the two movements associated with each of the three components; thus, a pattern includes a component of flexion or extension for sagittal plane movement, abduction or adduction for frontal plane movement, and medial or lateral rotation for transverse plane movement.

The intermediate joints—the elbow and knee—each having essentially one degree of freedom, may be in a position of flexion or extension. The distal joints are in a position consistent with the proximal joints regardless of the action at the intermediate joints.

The first diagonal pattern is the same for both the upper and lower extremities. The proximal joint—shoulder or hip—is flexed, adducted, and laterally rotated (Figs. 1.42(a) 1.43(a)). The intermediate joints are either flexed or extended. The

Figure 1.44. Skilled patterns of human movement. (a), gymnastic forward needle: one leg balance with left upper extremity supporting left lower extremity. (b), gymnastic gran battement: one-leg kick to momentary balance with left lower extremity unsupported.

distal joints show patterns similar to the proximal joints. The reciprocal actions at the shoulder and hip are extension, abduction, and medial rotation (Figs. 1.42(b) 1.43(b)).

The second diagonal pattern differs between upper and lower extremities. The upper extremity second diagonal pattern is composed of shoulder flexion, abduction, and lateral rotation with consistent patterning at the distal joints (Fig. 1.42(c)). The reciprocal pattern consists of shoulder extension, adduction, and medial rotation (Fig. 1.42(d)). The elbow is flexed or extended.

The second diagonal pattern for the lower extremity consists of hip flexion, abduction, and medial rotation (Fig. 1.43(c)). The reversed actions are extension, adduction, and lateral rotation (Fig. 1.43(d)). Again, the distal joint actions are similar in pattern to those of the proximal joints and the knee is either flexed or extended.

Skilled Patterns of Movement

After the student has equipped himself (herself) with the necessary verbiage for describing human movement, he is prepared to communicate with others even when pictures are not available. Figure 1.44 depicts a skilled gymnast in two different poses of a one-leg balance. With the exception of the right upper extremity, which was used to brace the body mechanically as the still picture was taken, all major joint positions are visible. Assuming that the movement started from the anatomical position in each of the two balances depicted, the reader is invited to describe all of the major joint actions and to make note of the planes of the movements, even when they deviate from the cardinal planes of the body. When comparing Fig. 1.44, (a) and (b), particular attention should be given to the position of the left upper extremity, the amplitude of the left lower extremity movement, and the position of the support of the right lower extremity. Some of the features which distinguish this gymnast's style are also noticeable, particularly in the more dynamic pose depicted in Fig. 1.44(b). One might also note whether any of the diagonal patterns are observable. The authors' description of the movements depicted in these illustrations should be consulted after the reader has attempted to identify all movements, grouping them according to the joints at which the actions occur; planes of movements and/or positions of body parts with respect to planes are also listed (see Appendix 1). After the reader

has experienced movement descriptions of this type, he (she) should be prepared to communicate movement completely and accurately to another person even without benefit of the visual picture. While the verbal experience might not be as aesthetically pleasing, the skill of movement description is essential to the human movement specialist.

THE CHALLENGE TO FOLLOW

The skeletal system, its series of links, and their interactions with each other have been discussed. Common descriptors have been provided as a kind of motor alphabet to ease the task of communication about human movement. Still photographs have been used to illustrate the movement descriptions and readers have been urged to imagine movement occurring before and following the instants in time when the photographs were taken. It would appear that movement description is easily rendered if the nomenclature is studiously followed. Just how easy remains to be seen in ensuing chapters where still photographs are often not provided and the reader's attention is directed to motion and complex movement skills.

CHAPTER 2

Motion

INTRODUCTION

In Chapter 1 the skeletal system was highlighted and a wide variety of static body positions have been indicated. To achieve these positions, muscular forces have invariably had to act upon the skeletal system. At this juncture, however, we are not yet concerned with such forces. Rather, we are concerned with describing the time sequences of events which would be noted by an external observer of body movements. Recordings made by such an observer may or may not be aided by technological devices such as photographic cameras, video recorders, or even more sophisticated devices which might be interfaced directly with a digital computer. To facilitate our descriptions it is of value to classify the various basic types of motion of relevance and to define important terms including velocity and acceleration. Following on these aspects, we introduce facets of applied mathematics which are generally helpful with regard to human movement studies.

TYPES OF MOTION

Motion *per se* always involves a changing of place or position of a body.

This change can be described by any one or more of four different classes of motion:

1. Translatory motion can be most simply stated as movement of a body from one place to another. Engineers specify that in translatory motion all points in a body describe parallel lines which may be either curved or straight, and that all points travel in the same direction at the same velocity or acceleration (Timoshenko and Young (1956); Pletta and Frederick (1964); LeVeau (1977)). Translatory motion may also involve one or more of the other three specific types of motion described as follows.

2. Linear or rectilinear motion occurs when all parts of a body move the same distance in the same direction in the same amount of time. It is always a translatory form of motion.

3. Angular or rotatory motion involves movement of a body about a fixed point so that all parts of the body travel in arcs as they move through the same angular displacement.

4. Curvilinear motion describes the movement of a body along a curved path such as the trajectory of a projectile.

Some Examples

From the foregoing it can be seen that translatory motion of a body part must

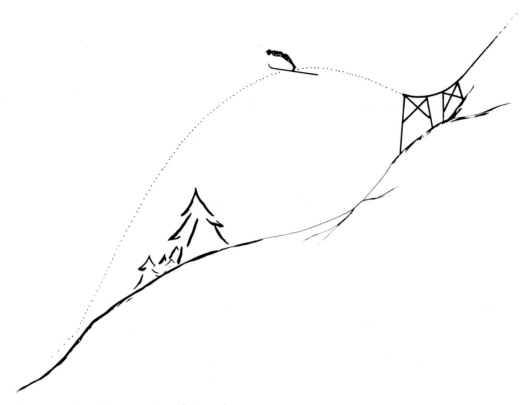

Figure 2.1. Curvilinear motion: ski jumping.

involve movement at more than one joint as the parallel lines must be straight or curved rather than circular. Thus, reciprocating actions of three or more segments at two or more joints can displace the distal end of the chain in translatory motion. (The reader is invited to try moving a finger along a straight or curved line of any length without motion in the wrist, forearm, elbow, or shoulder.)

A passenger in an automobile traveling along a straight highway is undergoing linear motion. When the highway curves, the motion of the car and passenger becomes curvilinear. A shussing skier, or one sliding down a track of fixed slope before taking off from a ski jump, would describe linear motions.

When the skier swerves to avoid a tree or go through a slalom gate there is curvilinear motion of at least some parts of his body (some movement will be rotatory as he shifts weight and swings arms and poles). On the other hand, when the skier makes his leap up and out from the jump track and assumes his forward lean he is undergoing essentially curvilinear motion as he travels outward and down to his landing on the slope below the jump (Fig. 2.1).

Human locomotion is translatory motion which has attributes of rotatory, linear and curvilinear motions. During the swing phase of the walking gait depicted in Figure 2.2, a lower extremity is swung forward, moving about an axis in the head of the femur (proximal end). At heel contact, that foot becomes weight-bearing as the body rides over it during the support or stance phase. During this portion of the step there is movement at a number of axes at the distal end of the lower extremity (in the toes, at the metatarsophalangeal joints, and at the ankle) as the thigh and

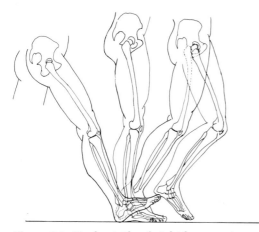

Figure 2.2. Single stride of right lower extremity: walking.

leg carry the rest of the body forward while they rotate over these lower axes. As the lower limbs are undergoing rotatory motions at alternating ends, the trunk is progressing curvilinearly. This latter motion is particularly clearly exhibited in the sagittal plane.

To appreciate the different types of motions and their areas of applicability in human movement it is necessary to become familiar with the fundamental principles involved, as well as with the anatomical and physiological factors presented in this text.

MECHANICS OF MOTION

Vectors and Scalars

Important in the description of motion characteristics are vectors which are defined to have both magnitude and direction. A scalar quantity is one which has magnitude only. Thus, a vector is described by two scalars: one reflecting magnitude, the other direction.

Scalar quantities can be added arithmetically and include such quantities as distance, mass, volume, area, time, temperature, energy, and speed. Speed *per se* has no direction and is the magnitude of a vector quantity, velocity.

Vector quantities, since they possess both magnitude and direction, can be represented graphically by segments of straight lines. The length of the line segment can be scaled to represent magnitude; the orientation of the line together with an arrowhead specify direction. Such quantities include *velocity* (where speed is the magnitude and direction is known), *displacement* from one point to another (*e.g.,* from point A to point B, from Boston to New York, etc.), and *force,* which might be a push, or a pull, or a weight.

The following discussion, illustrated by an example, will serve to distinguish some important differences in the properties of scalars and vectors.

If an individual (a) drives 5 miles, (b) walks 2 miles, and (c) rides a bicycle 3.3 miles, he has traveled 10.3 miles. These distances have magnitude only and so are scalar quantities and can be added as above. We have no idea where the traveler found himself at the end of his jaunt as no *directions* of travel were given. If we are

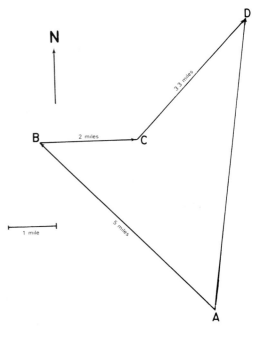

Figure 2.3. Vector diagram of distances traveled: $\vec{AD} = \vec{AB} + \vec{BC} + \vec{CD}$.

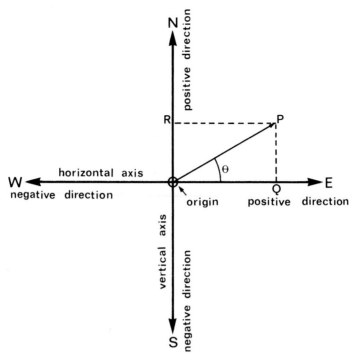

Figure 2.4. Vector components: OQ and OR are, respectively, the horizontal and vertical components of OP.

$$OQ = OP \cos \theta$$

$$OR = OP \sin \theta$$

Note:

$$OP^2 = OQ^2 + OR^2$$

$$OP = \sqrt{OQ^2 + OR^2}$$

$$\tan \theta = PQ/OQ = OR/OQ$$

given the directions for each of the above magnitudes we are dealing with vector quantities, *i.e.*, displacement, and we can determine where the traveler ended his tour.

Given he traveled (a) 5 miles northwest, (b) 2 miles east, and (c) 3.3 miles northeast: vectors can be added graphically as in Figure 2.3 by laying out vectors to scale and in proper directions. In this figure, vector $\vec{AB}$ represents the 5 miles northwest, $\vec{BC}$ the 2 miles east, and $\vec{CD}$ the 3.3 miles northeast. Note that the arrows

above $\vec{AB}$, $\vec{BC}$, and $\vec{CD}$ indicate that these are vector quantities. For $\vec{AB}$, the length AB indicates the magnitude of the vector which is directed from A to B. By measuring the vector $\vec{AD}$ we find that the traveler has ended his trip 5.9 miles north-northeast of his starting point. By drawing vector $\vec{AD}$ we have closed the polygon of displacements to produce the resultant of the summation of vectors $\vec{AB}$, $\vec{BC}$, and $\vec{CD}$.

A graphical solution for the summation of distance vectors has been provided.

It is also possible to solve this type of problem using trigonometry. In this, each vector is resolved into its horizontal and vertical components with regard to a set of mutually perpendicular reference axes. In choosing the horizontal axis to be positive from west to east and the vertical axis to be positive from south to north, the requisite components can be resolved by recognizing that horizontal components are the product of the vector magnitude and the cosine of the angle between the vector and the horizontal axis, and vertical components are the product of the vector magnitude and the sine of the angle between the vector and the horizontal axis (Fig. 2.4).

Each vector and its respective components can be tabulated as follows:

Vector	Magnitude (miles)	Direction θ*	Cos θ	Horizontal Component	Sin θ	Vertical Component
$\vec{AB}$	5	135°	−0.70711	−3.54	+0.70711	+3.54
$\vec{BC}$	2	0°	1	+2.0	0	0.0
$\vec{CD}$	3.3	45°	+0.70711	+2.33	+0.70711	+2.33

Sum of horizontal components = 0.79
Sum of vertical components = 5.87
Resultant = $\sqrt{0.79^2 + 5.87^2}$
$\quad\quad\quad = \sqrt{34.4569 + 0.6421}$
$\quad\quad\quad = \sqrt{35.099} = 5.92$ miles

The angular disposition, θ_R, of the resultant can be determined since

$$\tan \theta_R = \frac{5.87}{0.79} = 7.4304$$

$$\theta_R = \arctan 7.4304 = 82°\ 20'$$

The traveler ended his tour 5.92 miles and 82° 20′ north of east of his starting point.

Vector $\vec{AD}$ can be completely specified by writing

$$\vec{AD} = 5.92 \text{ miles} \lfloor 82°\ 20'$$

where 5.92 miles is the magnitude of $\vec{V}_{VA}$ and $\lfloor 82°\ 20'$ signifies that the vector is disposed at 82° 20′ with respect to the chosen horizontal reference axis.

---

* Angle with respect to horizontal axis.

To solve this kind of problem the arithmetic can be greatly facilitated by using a hand-held calculator having trigonometric capabilities. In the absence of the latter capabilities, resort must be made to pertinent trigonometric tables.

Linear Motion

Velocity

Suppose that in the foregoing example the 5 miles were driven in 10 minutes, the 2-mile walk took 40 minutes, and the bicycle ride took 10 minutes, so that the time elapsed was 1 hour. Cognizance must now be taken of time which is a scalar quantity.

The velocity in traversing BC (Fig. 2.3) follows from the definition of velocity:

$$\vec{V}_{BC} = \vec{BC}/t \quad\quad\quad (2.1)$$

where $\vec{BC}$ is the displacement of magnitude, 2 miles, that occurs in an easterly direction in a time t = 40 minutes. This displacement occurs at a speed of $(2 \times 60)/40$ = 3 mph also in an easterly direction. Thus, velocity $\vec{V}_{BC}$ is completely specified in magnitude and direction. Its magnitude is the scalar, speed, and its direction follows from the displacement vector of concern (BC). Its units (e.g., mph) reflect a time rate of change of displacement. As defined in Equation 2.1, velocity is the quotient of a vector quantity and a scalar quantity and thus retains the vector property.

The average velocities of each of the three aforementioned modes of transport are $\vec{V}_{AB}$ = 30 mph $\lfloor 135°$, $\vec{V}_{BC}$ = 3 mph $\lfloor 0°$, and $\vec{V}_{CD}$ = 20 mph $\lfloor 45°$. The average velocity for the complete trip $\vec{V}_{AD}$ = 5.92 mph $\lfloor 82°\ 20'$.

Velocity as a vector is always in a

straight line. The example given placed each velocity in its own line but, as each was in a different direction, they could only be added by using the quantitive displacement as calculated by vector summation.

When all velocities (or any vectors) are in the same direction or straight line they can simply be added arithmetically.

In human movement studies the velocity units usually used are feet per second (ft/sec) and meters or centimeters per second (m/sec or cm/sec). These units are more practical in such studies.

Average and Instantaneous Velocities

In the foregoing discussion the *average velocities* for each segment of the trip were calculated. Only the average value could be obtained, since the details of each segment from instant to instant were not given. Thus, we might speculate on how the trip might have been made. For example, the 5 miles driven in 10 minutes could have been done by driving steadily at 30 mph all the way, or at 45 mph for 6 minutes and at 7.5 mph for the remaining 4 minutes. Clearly there are an infinite number of possibilities unless we know the record of velocities from instant to instant.

The instantaneous velocity-time graph for a jogger might appear as in Figure 2.5.

The jogger spends the first minute getting up (accelerating) to the steady velocity level of 10 km/hr. He maintains this for 6 minutes, then slows down over a period of a minute (decelerates) to a standstill when he rests for 2 minutes and then repeats the previous performance. The average velocity for the 18 minutes recorded is obtained by taking the area under the instantaneous velocity-time graph and dividing by the base-line length of 18 minutes, i.e.

$$v_{av} = [(\tfrac{1}{2} \times 1 \times 10) + (6 \times 10)$$
$$+ (\tfrac{1}{2} \times 1 \times 10)$$
$$+ (2 \times 0) + (\tfrac{1}{2} \times 1 \times 10)$$
$$+ (6 \times 10)$$
$$+ (\tfrac{1}{2} \times 1 \times 10)]/18$$
$$= 140/18 = 7.77 \text{ km/hr.}$$

So, the average velocity is 77.7% of the maximum steady velocity. This value

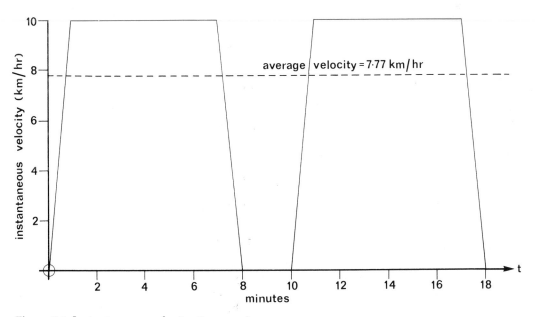

Figure 2.5. Instantaneous velocity-time graph.

would be reduced further by a longer rest period, slower rises to, and rapid declines from, the steady level.

Acceleration

Accleration is defined as the time rate of change of velocity. It should be borne in mind that velocity depends upon both speed and direction and therefore acceleration could be affected by changes in either or both of these parameters.

If the direction of the velocity vector is maintained fixed, then acceleration

$$a = \frac{\text{change in velocity}}{\text{time taken for change}} \quad (2.2)$$

Let the velocity of a body moving along a straight line be v_1 m/sec at a time t_1 sec and consider that the velocity undergoes a change to velocity v_2 m/sec in a time interval Δt sec such that $\Delta t = t_2 - t_1$. Then

$$a = \frac{(v_2 - v_1)}{(t_2 - t_1)} \frac{\text{m/sec}}{\text{sec}} \quad (2.3)$$

The dimensions of a in this case are (m/sec)/sec which can be written m/sec/sec, m/sec^2 or $m \cdot sec^{-2}$.

Let

$$\Delta v = v_2 - v_1 \quad (2.4)$$

and

$$\Delta t = t_2 - t_1$$

where the prefix Δ (Greek letter delta) is read as "the change in"

then

$$a = \frac{\Delta v}{\Delta t} \quad (2.5)$$

If v_2 exceeds v_1 for positive values of t, a will be positive; if v_1 exceeds v_2 for positive values of t, v and hence a will be negative. A *negative acceleration is deceleration*.

Turning to the example illustrated in Figure 2.5, it will be seen that the jogger has uniform accelerations and decelerations of 0.167 km/min^2, i.e., 0.0464 m/sec^2.

An Important Case

In cases of constant acceleration in a straight line (the situation encountered by bodies acted upon by gravity) the velocity after any elapsed time t is given by

$$v = v_0 + at. \quad (2.6)$$

where v_0 is the initial velocity or the velocity at $t = 0$, and a is the acceleration.

It should be evident that for acceleration in a straight line, the velocity will be directed along that same line as will the displacement, Displacement s can be calculated by realizing that

$$s = s_0 + vt \quad (2.7)$$

where s_0 is the displacement measured from the origin at time $t = 0$, and v is the velocity that obtains for time t.

The following example illustrated by a graphical solution shows the determination of displacement as a function of time from a knowledge of acceleration. Reference should be made to Figure 2.6 in which there are graphs (top to bottom) of acceleration, velocity, and displacement. Given that acceleration a is maintained constant for a body initially at rest (i.e., $v = 0$) at the origin (i.e., $s = 0$) at time $t = 0$, the graph of v against time can be constructed by recalling Equation 2.6. Since $v_0 = 0$, $v = at$. Substituting values of a (2) for the range of values of t of concern leads to a number of discrete points through which the straight line can be drawn as shown. As expected, for a constant acceleration the velocity of a body increases linearly.

From Equation 2.7, $s = s_0 + vt$. Initially, $s = 0$ (i.e., $s = 0$ for $t = 0$) and therefore $s = vt$.

For $t = 1$, $v = 2$ and the average speed in the time interval $t = 0$ to $t = 1$ is $v_{av} = 1$. Thus, in this interval $s = 1 \times 1 = 1$. The point (A) can therefore be plotted. In the interval between $t = 1$ and $t = 2$ we expect s to have increased by twice as much as in the preceding 1-unit interval. This increase must, according to Equation 2.7, be added to the displacement or distance already traversed. The average velocity in the in-

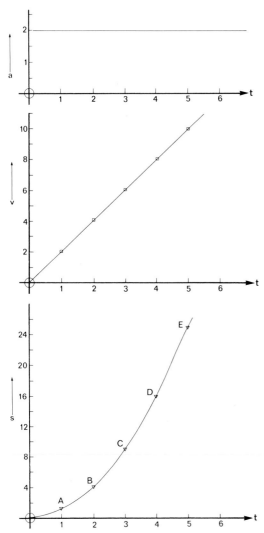

ing curve turns out to be a parabola as shown. The equation of the parobola is

$$s = t^2$$

If a had initially been chosen to be 1 unit instead of 2 all of the s values would be reduced by a factor of 2. If a had been doubled, all of the values of s would be doubled. Thus, we can write s in terms of a and t as follows

$$s = \tfrac{1}{2} at^2 \qquad (2.8)$$

For a body with initial velocity v_0 at $t = 0$, subjected to an acceleration a thereafter,

$$s = v_0 + \tfrac{1}{2} at^2 \qquad (2.9)$$

The simplest example of constant acceleration for a body is that due to the gravitational pull of the earth which accelerates a freely falling body at $g = 9.81$ m/sec^2 (32.2 ft/sec^2). The direction of g is always vertically downward (being directed toward the center of the earth).

Rotational Movement

Rotation of a body about a fixed point or axis causes any portion of that body to travel in a circular path as it undergoes angular displacement. As this occurs when any body segment moves on another (rotatory motion), it behooves the kinesiologist to have some understanding of the laws governing such action.

Considerations relating to circular motion are of fundamental importance. In what follows concern is for the angular movements, velocities, and accelerations relating to a body moving in a circular path, whether it is a ball on a string or a space capsule in orbit. Results indicated can also be applied to the movement of a distal end of a body kinematic chain moving in an arc about an axis through a proximal joint.

Angular Displacement and Velocity

Imagine a weightless object OP pivoted at and rotating about O (Fig. 2.7). O

Figure 2.6. Acceleration, a, velocity, v, and displacement, s, as functions of time t for a body initially at rest at the origin of measurement.

terval of concern is 3 units and the distance traversed in the interval is 3 units. The accumulated distance is therefore 4 units corresponding to point (B).

Points (C), (D), and (E) follow in a similar fashion. If a smooth graph is drawn through these points (and others which can be constructed as desired), the result-

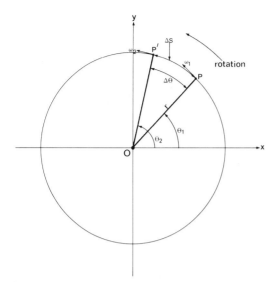

Figure 2.7. Circular motion:

$$OP = r$$

$$PP' = \Delta s$$

is the origin of the horizontal (X) and vertical (Y) axes shown. At time t_1 sec OP makes an angle θ_1 radians with the X-axis and at t_2 sec this angle is θ_2 radians.

Now $\Delta\theta = \theta_2 - \theta_1$ is the change in angular displacement and $\Delta t = t_2 - t_1$ is the time over which that change takes place.

Angular velocity ω $\qquad\qquad$ (2.10)
$$= \frac{\Delta\theta}{\Delta t} \text{ (radians/sec)}$$

and the velocity is directed tangentially at the instantaneous position occupied by P.

If the value of ω is constant the angular displacement in a time t is $\theta = \omega t$. This angular displacement corresponds to a linear distance s traversed by P.

It is well-known that

$$s = r\theta \qquad\qquad (2.11)$$

where r is the radius of the movement. (For a complete traverse of the circle, $\theta = 2\pi$ radians, and $s = 2\pi r$.)

It follows from Equation 2.11 that since r is a constant

$$\Delta s = r\Delta\theta \qquad\qquad (2.11(a))$$

Dividing both sides by Δt

$$\frac{\Delta s}{\Delta t} = r\frac{\Delta\theta}{\Delta t}$$

recalling Equation 2.10 and realizing that $v = \Delta s/\Delta t$ leads to

$$v = \omega r \qquad\qquad (2.12)$$

(It will be helpful to remember that 2π radians $= 360°$.)

Angular Acceleration

As indicated in Figure 2.7 the angular velocity at any point on the circular path is in a straight line tangential to the path (perpendicular to the radius). While the magnitude of ω may be fixed, its line of action changes from moment to moment and position to position. Angular acceleration α is defined by

$$\alpha = \frac{\omega_2 - \omega_1}{t_2 - t_1} = \frac{\Delta\omega}{\Delta t} \qquad (2.13)$$

(This result is analogous to that for linear acceleration (see Equations 2.3 and 2.5).)

Radial Acceleration

Figure 2.8 shows a graphic method for deriving instantaneous acceleration in magnitude and direction when a body OP is rotated at constant angular velocity ω in a clockwise direction.

OP is the initial position occupied by the body and OP' its final position. The line PY is constructed such that PY is perpendicular to OP and therefore in the direction of the instantaneous ω. Similarly line $XP'Z$ is constructed tangential to OP' at P' to represent the direction of ω at P'. X is the intercept with PY; XY is made the magnitude of ω to a suitably chosen scale, and likewise for XZ.

Thus, XY represents $v_1 = \omega r$ and XZ represents

$$v_2 = \omega r$$

$$v_2 - v_1 = \vec{XZ} - \vec{XY} = \vec{XZ} + \vec{YX} = \vec{YZ}$$

It will be seen that YZ is parallel to OX.

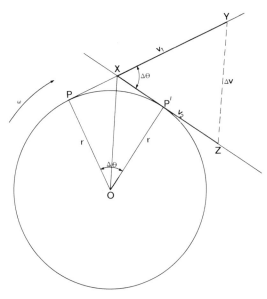

Figure 2.8. Graphical determination of radial acceleration.

Thus, in the angle $\Delta\theta$ which occurs in time $(\Delta\theta/\omega)$, v is directed radially toward the origin. This renders the direction for an acceleration component a_R, its magnitude is

$$a_R = \omega^2 r \qquad (2.14)$$

(This follows by considering that $\Delta v = \omega r \Delta\theta$ (refer to Fig. 2.8 and Equations 2.11 (a) and 2.12). In Figure 2.8

$$\Delta v = XY.$$

$$\Delta v = \omega r \Delta\theta$$

Since

$$a_R = \frac{\Delta v}{\Delta t},$$

which follows from the definition for linear acceleration, and

$$\Delta t = \frac{\Delta\theta}{\omega},$$

then by substituting in the expression for a_R, Δv and Δt just described, it follows that

$$a_R = \frac{\omega r \Delta\theta}{\Delta\theta} \cdot \omega = \omega^2 r.)$$

Since

$$v = \omega r$$

$$a_R = \frac{v^2}{r} \qquad (2.15)$$

a_R is a radial or *centripetal* acceleration.

Tangential Acceleration

When motion in a circular path is not constant, the velocity vectors will have different lengths, as in Figure 2.9 (a). If the vectors of v_1 and v_2 are drawn as before (Fig. 2.9 (b)) with the angle $\Delta\theta$ between them, the vector Δv represents the resultant change in velocity. In Figure 2.9 (c) this change of velocity has been resolved into two components, the change in radial velocity Δv_R which results from the *change in direction*, and the change in tangential velocity Δv_T which results from the *change in the magnitude* of the velocity. The tangential acceleration a_T is:

$$a_T = \frac{\Delta v_T}{\Delta t} \qquad (2.16)$$

where tangential acceleration is equal to the change in tangential velocity divided by the change in time, and it will be in the direction of the changing velocity (Fig. 2.9 (d)).

As a_R and a_T form two sides of a right triangle, when $\Delta\theta$ is very small (practically zero) the actual resultant acceleration is:

$$a = \sqrt{a_T^2 + a_R^2} \qquad (2.17)$$

EXAMPLE—Rotational Velocity and Acceleration Components

A figure skater initially at rest uniformly increases her rotational speed about a vertical axis until after 15 seconds she is spinning at 300 rpm. Find the average angular acceleration of the skater and the final linear velocity and linear acceleration of her shoulder tip if it is 20 cm from the axis of spin. Then discuss the

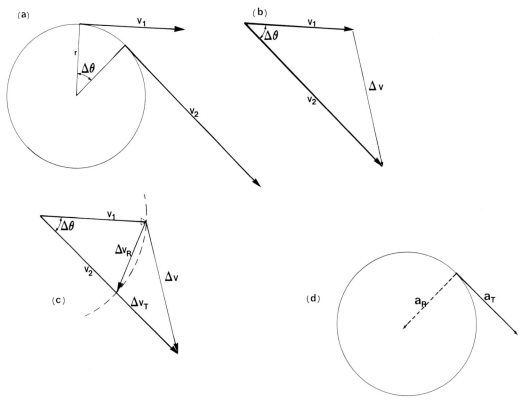

Figure 2.9. Graphic method for deriving instantaneous acceleration; (a) and (b), tangential acceleration; (c) and (d) illustrate the directions of a_R and a_T.

relative values of radial and tangential acceleration components as a function of speed.

A speed of 300 rpm corresponds to

$$2\pi \times 300/60 \text{ rad/sec} = 10\pi \text{ rad/sec}$$

Recalling Equation 2.13:

$$\alpha_{av} = (\omega_2 - \omega_1)/\Delta t = (10\pi - 0)/15$$

$$= 2\pi/3 \text{ rad/sec}^2 = 2.105 \text{ rad/sec}^2$$

From Equation 2.12 the instantaneous linear velocity at the shoulder has a magnitude

$$v = \omega r = 10\pi \times 20 = 200\pi \text{ cm/sec}$$

$$= 628.2 \text{ cm/sec}$$

It is directed at right angles to the line drawn from shoulder to shoulder, and in the direction of spinning.

The centripetal acceleration a_R follows from Equation 2.15:

$$a_R = v^2/r = \omega^2 r$$
$$= 100\pi^2 \cdot 20 = 2000\pi^2$$
$$= 19739.3 \text{ cm/sec}^2$$

The tangential acceleration a_T is 0 since a constant angular velocity is maintained when the final spinning speed is reached.

In progressing from rest and accelerating to 300 rpm there will, of course, be both radial and tangential acceleration components. At 150 rpm, the radial component

$$a_R = 19739.3/4 = 4934.83 \text{ cm/sec}^2$$

(since a_R depends on v^2).

To find a_T, recourse must be made to

Equation 2.16 and Figure 2.9. At 150 rpm

$$\omega_1 = 5\pi \text{ rad/sec}$$

At a time Δt later, a new angular speed ω_2 will be reached because of the average angular acceleration $\dfrac{2\pi}{3}$ rad/sec

$$\omega_2 = 5 + \frac{2\pi}{3} \Delta t$$

$$\omega_2 - \omega_1 = \Delta \omega$$

i.e.

$$\Delta \omega = \frac{2\pi}{3} \Delta t$$

$$\Delta v_T = \Delta \omega = \frac{2\pi r}{3} \Delta t$$

and

$$a_T = \frac{\Delta v_T}{\Delta t} = \frac{2\pi r}{3} = \frac{2\pi \times 20}{3}$$

$$= 41.95 \text{ cm/sec}^2$$

On reviewing the steps to acquire this result, it will be seen that a_T is independent of ω, being constant at 41.95 cm/sec² until the steady speed of 300 rpm is reached when it drops to 0.

$$a_R = \omega^2 r = (2\pi n)^2 .20 = 80\pi^2 n \text{ cm/sec}^2$$

(where n is the number of rps). For all but very low speeds, less than about ½₀ rps, a_R is much greater than a_T.

Analogies between Equations for Linear and Rotational Quantities

The reader should note the similarities between the two sets of equations indicated in Table 2.1.

SOME NOTES ON CALCULUS

The differential and integral calculus provides a number of mathematical tools of value in kinesiology and biomechanics. In the following discussion the essence of

Table 2.1

Analogies between Equations for Linear and Rotational Quantities

Linear	Angular
$a = \dfrac{\Delta v}{\Delta t}$	$\alpha = \dfrac{\Delta \omega}{\Delta t}$
$v = v_0 + at$	$\omega = \omega_0 + \omega t$
$s = s_0 + vt$	$\theta = \theta_0 + \omega t$
$s = v_0 t + 1/2\, at^2$	$\theta = \omega_0 t + 1/2\, \alpha t^2$

differentiation and integration as useful tools will be provided.

Essentially, differentiation helps to distinguish special features of the variations of one function as a consequence of related variations which occur in a variable which influences the function of concern. For example, a distance-time graph (s vs t) for a runner would show the interrelationship between the parameters distance (s) and time (t). When the runner goes faster, more distance is covered in a selected time interval Δt. "Differentiating" the distance-time characteristic by examining the distances covered in the time interval Δt we can perceive changes in the runner's characteristics. In fact, what we are doing is to create a plot of the speed-time (v vs t) characteristic for a selected time epoch Δt. As Δt is made smaller and smaller for each (s, t) value on the graph, the closer we come to generating the actual instantaneous speed-time characteristic.

Integration is the reverse process of differentiation. Given the speed-time (v-t) relationship for a runner, the distance (s) traveled can be found by integrating (or *adding together*) all of the contributions of distance (s) traveled in each small increment of time Δt, which make up the time span of concern. The distance traveled in each increment is $\Delta s = v \times \Delta t$ (and v will take on a distinct value depending on t). Again, the smaller we make Δt for each (v, t) value on the graph depicting their relationship, the closer we come to generating the actual instantaneous distance-time characteristic.

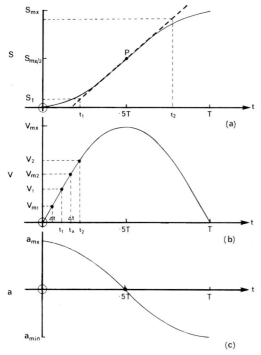

Figure 2.10. Differentiation and integration. Relationships between distance (s), speed (v), and acceleration (a) as functions of time (t).

By mathematical definition, as Δt is made to approach zero we obtain the derivative of s at point P, i.e.,

$$\frac{ds}{dt} = \frac{\Delta s}{\Delta t}, \ \Delta t \to 0$$

is the derivative of s with respect to t and we have differentiated the function s.

It will be recognized that speed

$$v = (s_2 - s_1)/(t_2 - t_1)$$

and thus at any point on the s-t curve

$$v = \frac{ds}{dt}$$

Thus, Figure 2.10 (b) which depicts v as a function of t can be generated. If we construct the closest approximating slope at each point on the s-t curve, the corresponding v values can be determined and plotted. Note that when the slope of s-t is 0 at $t = 0$ and $t = T$, so is $v = 0$.

By similar arguments to those just given it can be shown that $a = dv/dt$. Hence the curve of a against t (Fig. 2.10 (c)) can be derived. Again note that when $v = v_{mx}$ at $t = T/2$, the slope of v, i.e., $dv/dt = 0$ and so is $a = 0$.

Differentiation

Figure 2.10(a) shows a plot of s against t for a runner who runs for a total time T. Through point P corresponding to $T/2$ and $S_{mx}/2$, i.e. $(T/2, S_{mx}/2)$, a straight line (dotted) is drawn to approximate as closely as possible to the s-t curve. Proceeding further away from P either to the left or the right results in more deviation of the dotted line from the s-t curve. Clearly the closer one approaches P, the more precisely does the dotted line segment in the region of P represent the variation of s with t. The slope of the curve at P can be described by the ratio

$$(s_2 - s_1)/(t_2 - t_1).$$

Let

$$\Delta s = s_2 - s_1 \quad \text{and} \quad \Delta t = t_2 - t_1$$

then slope $= \Delta s/\Delta t$.

Integration

If we wish to determine the s-t curve from a knowledge of the v-t curve we would use the process of integration. Starting at $t = 0$ on the v-t curve and realizing that $s = 0$ we know that for $t = 0$, $s = 0$ on the s-t curve. For a time increment Δt commencing at $t = 0$, i.e., up to a time t_1 as shown in Figure 2.10 (b) the speed would have attained a value v_1, but the average speed during Δt would be the value nearly midway between 0 and v_1, i.e., $v_{m1} = v_1/2$.

The distance Δs that would be covered in Δt at an average speed v_{m1} would be $\Delta s_1 = v_{m1} \times \Delta t$. Likewise between t_1 and t_2 the distance $\Delta s_2 = v_{m2}\,\Delta t$.

Clearly, as t values about t_a (corresponding to v_{m2}) are made smaller and smaller, the values such as v_1 and v_2 bounding v_{m2} would more closely approximate v_{m2} until for $\Delta t = 0$ they would all be equal. The contribution to distance in these general circumstances would grow closer to

$$\Delta s = v \, \Delta t$$

where v is the speed value at a particular value of t.

(Note that this can be transposed and written $v = \Delta s/\Delta t$ and as $\Delta t \to 0$, $v = ds/dt$ as before.)

Summing all the Δs contributions as t proceeds we can write

$$\Sigma \Delta s = \Sigma v \cdot \Delta t$$

for $\Delta t \to 0$ we define the mathematical relationship

$$\Sigma \Delta s = s = \int_{t_i}^{t_f} v \, dt$$

where s is the total distance traveled in the interval t_i (initial time) to t_f (final time) arising from the integrated values of v for all of the time epochs Δt as $t \to 0$. $\int$ is the mathematical sign for integration and the range or limits of integration are often depicted with it as

$$\int_{t_i}^{t_f}$$

From a graphical standpoint, so long as we keep Δt small enough to ensure that $v = (v_1 + v_2)/2$, we can progressively increment s at each step in Δt by an amount $v\Delta t$. In this way, the s-t curve can be generated from the v-t curve. Likewise, the v-t curve can be generated from the a-t curve. Also, it is worth noting that

$$v = \int_{t_i}^{t_f} a \cdot dt$$

General Comments

In general any variables that are mathematically related can be differentiated or integrated with respect to each other, e.g., $y = f(x)$ can be differentiated to give

$$\frac{dy}{dx} = \frac{df(x)}{dx}$$

or integrated according to

$$\int_{x_i}^{x_f} y \, dx = \int_{x_i}^{x_f} f(x) \, dx$$

For many mathematical functions there are clear-cut relationships for determining differentials or integrals. Often for the purposes of the kinesiologist or biomechanician, graphical methods or numerical techniques implemented on computers, but fundamentally along the lines discussed in this section, are used.

CONCLUSION

The material presented in this chapter should provide the student with a background sufficient to pursue the substance presented in subsequent chapters. The student whose mathematical background is lacking may wish to consult some of the applied mathematics or physics texts indicated in the list of references at the end of the text.

CHAPTER 3

Fundamentals of Applied Mechanics

INTRODUCTION

This chapter presents a review of the fundamentals of applied mechanics necessary to understand and apply important concepts in describing relevant force, motion, and energy relationships in human movement.

FORCE

The one universal force, that of gravitation, was first defined by Sir Isaac Newton in the 17th century.

This **Law of Universal Gravitation** states that:

ANY TWO BODIES IN THE UNIVERSE HAVE A GRAVITATIONAL ATTRACTION FOR ONE ANOTHER. IF THEIR MASSES ARE m_1 AND m_2 AND THEIR DISTANCE APART, r, IS LARGE COMPARED WITH THE SIZE OF EITHER, THEN THE FORCE ON EITHER BODY POINTS DIRECTLY TOWARD THE OTHER BODY AND HAS THE MAGNITUDE

$$F = G\, m_1\, m_2/r^2 \qquad (3.1)$$

where F is the gravitational force, m_1 and m_2 the respective masses of the two bodies, r the distance between them, and G is a gravitational constant having the same value for all bodies. Evidently, force is a vector quantity since it has magnitude and direction. For an object on or close to a planet such as the earth the resultant force acting between the planetary mass and the object is directed to the center of the planet. Thus, at any place on earth, gravitational force is observed to be directed vertically downward.

The force described by Equation 3.1 depends on the masses m_1 and m_2. Mass is defined as a quantity of matter and as a measure of the quantity to which inertia is ascribed. It is related to volume V and density ρ of the mass by the relation

$$m = \rho\, V \qquad (3.2)$$

Weight, on the other hand, is defined as a force with which a body is attracted toward the earth (or planetary or lunar type of mass). Therefore, weight is a result of the gravitational attraction which draws people or objects on or near the surface of a planet toward its center.

An examination of the relationship between weight and mass follows logically after a consideration of Newton's Laws of Motion.

Newton's Laws of Motion

The bases for the modern study of motion were laid by Sir Isaac Newton in the 17th century when he formulated his three laws of motion. The early translation from the original Latin is in language which is somewhat archaic, but beautifully expressive:

First Law. "Every body persists in its state of rest or of uniform motion in a straight line unless it is compelled to change that state by forces impressed on it."

Second Law. "The change of motion is proportional to the motive power impressed, and is made in the direction of the right (straight) line in which the force is impressed."

Third Law. "To every action there is always opposed an equal reaction; or, the mutual actions of two bodies upon each other are always equal, and directed to contrary parts" (Resnick and Halliday (1960)).

During the years since Newton's lifetime, these statements have been worked over and reworded to make their meaning clear to each generation of students.

The first law is cited today as follows: A BODY REMAINS AT REST, OR IN A STATE OF UNIFORM MOTION IN A STRAIGHT LINE UNLESS ACTED ON BY AN APPLIED FORCE. Today this has a far more explicit meaning to the average individual than it did a few years ago before the space program. Apollo moon rockets remained at rest on their launching pads at Cape Kennedy until ignition of their rocket fuel. They continued to accelerate as long as the "burn" lasted and, when the rocket motors cut off, they had reached escape velocity. The command capsule continued *in the same straight line* until (1) it was acted on by a short burst of rocket fire for a course correction, and (2) it came into the pull of the gravitational force of the moon.

This law is also described as the **law of inertia** because it describes the quality of needing a force to change the state of rest or of motion of a body, and it implies as a consequence a resistance to such a change. The concept of inertia is sometimes applied interchangeably with that of mass. A body in space or on the moon will have the same mass or resistance to change in its state of movement or of rest, i.e., *inertia,* as it had on earth. It will take the same amount of force to put it in motion (accelerate it), to stop it moving, or to change its direction as it did on earth. The weightlessness of space does not change this, nor does the one-sixth gravity force on the surface of the moon relative to the earth force at its surface.

The only difference is the absence of or the decrease in the force pulling an object toward the center of a planet or satellite, gravity.

The first law leads into the second, which is concerned with force, mass, and acceleration. Actually, this law as Newton stated it used the term motion as the product of mass and velocity, mv (Resnik and Halliday (1960), Tricker and Tricker (1967)). Taking this law as Newton stated it, the equation is:

$$F = \frac{mv_2 - mv_1}{\Delta t} \qquad (3.3)$$

where m is the body mass, v_1 the velocity at time t_1, v_2 the velocity at time t_2, $\Delta t = t_2 - t_1$, and F is the vector sum of all the forces acting on the body. From Equation 3.3 it will be noted that since v_1 and v_2 are vectors, F must also be a vector. This equation can be treated algebraically to yield:

$$F = m \frac{(v_2 - v_1)}{(t_2 - t_1)} \qquad (3.4)$$

It will be recognized that

$$(v_2 - v_1)/(t_2 - t_1) = a$$

the acceleration in the interval Δt. Hence

$$F = ma \qquad (3.5)$$

From this there follows the more modern statement of *Newton's Second Law:* IF A

BODY OF MASS, m, HAS AN ACCELERATION, a, THE FORCE ACTING ON IT IS F, DEFINED AS THE PRODUCT OF ITS MASS AND ACCELERATION ($F = ma$).

Newton's *Third Law* is very simply expressed nowadays as follows: TO EVERY ACTION THERE IS AN EQUAL AND OPPOSITE REACTION. Thus, forces work in pairs. When a man's foot presses on the ground as he walks, the ground pushes back with an equal but opposite force. The ground is acted upon as the foot strikes it and it reacts with an equal force in the opposite direction, causing motion in that direction when possible. (Fig. 3.1) As the running long jumper takes off, his take-off foot thrusts against the board, the board pushes back at him, and he is propelled through the air by a force equal to the thrust of his take-off foot.

One of the problems met with in the space program was this same law of action and reaction. When there is no large mass such as the earth or moon to react against the thrust of a foot or a hand, the body itself responds by turning or moving in the opposite direction. This can be illustrated in the laboratory by someone standing on a freely movable turntable with one arm abducted 90°. Regardless of whether he swings the arm to the left or right in horizontal flexion or extension, the table reacts by turning him in the opposite direction.

Acceleration g due to Gravity

Following from Equation 3.5, mass is defined mathematically by

$$m = F/a \qquad (3.6)$$

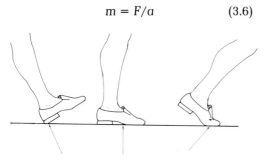

Figure 3.1. Action and reaction. The foot lands and exerts pressure on the ground; the ground reacts with an equal but opposite force.

The force of gravity on earth gives rise to an acceleration g; thus, mass can be derived from a knowledge of F and a with $a = g$ in Equation 3.6.

From an earlier discussion it will be evident that weight

$$W = mg \qquad (3.7)$$

The system of units used and conventions play roles in describing weight (force) and mass:

If mass is in grams and g is in centimeters per second^2, weight is in dynes, e.g.:

$$W = 10 \text{ grams} \times 980.6 \text{ cm/sec}^2$$
$$= 9806 \text{ dynes}$$

If mass is in kilograms and g in meters per second^2, weight is in newtons, e.g.:

$$W = 10 \text{ kg} \times 9.806 \text{ m/sec}^2$$
$$= 98.06 \text{ newtons}$$

If mass is in pounds and g in feet per second^2, weight is in poundals, e.g.:

$$W = 10 \text{ lb (mass)} \times 32.17 \text{ ft/sec}^2$$
$$= 321.7 \text{ poundals}$$

If mass is in slugs and g in feet per second^2, weight is in pounds, e.g.:

$$W = 10 \text{ slugs} \times 32.17 \text{ ft/sec}^2$$
$$= 321.7 \text{ pounds}$$

The last two items above illustrate the confusion existing in the American and English systems of nomenclature. This confusion extends somewhat to the metric system also, as it is common practice to use the terms gram and kilogram as well as pound to define both mass and weight. However, although the difference between mass and weight has been noted above, this text follows the common practice and uses the terms grams, kilograms, and pounds as units of both mass and weight. Current trends, internationally, are toward use of the SI system of units (See Appendix 2).

Variations of g

As the amount of gravitational attraction is inversely proportional to the square

Table 3.1
Variation of g with Altitude at 45° latitude

Altitude	g	Altitude	g
ft	ft/sec²	m	m/sec²
0	32.174	0	9.806
1,000	32.170	1,000	9.803
4,000	32.161	4,000	9.794
16,000	32.124	8,000	9.782
60,000	31.988	16,000	9.757
100,000	31.865	32,000	9.708
500,000	30.631	100,000	9.598

of the planet's radius (refer to Equation 3.1), the shape of the planet affects the amount of force exerted at any given point on its surface. For example, on earth this gravitational force causes an object to be accelerated 9.806 m/sec² (SI system) or 32.17 ft/sec² (English system) when it is falling at 45° latitude, i.e., halfway between the equator and either pole. However, because the earth is flattened at its poles, making the distance to its center slightly less at these two points, a falling object would be acclerated 9.832 m/sec², or 32.257 ft/sec², in the polar regions. When falling at the equator, where the distance to the center is greatest, the same object falling at sea level would undergo less acceleration; i.e., 9.78 m/sec² or 32.08 ft/sec². In other words, the object would be heaviest when weighed at either pole, lighter when weighed at 45° latitude, lighter still at equator sea level, and lightest of all on a mountain peak along the equator, i.e., in Bolivia, Kenya, or Sumatra (Tables 3.1 and 3.2) (Resnick and Halliday, (1960)).

Force Relations

The forces acting upon a body are usually summed using vector methods to derive their resultant values. Depending on how forces are applied to given structures that are able to move under their influence, various kinds of motions can result. It is of value to examine the types of relationships that exist for various fundamental ways in which forces might act.

Linear Forces

When forces act in the same straight line they are said to be linear forces. This is the simplest combination of forces. Figure 3.2 shows two pairs of linear forces. Pair A, R are opposite forces acting in the same line; pair X, F are equal and opposite forces acting in the same line. (Action and reaction are equal and opposite.)

If the weight is initially at rest, when the arm force A exceeds the rope pull R which is equal to the suspended weight (for a frictionless pulling system), the weight will be lifted and the subject's stiffened body will move backwards while pivoting about the feet. When R exceeds A, the body will tend to move forward.

Consider a situation in which forces of 10, 3, and 5 lb are pushing a body to the right, while at the same time forces of 2 and 4 lb are pushing the same body to the left. These forces have direction and magnitude and so are vector forces which can be graphically illustrated by arrows whose length is scaled to represent the amount of force (Fig. 3.3). Vectors pointing to the right or upward are by convention considered as positive, while those pointing in the opposite directions are considered negative. Adding graphically in Figure 3.3(b) we end up with a positive vector 12 units long (Figure 3.3(d)) which is equivalent to 12 lb. Mathematically:

$$R = \sum_{i=1}^{n} F_i \qquad (3.8)$$

Table 3.2
Variation of g with Latitude at Sea Level

Latitude		Feet/sec²	Meters/sec²
Equator	0°	32.0878	9.78039
	10°	32.0929	9.78195
	20°	32.1076	9.78641
	30°	32.1302	9.79329
	40°	32.1578	9.80171
	50°	32.1873	9.81071
	60°	32.2151	9.81918
	70°	32.2377	9.82608
	80°	32.2525	9.83059
Pole	90°	32.2577	9.83217

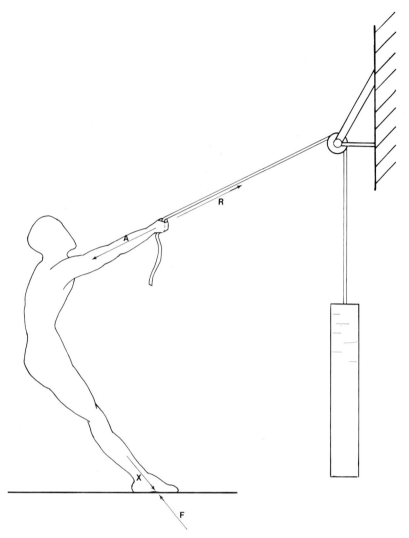

Figure 3.2. Linear force system; two pairs of forces. *A*, force exerted by the arms; *R*, the force exerted by the rope; *X*, the force exerted by the feet; *F*, the force exerted by the floor.

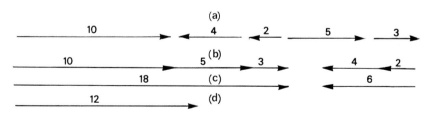

Figure 3.3 Vector diagram of forces described in text. (a), vectors representing each of the different forces (see text); (b), (c), and (d), vectors added graphically.

where R is the resultant of the forces, F_1, $F_2 \cdots F_n$ involved:

$$R = +10 - 4 + 5 + 3 - 2 = 18 - 6$$
$$= +12 \text{ lb}$$

Parallel Forces

In this case the action lines of the forces under consideration are parallel to one another. As the forces are applied at some distance from each other a different approach is necessary for the analysis and calculations relevant to situations of this kind. If the body upon which parallel forces are acting is to be in equilibrium, the sum of the forces acting in one direction must be completely nullified by the sum of the forces acting in the opposite direction. Furthermore, cognizance must be taken of the rotational influences that can occur when parallel forces act on a body. This aspect will become more readily apparent when attention is paid to levers (see later). At this juncture, it will be sufficient to note that the two aforementioned facets must be considered.

Force Couples

A special case of parallel forces occurs when there are two forces of equal magnitude acting at a distance from each other and in opposite directions. Under these circumstances they produce a turning action as in Figure 3.4 in which two boys cooperate to turn the boat end for end. As long as the force exerted by boy A is equal and opposite to that exerted by boy B, the boat will not go anywhere; there will be no linear displacement or acceleration, and the resultant of the two forces will be zero.

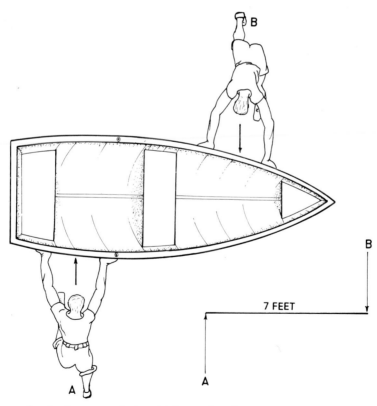

Figure 3.4. Example of a force couple in which equal and opposite forces are applied to each end of the boat. The torque or moment of this couple (T_c) is the product of one force and the distance between them.

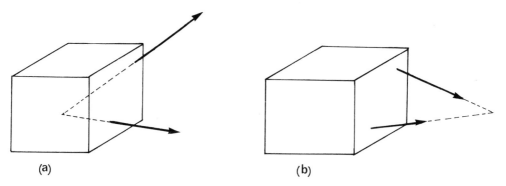

(a)   (b)

Figure 3.5. Concurrent force. (a), lines of force intersect inside the body; (b), lines of force intersect outside the body.

But the boat is turned. The forces acting in this situation are known as a couple and the moment or torque is the product of one of the equal and opposite forces multiplied by the distance between them:

$$T_c = Fd \qquad (3.9)$$

where the subscript c refers to the couple.

In Figure 3.4 each boy is exerting a 30-lb force and they are 7 ft apart. Boy A is exerting a clockwise force, thus the resultant of the two forces will be 0:

$$R = \Sigma F = 0$$

$$R = +30 - 30 = 0$$

but the moment or torque:

$$T_c = 30 \text{ lb} \times 7 \text{ ft}$$

$$= 210 \text{ lb-ft}$$

Concurrent forces

Forces whose action lines meet at a point are concurrent. Such forces may be applied to a body from two or more different angles so that projections of their action lines will cross. This intersection need not be inside the body. Figure 3.5(a) presents a simple situation where the action lines intersect within the body, while (b) illustrates an intersection outside the body. Another example occurs in normal quiet standing where the line of gravity falls slightly anterior to the ankle joint so that gravity has a clockwise action on the body and only the counterclockwise tension in

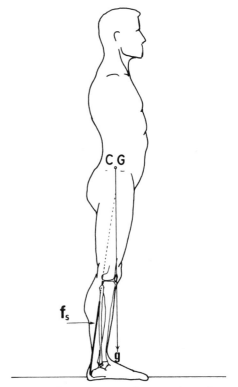

Figure 3.6. The concurrent force system of gravity g and soleus muscle force f_s, maintaining the upright posture. CG, center of gravity.

the calf muscle maintains erect posture (Figure 3.6).

With all of these concepts as a basis, it is possible to proceed to solving problems that arise in the study of human

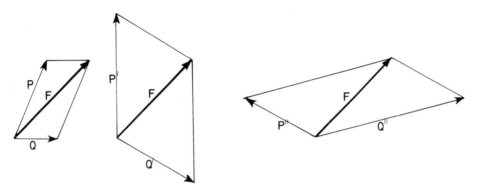

Figure 3.7. Resolution of force F into two components: P and Q, P' and Q', and P'' and Q'' by constructing parallelograms of force around force F.

movement, in particular those involving kinetics. Kinetics itself is concerned with the forces that either produce or change the state of rest or motion of a mass, living or inert. When studying the kinetics of the human body we must be aware of two kinds of forces which act on it, internal and external. Internal force is exerted by muscular *tension*, either while shortening or being lengthened by an outside force. The major external force is, of course, gravity. As gravitational force pulls all bodies toward the center of the earth, its action line is always vertically downward. Other external forces include those of impact as in catching or striking, falling or contact in sports, water resistance when swimming, the force exerted by a fiberglass pole in vaulting, etc. The action line of these external forces other than gravity depends upon the situation being analyzed. As all of these forces, both internal and external, have direction and magnitude, they are vector quantities and subject to vector analysis.

Composition and Resolution of Forces

Many times when a number of forces are acting on a body it is desirable to find a single force or resultant that will have the same effect on the body as that of the combined forces that it replaces. This process is known as the composition of forces. This problem of finding a resultant

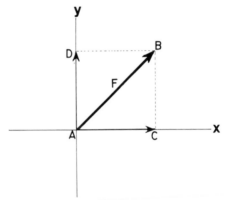

Figure 3.8. Resolution of force F into vertical and horizontal components. See text.

of two or more forces has as its concomitant the problem of resolving a single force into two or more components such that the combined action of the two new forces will be equivalent to that of the original force. The most common situation that the student of human movement will encounter is the necessity to resolve a single force into two components.

Graphical Resolution of Forces

Any single force can be replaced by two or more components whose combined action will be the same as that of the original force. The simplest situation of resolving a force into two components involves the construction of a parallelogram of forces such that the original force forms the diagonal (Fig. 3.7). In each of the above cases the pairs of force components of the force F are indicated by P and Q, P' and Q', and P" and Q". These forces, acting in unison on a body, will produce the same effect as the single force F. However, it is more desirable to specify the direction that each of the two replacements must take, and the most frequent procedure is to resolve the force F into two components at right angles to each other. These directions are normally horizontal and vertical, forming an X component on the horizontal or x axis, and a Y component along the vertical or y axis (Fig. 3.8). The force F is at 45° with the horizontal, and the vector AB represents the force. The x and y axes are drawn through point A, the origin of the force. Perpendiculars from point B to the two axes define the two components, vectors AC (the X component) and AD (the Y component).

The two component force vectors form the two legs of a right angle triangle, with the force as the hypotenuse, so the Pythagorean theorem is applicable: "The square of the hypotenuse is equal to the sum of the squares of the two sides." Consequently,

$$\vec{AB}^2 = \vec{AC}^2 + \vec{AD}^2$$

$$\vec{AB} = \sqrt{\vec{AC}^2 + \vec{AD}^2}$$

Graphical Composition of Forces

A number of forces acting on a single body can be reduced to a single force whose action will be the equivalent of the combined action of all of the original forces. When it is desirable to manipulate forces in this manner there are two general approaches, graphical and mathematical. Parallel forces can best be composed mathematically, while linear and concurrent forces can be treated by both methods. A graphical solution of a linear force problem is presented above under "Linear Forces." Finding the resultant of a pair of concurrent forces is the reverse of resolving a force into two components as in Figures 3.7 and 3.8. The pair of concurrent forces becomes the sides of a parallelogram, and the single resultant force R is the diagonal of the parallelogram which

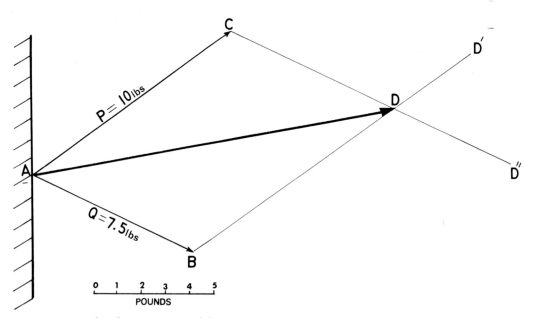

Figure 3.9. Graphical composition of forces P and Q. See text.

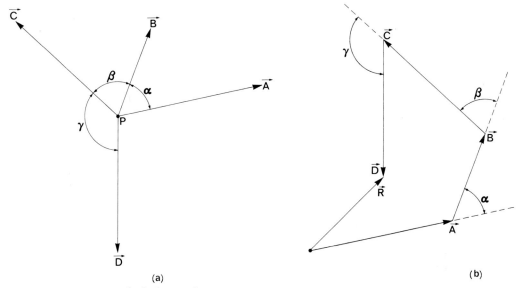

Figure 3.10. (a), forces $\vec{A}$, $\vec{B}$, $\vec{C}$, and $\vec{D}$ at a point P. (b), constructing the polygon as shown leads to the resultant force $\vec{R}$ in magnitude and direction.

originates from the junction of the two forces.

In Figure 3.9 two forces, P of 10 lb, and Q of 7.5 lb, are acting on a single body.

Problem. Find the magnitude and the direction of the resultant (the single force that can replace P and Q).

Solution. Construct a parallelogram of forces by drawing line BD' parallel to vector $\vec{AC}$, and line CD'' parallel to vector $\vec{AB}$. D is the point of intersection of BD' and CD''. The vector $\vec{AD}$ is the resultant of forces P and Q.

More than Two Forces

With three forces $\vec{A}$, $\vec{B}$, and $\vec{C}$ it is possible to construct two parallelograms. The three vectors $\vec{A}$, $\vec{B}$, and $\vec{C}$ are drawn from a common origin and appropriately scaled along their lines of action. The resultant $\vec{D}$ of two components, say $\vec{A}$ and $\vec{B}$, is determined by a parallelogram of forces and then another parallelogram constructed in which $\vec{C}$ and $\vec{D}$ are the vectors to be summed. In this way the resultant of the three forces is arrived at. Clearly, this method can be extended to cover as many forces as required. For many vectors it is a laborious method. Another useful graphical method is the solution which derives from the construction of a polygon of forces.

Polygon of Forces

In this construction, vectors are added graphically to each other by drawing them sequentially (and in appropriate lines of action with pertinent scaling) such that the termination of one vector is the starting point for drawing the next one. Figure 3.10 illustrates the situation for four vectors $\vec{A}$, $\vec{B}$, $\vec{C}$, and $\vec{D}$.

The vector $\vec{R}$ that closes the polygon shown in Figure 3.10(b) is the resultant.

Mathematical Composition and Resolution of Concurrent Forces

This technique has already been illustrated by way of an example in Chapter 2 and it relies on the essential trigonometric functions sin, cos, tan, etc. (See Appendix 3 for a review of these relationships.) Unless the angles that each force makes with the horizontal are known, a force diagram is drawn, reproducing the action lines of the forces in question, and a horizontal x

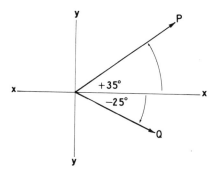

Figure 3.11. Force diagram from problem solved graphically in Figure 3.9 with horizontal x and vertical y axes added. See text.

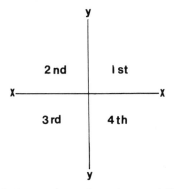

Figure 3.12. x and y axis system and the four quadrants.

axis is added, intersecting the action lines. (The y axis is not always needed but can be drawn in if desired.) Figure 3.11 is a force diagram of the problem solved graphically in Figure 3.9. Both the x and y axes have been added at the origin of the two forces. The action line of force P lies in the first quadrant of the x-y system (Fig. 3.12), so both the X and Y components are positive. On the other hand, the action line of force Q lies in the fourth quadrant and, while the X component is positive, the Y component is downward and therefore negative. By measuring the angles on the diagram, force P is at 35° with the x axis and force Q forms a downward angle of 25° with the same axis.

The X component, R_X, of the resultant force will equal the sum of the X compo-

nents, ΣF_X, of the forces involved, and similarly the Y component, R_Y, of the resultant will equal the sum of the Y components, ΔF_Y, of the force involved, i.e.

$$R_X = \Sigma F_X$$

and (3.10)

$$R_Y = \Sigma F_Y$$

To find the resultant (in magnitude and direction) of forces P and Q (Fig. 3.11) we proceed as follows:

Table of known data

Force	$P = 10$ lb	$Q = 7.5$ lb
Angle with horizontal	+35°	−25°
Sine	0.57358	0.42262
Cosine	0.81915	0.90631

$R_X = P \cos 35° + Q \cos 25°$

$\quad = (10 \text{ lb} \times 0.81915) + (7.5 \text{ lb} \times 0.90631)$

$\quad = 8.195 \text{ lb} + 6.7973 \text{ lb}$

$\quad = 8.2 \text{ lb} + 6.8 \text{ lb}$

$\quad = 15.0 \text{ lb}$

The horizontal component of forces P and Q is 15 lb.

$R_Y = P \sin 35° - Q \sin 25°$

$\quad = (10 \text{ lb} \times 0.57358) - (7.5 \text{ lb} \times 0.42262)$

$\quad = 5.7358 \text{ lb} - 3.16965 \text{ lb}$

$\quad = 5.7 \text{ lb} - 3.2 \text{ lb}$

$\quad = 2.5 \text{ lb}$

The vertical component of forces P and Q is 2.5 lb.

The resultant being sought is the hypotenuse of the right triangle whose two sides are represented, respectively, by vectors 15.0 and 2.5 units long.

By using the Pythagorean theorem the magnitude of the resultant can be determined:

$$R^2 = 15^2 + 2.5^2$$

$$= 225 + 6.25$$

$$= 231.25$$

$$R = \sqrt{231.25}$$

$$= 15.2 \text{ lb}$$

The resultant equals 15.2 lb, but the direction has yet to be located, i.e., the angle that the resultant makes with the horizontal. Angle θ is the angle whose sine is Y divided by R. This is expressed as:

$$\theta = \arcsin Y/R$$

$$\theta = \arcsin 2.5/15.2$$

$$= \arcsin 0.16453$$

$$= 9° 28'$$

Alternative Solution. The magnitude and direction can also be determined as follows:

$$\theta = \arctan Y/X$$

$$= \arctan 2.5/15.0$$

$$= \arctan 1.66666$$

$$= 9° 28'$$

This gives the direction of the resultant but not its magnitude, which is equivalent to the hypotenuse of the right triangle. Cos $\theta = X/R$ and sin $\theta = Y/R$ so either can be used to find R, the hypotenuse, as $R = Y/\sin \theta$ or $X/\cos \theta$.

$$R = Y/\sin \theta$$

$$= 2.50/0.16447$$

$$= 15.2 \text{ lb}$$

The resultant is 15.2 lb upward at an angle of 9° 28' (cf. results found by the Pythagorean theorem).

Positive and Negative Components

Whether or not a force component is positive or negative or zero depends upon its direction in relation to the x-y axis system which divides a plane into four quadrants (Fig. 3.12).

A force whose direction upward and to the right places it in the first quadrant will have both X and Y components positive. When the direction is still upward but to the left, the force will be in the second quadrant, and the Y component remains positive but the X component becomes negative. Any force directed downward and to the left will be in the third quadrant and both X and Y components will be negative, while forces directed downward and to the right are in the fourth quadrant where the Y component is still negative but the X component is again positive.

To summarize:

All forces directed upward have positive Y components.

All forces directed to the right have positive X components.

All forces directed downward have negative Y components.

All forces directed to the left have negative X components.

The student should bear these facts in mind when calculating mathematical solutions in the composition of force systems.

Equilibrium Forces

For a body to be in a state of equilibrium it is necessary for the total forces acting upon it to be equal to 0. Mathematically:

$$\Sigma F = 0$$

MOMENTUM

Momentum is the product of the mass of a body and its velocity, and as such it is a vector quantity, a *quantity of motion* possessed by a body: $P = mv$. If mass is in pounds and velocity in feet per second, momentum P is expressed as pound feet per second; similarly if mass is in kilograms and velocity in meters per second, P is expressed as kilogram meters per second.

Application of Force and Changes in Momentum

The equation of Newton's original statement of his Second Law

$$F = \frac{mv_2 - mv_1}{\Delta t} \tag{3.3}$$

involving the time rate of change of momentum can be used for determining the force involved in striking a ball if the necessary data are available: the duration of the contact of the implement with the ball; the velocities of the ball before and after it was struck; and the mass of the ball. It is only in recent years, since ultra-high speed photography has made the acquisition of such data possible, that the use of this formula has become practicable in sports analysis. Unfortunately, as the speed of the struck ball also is influenced by the elastic qualities of the ball and possibly those of the striking implement as well, the estimate of the force involved will not be entirely accurate.

Equation 3.3 can be rearranged as follows:

$$F \times \Delta t = mv_2 - mv_1 \qquad (3.11)$$

where the product $F \times \Delta t$ is referred to as the *impulse* that causes the *change in momentum* $(mv_2 - mv_1)$. The impulse consists of a force F applied for a short duration Δt.

It is evident that a small force applied for a long enough period of time can produce momentum changes comparable to a far larger force applied for a briefer interval.

To achieve a large change of momentum in a given period Δt, a large force F is required. Thus, in projectile skills such as throwing, greater body forces are needed to attain higher projectile velocities. This explains why baseball pitchers may hurt their arms when they attempt to achieve maximal velocity of the ball by applying a maximal amount of force over as brief an interval as possible. The resulting tangential and radial accelerations allied with upper extremity movements create forces which are responsible for the strained ligaments, damaged bursae, and possibly torn muscles that pitchers occasionally suffer.

Receiving a Force

When considering Equations 3.3 and 3.11 from another viewpoint, that of catching a fast ball or landing from a high jump, the greater the increase in the amount of time consumed in catching the ball or for completing the landing, the less will be the force felt by the catcher or jumper at any one instant while his body is receiving the force and the less will be the damage to body tissues. In catching a fast ball, even with a catching mitt, the hands are stretched forward to meet the ball and are drawn toward the body as the ball contacts the glove. All of this increases the time during which the momentum of the ball is absorbed by the catcher. Landing pits for long and high jumps were originally filled with sand, then with sawdust, wood shavings, or tanbark. Today most of them are filled with chunks or sheets of foam rubber. The jumper, if he lands on his feet (this is chiefly in the case of the long jump), dorsiflexes at the ankles, flexes the knees, hips, and spine, and may go into a roll. The high jumper or pole vaulter may land on his shoulders and back or with a roll or both. In any case the force generated by his momentum is spread over as large an area of his body as possible so that any one area is impacted by a comparatively small amount of force. This type of landing also increases the duration of the landing process so that there is less force for the body to absorb from instant to instant. The landing medium also absorbs some of the momentum, as the particles of sand, shavings, tanbark, or foam rubber fly in all directions when some of the jumper's momentum is transferred to the landing medium. The ultimate result is that the force felt by the jumper at any one instant over any one area of his body is decreased to the point where there is little if any damage to the tissues. The boxer rolls with the punch for the same purpose: to decrease the force with which he is hit by increasing the time of contact, which in turn decreases the change in momentum per unit of time. It is the abrupt change in momentum from a large quantity to little or none that is harmful to the body (e.g., an automobile collision). This corresponds to the condition of a large impulsive force (see Equation 3.11).

Conservation of Momentum

The above illustrations all have spoken of momentum as being "absorbed" over as long a period of time as is conveniently possible, or that some of the momentum has been transferred under each of the circumstances. This slow deceleration, as has been pointed out, is accomplished so that the amount of force applied to the body at any given instant is as small as it can be made while the momentum is being altered. These facts imply that momentum is not lost or destroyed, and it is not.

If one observes the interaction of two or more bodies with m_1v_1, m_2v_2, m_3v_3 ... etc. in a closed system in which there are no resultant or outside forces involved and records the sum of the various momenta both before and after the interaction, he will find that they are identical in value. Mathematically:

$$\sum_{i=1}^{n} m_i v_i = K \qquad (3.12)$$

where $\sum_{i=1}^{n}$ means the summation of the product following, using all of the values of $i = 1$ through n. K is a constant value. This is the *law of conservation of momentum*.

At the same time it must not be forgotten that the law of conservation of mass also applies to the aforementioned interaction so that:

$$\sum_{i=1}^{n} m_i = M \qquad (3.13)$$

where M is a constant value.

Whatever happens, the momentum lost by one object or body is gained by another. The chunks of rubber in the landing pit fly around as the jumper lands, and whatever momentum is not transferred to the rubber is eventually absorbed by the earth, whose mass is so enormous that we cannot measure any change that is caused. Similarly, when the outfielder catches a line drive, the momentum of the ball (whose mass is extremely small) is transferred to his body and then, as with the jumper, to the earth.

ENERGY

Energy is a scalar quantity and is often described as the capacity to do work. It appears in many forms; electrical, chemical, nuclear, mechanical.

In a mechanical context, when a force F is applied to a body so that it moves through a distance x, the work done on the body or energy expended is

$$W = Fx \qquad (3.14)$$

Work makes use of the same units as energy. In the SI system force is expressed in newtons, distance in meters, and work or energy in joules.

The breakdown of gasoline provides the energy to drive the internal combustion engine; the splitting of the high energy phosphate bond ($\sim$ (P)) of adenosine triphosphate (AT-(P) $\sim$ (P) $\sim$ (P)) provides the energy for muscular contraction. These are examples of chemical energy. Mechanical energy, which is our concern at the moment, involves both potential and kinetic energy.

Potential Energy

Potential energy (PE) is also spoken of as a gravitational energy or energy of position. Accordingly, it is defined by

PE = weight $\times$ height

(i.e., height relative to reference position)

or

$$PE = mgh \qquad (3.15)$$

where mg is the product of mass m and gravitational acceleration g to yield gravitational force or weight, and h is the height of the mass m relative to a chosen reference level (usually taken to be the surface of the earth).

A 20-lb boulder balanced on the edge of a 70-ft cliff has a potential energy of 1400 ft-lb relative to the foot of the cliff. It could do considerable damage to a highway at the foot of the cliff.

A gymnast hanging from the stationary rings has a certain amount of potential energy. If he weighs 140 lb and his center of gravity is raised 0.5 ft when he leaps upward to grasp the rings, he will increase his potential energy by 70 ft-lb in relation to the floor.

If the gymnast changes his position to the front uprise causing his center of gravity to rise approximately another 4 ft, his potential energy will increase further by some 560 ft-lb (140 × 4).

Thus, the total increase in potential energy relative to the starting position is 630 ft-lb.

Kinetic Energy

Kinetic energy (KE) as its name implies, is the energy of motion. The amount of kinetic energy in a moving body is determined by:

$$KE = \frac{1}{2} mv^2 \qquad (3.16)$$

(Kinetic energy equals one-half the mass times the square of the velocity.)

Thus, the quantity of kinetic energy of any given mass depends solely on its velocity, while the potential energy of that mass depends solely upon its position.

Conservation of Energy

The law of conservation of mechanical energy states that, in a closed system where there are no outside forces present, the sum of the kinetic energy and the potential energy is equal to a constant for that system:

$$PE + KE = a\ constant \qquad (3.17)$$

so that there is no change in the total amount of energy in the system. Thus, there is no change in the sum of the potential and kinetic energies for that system at any one instant in any given situation.

Suppose the gymnast depicted on the flying rings in Figure 3.13 swings through the arc shown. His overall body motion would be like that of a pendulum. At each high point of his swing he has only potential energy since his velocity momentarily falls to 0. With respect to the lower point

that he passes through he has only kinetic energy. In general, $PE = mgy$ where y is the level of the mass m above the reference location. Thus

$$PE + KE = mgy + \frac{1}{2} mv^2.$$

At the top of the swing obviously

$$y = h \quad \text{and} \quad v = 0;$$

at the bottom of the swing

$$y = 0 \quad \text{and} \quad v = v_{max}$$

where v_{max} is the maximum velocity attained.

For the top of the swing

$$PE + KE = mgh + 0 = mgh$$

which by the law of conservation of energy must be fixed for the system.

Therefore, for the low point of the swing:

$$PE + KE = 0 + \frac{1}{2} mv^2 = mgh$$

i.e.

$$mgh = \frac{1}{2} mv^2$$

or

$$h = v^2/2g \qquad (3.18)$$

By rearranging and taking the square root of both sides of the equation

$$v = \sqrt{2gh} \qquad (3.19)$$

Thus, from a knowledge of h, v can be calculated.

It should be realized that the effects of frictional forces (air friction, ring-on-rope friction and the like) have been ignored in this treatment. In fact for any system or flight path where friction is negligible the potential energy and the highest point of the path will provide the constant for any other point in the path, and this potential energy will be equal to the maximal kinetic energy of the system.

If the gymnast's center of gravity rises 5 ft during the swing and he weighs 140 lb, the constant is most easily calculated at the top of the swing:

$$PE = 140 \times 5 = 700\ ft\text{-}lb$$

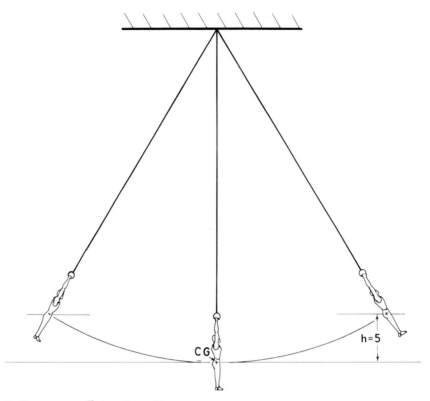

Figure 3.13. Gymnast on flying rings. See text.

As KE is 0 at the top of the swing, 700 lb-ft is the constant energy for the system. As PE + KE will always equal 700 ft-lb, it is a very simple matter to determine the kinetic energy for a drop of any given distance.

Firstly, at the lowest position velocity v follows from Equation 3.19.

$$v = \sqrt{2 \times 32.2 \times 5} = \sqrt{322}$$

$$= 17.94 \text{ ft/sec}$$

If there has been a drop of 1 ft from the highest position, then the kinetic energy at that position is determined as follows:

$$PE + KE = 700 \text{ ft-lb}$$

$$140 \times (5 - 1) + KE = 700$$

(The drop is 1 ft; therefore, the height above the reference position is $5 - 1 = 4$ ft.) That is, PE at this position is

$$140 \times 4 = 560 \text{ lb-ft}$$

Hence KE in this position

$$= 700 - 560$$

$$= 140 \text{ ft-lb}$$

The velocity at this position is obtained from

$$\tfrac{1}{2} mv^2 = 140$$

$$\tfrac{1}{2} \left(\frac{140}{32.2} \right) v^2 = 140$$

$$v^2 = 64.4$$

$$v = 8.025 \text{ ft/sec}$$

Work and Power

Power is defined as the time-rate of performing work: e.g., so many joules per

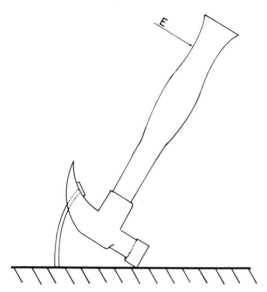

Figure 3.14. The hammer as a lever. E is the line of force of the effort used to pull the nail. The effort arm is made up of all parts of the lever between E and the fulcrum where the hammer head contacts the board.

(sometimes termed load) R is a second force which tends to rotate the lever in the direction opposite to the effort. The lever arm of the resistance or RA consists of all parts of the lever between the axis and the point where the resistance is applied.

In Figure 3.14 the fulcrum is at the point of contact of the hammer head and the board; the effort is applied as indicated in the figure so that the EA includes all of the handle of the hammer below the point where the effort is applied, plus that part of the hammer head from the shaft to the fulcrum. The resistance is the force required to withdraw the nail being pulled and is applied to the claws of the hammer by the nail head. The RA then is all of the hammer head between the fulcrum and the point where the claws are gripping the nail head.

Classification of Levers

Levers fall into three categories or classifications depending upon the relation-

second or so many foot-pounds per second. One horsepower is 550 foot-pounds of work per second.

LEVERAGE

A brief review of the principles of levers can perhaps assist the student in understanding how the laws of leverage apply to human movement. A lever is a device for transmitting force, and it is able to do work when work is done upon it. It is generally a rigid bar or mass (but need be no particular shape), which rotates around a fulcrum (fixed point) on an axis perpendicular to the plane of motion. The rotation is caused by a force applied to the rigid bar or mass. If this force or effort is used to overcome a resistance, it is designated as the effort E, and all parts of the lever between the axis and the point where the effort is applied are designated as the lever arm of the effort or simply the effort arm, EA. The resisting force or resistance

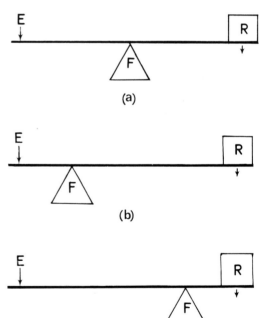

Figure 3.15. First class levers. E, effort; F, fulcrum; R, resistance, or load. See text.

ship between the positions of the fulcrum F, the effort E, and the resistance R.

First Class Levers. When the fulcrum on the axis is located between the effort and the resistance, the EA may be equal to, greater than, or less that the RA (Fig. 3.15). This type of lever may be used to gain speed if the fulcrum is nearer the point where the effort is applied (Fig. 3.15(b)) or to gain force if the fulcrum is closer to the resistance (Fig. 3.15(c)). Examples of first class levers in the body include the triceps extending the elbow while the motion of the hand and arm is resisted, and the free foot being plantarflexed at the ankle while the ball of the foot and toes act upon a resistance such as an accelerator pedal in an automobile.

Second Class Levers. When the resistance R lies between the axis and the effort so that consequently the effort arm length EA is always greater than the resistance arm length RA, the lever is a second class lever (Fig. 3.16). A man rolling a wheelbarrow is an example of such a system. The man supplies the effort at the handles, the load is in the barrow, and the fulcrum is in the axis of the wheel. Examples within the body are discussed later in the text.

Third Class Levers. When the positions of E (effort) and R (resistance) are reversed, the lever is a third class lever (Fig. 3.17). RA (resistance arm) length is always longer than EA (effort arm) length. There are many third class levers in the human body; among them are the biceps causing flexion of the forearm about the elbow and the iliopsoas flexing the thigh about the hip joint.

Moment of Force

The moment of force, M, allied with a force such as E (effort) or R (resistance)

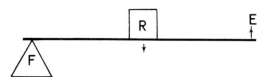

Figure 3.16. Second class lever. E, effort; F, fulcrum; R, resistance or load.

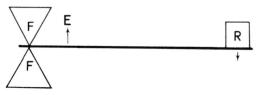

Figure 3.17. Third class lever. E, effort; F, fulcrum; R, resistance or load.

acting on a lever is defined as the product of the force and the distance perpendicular to its line of action from the fulcrum: the moment for resistive forces

$$M_R = R \times RA$$

the moment for effort forces

$$M_E = E \times EA$$

It will be realized that for all three classes of levers the resistive moments must be exactly counterbalanced by the effort moments to achieve equilibrium. For equilibrium

$$M_E = M_R$$

$$E \times EA = R \times RA$$

Taking ratios of forces to one side of the equation leads to

$$\frac{R}{E} = \frac{EA}{RA}$$

This ratio is defined as the mechanical advantage (M.Adv.) of the lever, i.e.

$$M.Adv. = \frac{R}{E} = \frac{EA}{RA} \qquad (3.20)$$

A large value of M.Adv. implies that little effort is required to balance a large resistance.

It should be noted that for a frictionless fulcrum, the energy expended in moving a resistance exactly equals that applied through the effort. Therefore, it follows that the ratio of distances moved by the resistance and effort will be the inverse of M.Adv. At this stage it will be helpful to refer to Figure 3.18. This shows a first class lever. If the effort E is increased or decreased about the equilibrium value resistance and effort extremities will respec-

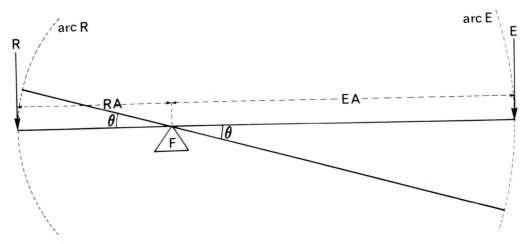

Figure 3.18. Relationships between distances moved for effort and resistance forces.

tively move along the arcs Arc_R and Arc_E.
Recalling Equation 2.11:

$$s = r\theta$$

The distances traveled along Arc_R and Arc_E for angle θ shown will be

$$s_R = RA\,\theta, \quad \text{and} \quad s_E = EA\,\theta.$$

If we consider that R and E are continually applied at right angles to the lever then work done on R (see Equation 3.14)

$$W_R = R \times RA\theta$$

Similarly, the work done by E

$$W_E = E \times EA\theta$$

From Equation 3.20

$$R \times RA = E \times EA$$

and therefore

$$W_E = W_R$$

as discussed earlier.
Also

$$s_R/s_E = RA/EA = E/R = \frac{1}{\text{M.Adv.}}$$

another result that was alluded to earlier.
In the case of a third class lever EA is always less than RA. It thus follows from Equation 3.20 and the foregoing discussion

that the mechanical advantage is less than 1, i.e., the effort E has to exceed the resistance R; but at the same time, when movement occurs, the distance traversed by R is greater than that traversed by E. Therefore, the velocity of movements of R is greater than that of E.

Force at the Fulcrum

The force acting at the fulcrum depends on the class of the lever and the effort E and resistance R forces. In the case of the first class lever (Fig. 3.15) the fulcrum must withstand a net force F_1 of;

$$F_1 = E + R$$

i.e., F_1 is a reaction force directed upward.
For the second class lever (Fig. 3.16) the fulcrum has to oppose a force F_2;

$$F_2 = R - E \quad (R > E)$$

i.e., F_2 is a reaction force directed upward. Turning to the third class lever (Fig. 3.17) the fulcrum for F_3:

$$F_3 = R - E \quad (E > R)$$

i.e., F_3 is a reaction force directed downward. (Hence the fulcrum arrangement shown.)
Some examples will be helpful at this

juncture. Note that the levers themselves are assumed to be weightless.

Example 1. Given a load of 12 lb and that EA and RA are both 3 ft. The moment of force of the resistance M_R (resistance multiplied by the perpendicular distance from the line of force of R to axis through the fulcrum) is:

$$M_R = R \times RA$$

$$= 12 \text{ lb} \times 3 \text{ ft}$$

$$= 36 \text{ lb ft}$$

This requires an equal but opposite moment of force to balance it:

$$E \times EA = R \times RA$$

$$E \times 3 \text{ ft} = 36 \text{ lb-ft}$$

$$E = \frac{36}{3 \text{ ft}}$$

$$E = 12 \text{ lb}$$

Fulcrum forces are:

Class 1: 24 lb upward

Class 2: 0 lb

Class 3: 0 lb

Example 2. Given that EA is equal to 6 ft while RA remains 3 ft and R is 12 lb (this could occur with either a first or second class lever (Fig. 3.19, (a) and (b)).

The moment of force of the resistance M_R is still unchanged at 36 lb-ft. Clearly

$$E \times 6 \text{ ft} = 12 \text{ lb} \times 3 \text{ ft}$$

$$E = \frac{36 \text{ lb-ft}}{6 \text{ ft}}$$

$$E = 6 \text{ lb}$$

Only 6 lb of force are needed to balance the resistance, which is only half as much force as required in the first situation. Thus we can say that a lever system with an EA twice as long as the RA has a mechanical advantage (M. Adv.) of 2 (see Equation 3.20).

Fulcrum forces are:

Class 1: 18 lb upward

Class 2: 6 lb upward

Example 3. Given the same 12-lb load as R but an RA of 6 ft and and an EA of 3 ft.

$$3 \text{ ft} \times E = 6 \text{ ft} \times 12 \text{ lb}$$

$$E = \frac{72 \text{ ft-lb}}{3 \text{ ft}}$$

$$= 24 \text{ lb}$$

So an effort of 24 lb is now needed to support the 12-lb load. In this case

$$\text{M. Adv.} = \frac{R}{E} = \frac{12}{24} = \frac{1}{2}$$

Fulcrum forces are:

Class 1: 24 + 12 = 36 lb upward

Class 3: 24 − 12 = 12 lb downward

Torque

In the simple levers used above to illustrate lever classes, EA and RA were always perpendicular to the line of force of E and R, respectively, and each ran directly to the axis passing through the fulcrum. Since the perpendicular distance from a line of force to an axis through the

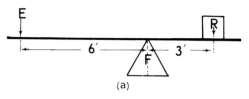

(a)

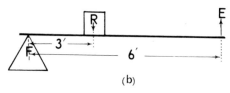

(b)

Figure 3.19. Examples of levers with 6-ft effort arms and 3-ft resistance arms.

fulcrum is the **moment arm** of that force (MA), we can write

$$E \times MA_E = R \times MA_R \qquad (3.21)$$

where MA_E is the moment arm of the effort and MA_R is the moment arm of the resistance. The product of $MA_R \times R$ (and of $MA_E \times E$) is a moment of force, sometimes called torque, where:

$$T = fd \qquad (3.22)$$

(the Greek letter tau) is the torque or force moment, f is the force (effort or resistance), and d is the moment (torque) arm, i.e., the perpendicular distance from the line of force to the axis. The units of torque are, therefore, force and distance, as pound-feet for example.

Both $E \times MA_E$ and $R \times MA_R$ are moments of force, more simply known as *moments*. As indicated earlier all moments are expressed in double units; e.g., pound-feet or pound-inches, gram-centimeters, or kilogram-meters, depending on the system of measures being used.

A torque or more simply a moment, tends to cause rotation about an axis. Torques causing clockwise rotations are considered positive; when they tend to rotate a body or lever in the opposite direction, counterclockwise, they are considered negative.* Thus, in an equilibrium situation the sum of all torques or moments must be equal to 0 and Equation 3.21 above is written:

$$(E \times MA_E) - (R \times MA_R) = 0 \qquad (3.23)$$

or in more general terms:

$$\Sigma M = 0 \text{ or } \Sigma T = 0 \qquad (3.24)$$

The sum of the moments is equal to 0.

Returning to the earlier example of the hammer as a lever (Fig. 3.14) the nail is exerting a counterclockwise or negative moment, while the effort to pull the nail is clockwise and this moment is positive.

---

* This is a matter of convention. The opposite convention, i.e., anticlockwise positive, clockwise negative, can be used. Whichever is adopted should be used consistently.

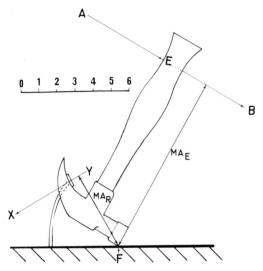

Figure 3.20. Force diagram of hammer. A-B, line of force of the effort; X-Y, line of force of the resistance (nail); F, fulcrum (on the axis); MA_R, moment arm of the resistance; MA_E, moment arm of the effort.

Example 4. The problem is to calculate the amount of force that must be applied at E to pull the nail, assuming that the nail is resisting with a force of 15 lb.

The solution involves determining the moment arms involved in both the resistance and the effort. One cannot simply draw a line from the nail head to the axis, or from the point of application of the effort E to the axis, and measure the distances involved. *A moment arm must be perpendicular to the line of force with which it is associated.* The first step then is to make a force diagram of the situation by adding the lines of force to Figure 3.14 as in Figure 3.20. The line of force of the nail, which is perpendicular to the hammer claws at the point at which the nail head is resting on them, is indicated by the line X—Y. The moment arm of the resistance, MA_R, is perpendicular to this line and runs directly to the axis where the hammer head is resting on the board. The line of force of the effort is indicated by line A-B (which is perpendicular to the handle). The MA_E runs from the axis and is perpendicular to A-B (Fig. 3.20). The MA_E is

10 in long and the MA_R is 4.3 in long. Since

$$\Sigma M = 0$$

$$E \times MA_E - R \times MA_R = 0$$

$$E \times 10'' - 15 \text{ lb} \times 4.3'' = 0$$

$$E = \frac{64.5 \text{ lb-in}}{10 \text{ in}}$$

$$E = 6.45 \text{ lb}$$

This 6.45 lb is just sufficient to balance the nail's resistance. To pull it will need more than this, possibly another 0.05 lb or, more likely, another ½ lb, or about 7 lb of force in all. The mechanical advantage of this system then is greater than 2, i.e.:

$$\text{M.Adv.} = \frac{R}{E} = \frac{15}{6.45} = 2.3$$

Likewise

$$\text{M.Adv.} = \frac{MA_E}{MA_R} = \frac{10}{4.3} = 2.3$$

EQUILIBRIUM

In the solution of equilibrium problems the sum of all the moments must be brought to 0 when an unknown force or moment arm is to be determined. Also, the sum of the forces must be equal to 0:

$$\Sigma M = 0$$
$$\Sigma F = 0 \tag{3.25}$$

and in equilibrium problems involving concurrent forces:

$$\Sigma X = 0$$
$$\Sigma Y = 0 \tag{3.26}$$
$$\Sigma Z = 0$$

where X, Y, and Z are the respective components of the forces involved. The Z component is needed only when the forces and resultants under study are in two or more different planes and a three dimensional approach is needed.

Equilibrium may be either passive, active, or dynamic. While passive equilib-rium is always static, static equilibrium need not always be passive. A body hold-ing a pose is static, but the equilibrium being maintained may be completely pas-sive, as occurs when an individual is re-clining and completely relaxed or, being relaxed, is wholly supported in some man-ner. On the other hand, if the pose is being held against gravity (or any outside force) as in Figure 3.21, (a) and (b), or even in quiet standing, muscular force is needed to hold the body segments in place. Under these circumstances the equilibrium is now active as muscular tension is creating force moments to balance those generated by gravity as the pose or stance was as-sumed. Dynamic equilibrium, on the other hand, is that maintained while the individ-ual is actively performing some form of motion or locomotion such that he keeps his center of gravity over a constantly changing base of support, as in Figure 3.22.

Equilibrium under Static Conditions

A body is in **stable equilibrium** when its *potential energy is at a minimum* and work must be done on it to cause a change in position. A living body is in stable equi-librium, then, only when it is lying down on a horizontal surface adequate to sup-port it. In order to change its position it must exert internal (muscular) force or have an outside force act upon it. When an individual maintains a given posture his center of gravity is held over his base of support and his line of gravity (LG) falls within that base (Fig. 3.21, (a) and (b)). If the individual is erect or nearly so, such that his center of gravity (CG) is some distance above his base, he has potential energy relative to the base. The larger the base, the greater is the amount of force that it takes to move the LG outside that base. Under this and other circumstances in which his *potential energy remains con-stant* (i.e., there is very little or no change in the height of his CG), he is in **neutral equilibrium**. The individual is in **unstable equilibrium** when his *potential energy is at a maximum* and his base of support is extremely small so that it takes very little force to move his LG outside his base. This

Figure 3.21. Static equilibrium. (a) and (b) show two positions called scales.

Figure 3.22. Illustration of dynamic equilibrium. Peggy Fleming (photography by W. Patriquin for the Boston *Herald-Traveler*).

situation may change very rapidly to a dynamic equilibrium or relapse into a neutral or even a stable equilibrium, depending on the individual's degree of motor control.

Equilibrium under Dynamic Conditions

A baby knows only stable equilibrium during the first months of his life; when he starts to creep he gains a small amount of potential energy but maintains an almost neutral equilibrium. When he starts to toddle his problems increase: he can only maintain neutral equilibrium through the aid of his parent's hand or by holding on to furniture. However, he soon develops dynamic equilibrium which varies between the neutral and the unstable as he learns to walk, run, jump, etc.

Over a Minimal Base. During all forms of locomotion any living body strives to maintain its balance, and all of its postural reflexes are geared to this end.

This equilibrium may be unstable at times, as when a ballerina dances on her toes or a figure skater performs as in Figure 3.22 or neutral as soon as she increases the size of her base and resorts to a less specialized form of locomotion.

The child learning a new motor skill or a new form of locomotion such as riding a bicycle has to learn to maintain as near neutral equilibrium as possible on a narrow, moving base. The inertia resulting from his linear motion contributes to his dynamic equilibrium and as long as the combined CG of body and machine continue to travel in the desired direction and their line of gravity falls on the line tangent to the front and rear wheels, balance is maintained. At the same time he must be prepared to fight this inertia when he makes a turn. Then he creates an inward force by steering into the turn and by an inward lean.

Over a Changing Base. When a gymnast or tumbler performs rolls, headsprings, and handsprings, cartwheels, etc., his base of support changes from his feet to other parts of his body. If his LG does not continue to travel in the direction of the next point of contact, dynamic equilibrium is lost and the performer collapses to the mat (stable equilibrium).

TORQUE AND ROTATIONAL MOTION

Torque or Moment of Force

In dealing with rotational movements or motion, the moment or torque fulfills the same function that force performs for motion in a straight line: it is the size of the moment that increases or decreases the angular velocity of a body and so produces acceleration or deceleration of the rotational movement.

Moment of Inertia

In discussing Newton's First Law earlier in this chapter it is mentioned that the terms mass and inertia could be used interchangeably. In rotational movement the moment of inertia I is the analog of mass in linear motion. If a body is divided into a large number of very small parts, and a typical particle has a mass m which is at a perpendicular distance r from the axis of rotation, then its contribution to the moment of inertia I is mr^2. Under these conditions the moment of inertia will be equal to the sum of all such contributions from all portions of the body (say n altogether) and:

$$I = \sum_{i=1}^{n} m_i r_i^2 \qquad (3.27)$$

where m_i is the mass of the ith portion and r_i its distance from the center of rotation.

The distance r for any particular portion is constant only for one given axis. As a body or body segment can rotate about different axes, each r_i will change with each change of axis. Thus, the moment of inertia is not a fixed constant for a body but is dependent on the distribution of the mass about the axis of rotation or the perpendicular distances. I can also be written as follows:

$$I = MK^2 \qquad (3.28)$$

The radius of gyration K is defined as the distance from the axis of rotation of a point at which the total mass m of a body might be concentrated without changing the moment of inertia of the body.

$$(M = \sum_{i=1}^{n} m_i)$$

The moment of inertia is the measure of the resistance of a body at rest to rotatory motion, or, if rotating, to change its state of rotation. As torque exerts a turning force on a body, it is analogous to the force in the equation $F = ma$. Therefore it can be stated that:

$$T = I\alpha \qquad (3.29)$$

Torque is equal to the product of the moment of inertia and the angular acceleration and so is analogous to force F as I is analogous to mass m and α to acceleration a in linear motion.

Angular Momentum

Linear momentum P equals the product of mass and velocity, so angular momentum A is the product of the moment of inertia and the angular velocity:

$$A = I\omega \qquad (3.30)$$

IF THE SYSTEM IS CLOSED AND THERE IS NO EXTERNAL TORQUE ACTING ON THE SYSTEM, THE TOTAL ANGULAR MOMENTUM REMAINS UNCHANGED,, EVEN THOUGH THE MOMENT OF INERTIA MAY BE ALTERED. This is the law of conservation of angular momentum. Note that while the quantity of mass remains unchanged, r or K (and so r^2 or K^2) can be changed.

An example involving the use of a "frictionless turntable" (Fig. 3.23) will serve to illustrate the use of the above relationships.

A man stands on a frictionless turntable holding a 15-lb barbell in each hand; his arms, with extended elbows, are abducted 90°. He is started spinning at a rate of 1 rps (revolution per second, 360°/sec). While he is spinning at this rate he pulls the barbells into his shoulders so that they are only 6 in from his axis of rotation. At the start of the experiment the barbells were held at 3 ft from the rotational axis. The problem is to determine the angular velocity with the barbells drawn in. Under the circumstances any change in the moment of inertia of his body will be so extremely small it can be ignored. Therefore as

$$I = mr^2 \qquad \text{(from 3.27)}$$

$$I = \frac{15 \text{ lb}}{g} \times 3\text{ft}^2$$

$$= \frac{135 \text{ lb-ft}^2}{g}$$

$$A = I\omega \qquad \text{(from 3.30)}$$

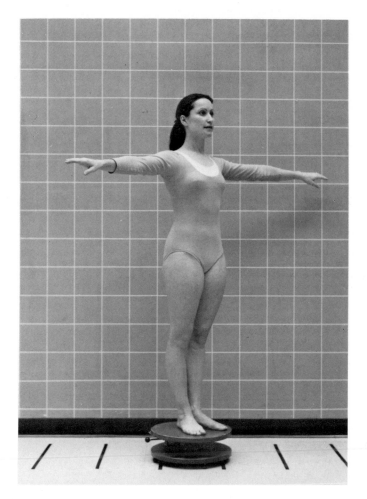

Figure 3.23. Frictionless turntable. Female subject can rotate her body relative to the floor using the "frictionless turntable" which is a mechanical bearing having very little friction.

$$A = \frac{135}{g}\,\omega_1$$

where ω_1 corresponds to 1 rps and $\omega_1 = 2\pi \times 1$ radians/sec.

When the hands draw the weights close to the body, their mass is unchanged but r has shrunk to 0.5 ft. Then

$$A = \frac{15\ \text{lb}}{g} \times 0.5\ \text{ft}^2 \times \omega_2$$

where ω_2 corresponds to the new angular velocity.

As A is unchanged (law of conservation of angular momentum), it follows that:

$$\frac{135\,\omega_1}{g} = \frac{3.75}{g} \times \omega_2$$

$$\omega_2 = \frac{135}{g} \times \frac{g}{3.75}\,\omega_1$$

i.e., $\omega_2 = 36\,\omega_1$. Hence, $\omega_2 = 36$ rps.

The greater the mass at a greater distance from the axis of rotation, the slower the man will rotate. When the mass is drawn closer to the axis, r or K is decreased so that the moment of inertia becomes smaller and, as the angular momentum remains unchanged, the increase in the angular velocity must balance the equation.

Use of these principles forms the basis of control in many skills, from figure skating to gymnastics to rebound tumbling to more elaborate competitive dives. In the air the diver's (or tumbler's) body rotates about its center of gravity and the performer can regulate the speed of his rotation by the posture or postures he assumes. If his position is a tuck, the radius of gyration mentioned earlier will be quite short, making the moment of inertia comparatively small, and he will spin rapidly about a transverse axis in the frontal plane. Thus, a good diver can complete one and a half somersaults from the low (1 m or 3 ft) diving board. If the performer feels that he is spinning too fast, he can decrease his angular velocity by opening his tuck slightly, increasing the radius of gyration and thus his moment of inertia. On the other hand, if he feels that, at his present speed of rotation he will not complete the number of somersaults that he planned, he can increase his rate of spin by tightening up his tuck (or pike).

In the twisting dives or leaps or the skating spins, the performer rotates about an axis running lengthwise of his body, and he controls his moment of inertia, and hence his angular velocity, during the twist or spin by the position of his arms just as the man did on the frictionless turntable. By these techniques, many a diver has "saved" a dive that started to go wrong.

Analogies between Linear and Rotational Motion

Throughout the preceding section the analogies occurring between rotational and linear motion have been indicated. Table 3.3 summarizes these analogies and serves as a reminder to the reader of the many important concepts and laws that have been presented.

Table 3.3

Analogies between Linear and Angular Motions

Linear Motion	Angular Motion
Distance (s)	Angle (θ)
Velocity (v)	Angular velocity (ω)
$v = \Delta s/\Delta t$	$\omega = \Delta\theta/\Delta t$
Acceleration (a)	Angular acceleration (α)
$a = \Delta v/\Delta t$	$\alpha = \Delta\omega/\Delta t$
Mass (m)	Moment of inertia (I)
	$I = \Sigma mr^2 = MK^2$
Force (F)	Torque (Υ)
$F = ma$	$\Upsilon = I\alpha$
Impulse	Angular impulse
$F\Delta t = m\Delta v$	$\Upsilon\Delta t = I\Delta\omega$
Momentum (P)	Angular momentum (L)
$P = mv$	$L = I\omega$
Kinetic energy (KE)	Kinetic energy (KE)
$KE = 1/2\ mv^2$	$KE = 1/2\ I\omega^2$
Potential energy (PE)	Potential energy (PE)
$PE = mgh$	$PE = mgh$
Work done (Fs)	Work done ($\Upsilon\theta$)
$Fs = 1/2\ mv^2 + mgh$	$\Upsilon\theta = 1/2\ I\omega^2 + mgh$
Power (Fs/t)	Power ($\Upsilon\theta/t$)

FRICTION

Frictional forces arise whenever one body moves over the surface of another. Thus, when human locomotion takes place, friction plays a role in many ways. For effective propulsion when each foot in turn reacts with the floor it is important that there be no slip between the floor surface and the contacting body member. In this case large frictional forces exist between the two interacting surfaces. With an oily floor surface, the friction would be reduced and both propulsion and stepping would be hindered. Different body movement skills would be demanded for effective balance and progression. Body joints have members which in general articulate by rolling and sliding over each other. To minimize frictional forces appropriate surface properties and lubrication are required. This accounts for the anatomical structure of joints which are

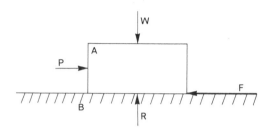

Figure 3.24. Diagrammatic representation of friction force.

bathed in synovial fluid. Frictional forces exist when tendons slide within guiding synovial sheaths.

The foregoing discussion suggests that frictional forces depend upon surface properties, lubrication, and the mode of movement between surfaces. Indeed, frictional forces depend on these factors and an understanding of the ways in which such factors can be controlled is advantageous in appreciating human movement aspects as well as designing devices of usefulness to man.

Figure 3.24 depicts a block of material A, having weight W upon a surface B. A horizontal force P is exerted in attempts to move A over B. W represents the weight of A. R is the reaction force of the surface B. It exactly counterbalances W. The force P required to move the block has to exceed slightly the frictional force F between A and B. By experimentally exploring the relationship between P and W for movement to occur it has been found that

$$P = F_{max} = \mu W = R$$

i.e.

$$F_{max} = \mu R \qquad (3.31)$$

which means that the maximum frictional force F_{max} depends on the reaction force R between the two surfaces and a factor which relates to the surface properties of A and B and is termed the coefficient of limiting friction.

From Equation 3.31 it is evident that F_{max} is independent of the surface area between A and B.

When relative motion or sliding between the surfaces occurs a value of μ_s, which is always lower than μ, pertains. Thus, the force for sliding

$$F_s = \mu_s R \qquad (3.32)$$

and

$$F_s < F_{max}$$

Coefficients of friction μ and μ_s are normally within a range from 0.1 to 1.0. For example, for wood on wood it is around 0.4; for metal on metal 0.2; for metal on metal but greased, 0.05.

These coefficients of friction describe the effects of different materials, the roughness of the contact surfaces, and the influence of lubricants.

When work is done by moving against frictional forces, such work generates heat between the surfaces.

Rolling friction results as a consequence of the deformations of both surfaces as one rolls over the other. Rolling friction is generally less than μ or μ_s by 100 to 1000 times, and depends on the reaction forces and the radius of curvature of the rolling member.

FLUID MECHANICS

The fluid medium in which the human body functions generally has an influence on performance. Most human activity takes place in the fluid medium of air and for most practical purposes the characteristic influences of this medium can be neglected. Air resistance or friction is usually not of consequence at the low speeds of movement involved in unaided human terrestrial activity.

However, a sky diver will rely on air-resistance properties and his interactions with them (for instance by changing the profile he presents normal to his path of flight) for control purposes.

As the density of the fluid medium increases fluid resistance and other properties will play a role in the mechanics of human performance. Water is a fluid medium in which most humans will function by floating and swimming. Walking in water up to the neckline is an energy-consuming endeavor relative to performing the same act in air. The fluid frictional forces resisting motion in water are far more substantial. Thus, resort is taken to other methods of propulsion with greater energy economy. Aspects pertinent to body equilibrium in fluids such as water include buoyancy and the body's center of buoyancy.

Buoyancy and Center of Buoyancy

Buoyancy is the upward force that any fluid exerts on an object or body which is partially or entirely immersed in it, and its force is equal to the weight of the volume of fluid which that object or body has displaced (derived from Archimedes' principle that a body immersed in a fluid experiences a buoyant force equal to the weight of the fluid it displaces). As gravity is considered as acting at the center of gravity (CG) of a body, so buoyancy may be considered as acting at the center of buoyancy (CB) of a body. If an object is of uniform density, CG and CB will coincide. However, neither boats, floats, fish, nor animals including man, have bodies of uniform density, so in them CB does not coincide with CG.

The center of buoyancy is defined as the center of gravity of the volume of the displaced fluid *before its displacement*. If one imagines a volume of water the same shape as the swimmer, the location of the CG of this volume corresponds exactly to that of the volume of a swimmer taken to have a uniform body density. The swimmer's CB, then, is located in the region which displaces the largest volume of water. Normally this is the thoracic region, although a very obese individual with large hips and thighs will have his CB nearer to the pelvic region.

Static Equilibrium in the Water

Static equilibrium in the water occurs only when the CB and CG are in the same vertical line. When the swimmer tries to float motionless his body will rotate until this condition is met (Fig. 3.25). The vertical floater (Fig. 3.25(c)) can, by altering the position of his extremities, move his CG cephalad in his body and thus change his floating angle (Fig. 3.25(d)). If an individual normally floats at a large angle with the horizontal, he would do well to start his float from the vertical position so that any rotation will result from the buoyant force of the water. If a swimmer who normally floats with his body close to the vertical starts his float from a horizontal position, his legs will drop with a gravitational acceleration that can well pull his head below water if he makes no effort to check the downward rotation.

Dynamic Equilibrium in the Water

As the water offers support to the swimmer and he makes progress by pushing against, or pulling his body along the surface of the water, dynamic equilibrium is not too much of a problem to him as long as he keeps to the usual horizontal or near horizontal mode of progress. However, when he attempts the various skills or stunts used in water ballet he may find it difficult to keep himself oriented to the stunt that he is trying to perform. Spears (1966) speaks of this as space orientation and points out that the ballet swimmer must have the complete body awareness which takes place through kinesthesia, i.e., through all parts of the proprioceptive system. The ballet swimmer is continually moving through three dimensions largely unfettered by gravity and, until he has been well-schooled to this condition in this medium, he can easily become disoriented.

Many stunts require the performer to keep the movement in a given plane and for his head to emerge either at the point of submersion (dolphins, somersaults, and related skills) or to emerge some distance away in a given direction, as in a porpoise. Sometimes the stunt involves a full or a half-twist so the swimmer should emerge facing in a different direction. If the performer depends on his inner ear mecha-

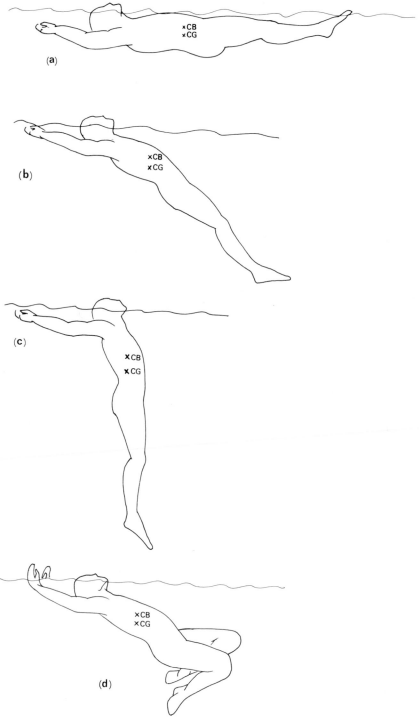

Figure 3.25. Floating positions dependent on the relationship between the swimmer's centers of buoyancy and gravity. (a), horizontal floater; (b), average floater; (c), vertical floater; (d), change of floating angle consequent on shift of body position relative to (c).

nisms to keep him oriented to the skill, he may find that they have betrayed him. This is particularly true in the early stages of learning ballet skills; thus, the beginning ballet swimmer orients himself by visual cues, checking his direction in relation to markings on the pool bottom or sides or in relation to nearby swimmers. In open, murky water this is, of course, more difficult. As he becomes more familiar with the performance, the ballet swimmer learns to depend on other proprioceptive cues, such as joint position and muscle tension, and he eventually achieves the total body awareness that enables him to perform an expressive movement.

It is difficult to account in detail for all of the dynamic forces acting upon all of the body members in an act as complex as swimming. Factors which can play a role are:

Surface drag. When water rushes over a propelled body there is a tendency for a resistance to be manifest between the body surface and the various layers of water adjacent to and then progressively farther from it. Surface drag in general depends on the body velocity, its surface area, its smoothness, and the nature of the fluid medium.

Form drag. This depends largely on the shape of the body, particularly its cross-sectional area perpendicular to the direction of fluid flow. The smoothness of the surface also plays a role in deciding this factor.

CONCLUSION

Having dealt with a number of important fundamental aspects of applied mechanics, we can now begin to consider the nature of the generation of forces by muscular action. This is the objective of Chapter 4. Chapter 5 deals with biomechanical examples which rely on the fundamental material embodied in this chapter.

CHAPTER **4**

Muscle—The Body's Force Generator

INTRODUCTION

Skeletal muscles are the prime movers of the human body. The forces they are able to produce under the control of the nervous system act on the bones to which they are attached to create the propulsive forces necessary for human movement. This chapter deals with important muscle properties and the factors which affect the magnitude of skeletal muscle contractile forces or tensions. Material on the mechanics of muscle action then follows. To gain insights to the structure and chemistry of skeletal muscle, the reader is referred to Chapter 7.

Man has about 640 skeletal muscles of many shapes and sizes, from the tiny stapedius muscle of the middle ear to the massive hip extensor, the gluteus maximus. Muscles are situated across joints and are attached at two or more points to bony levers. Forces giving rise to movements are usually generated with a concomitant shortening (or lengthening) in length and broadening (or narrowing) in width of the muscle.

Muscles differ in shape according to their functions. Some are long and slender for speed and range of movement, such as the biceps brachii; others are sheetlike to form supporting walls, such as the oblique abdominals; and some are multiple-headed to distribute and vary movement, such as the deltoid.

PROPERTIES OF SKELETAL MUSCLE

Muscle has four well-developed characteristic properties: irritability, contractility, distensibility, and elasticity. **Irritability** is the ability of muscle tissue to respond to stimulation. Muscle is our second most highly irritable tissue, being exceeded in this capacity only by nerve tissue.

The most distinguishing characteristic of muscle is **contractility.** By contractility, reference is made to the capacity of muscle to produce tension between its ends: to exert a pull. **Relaxation** is the opposite of contraction. It is entirely passive; the giv-

85

ing up of tension. Both relaxation and contraction progress from zero to maximal values over a finite time. Neither is instantaneous.

Muscles have a third property which is important in their function. They are distensible: they can be lengthened or stretched by a force outside the muscle itself. The stretching force can be the pull of an antagonistic muscle, of gravity, or of a force exerted by an opponent. Distensibility is a reversible process and the muscle suffers no harm so long as it is not stretched in excess of its physiological limits.

Finally, a muscle is **elastic:** unless it has been overstretched, it will recoil from a distended length. Distensibility and elasticity are separate and antagonistic properties whose coexistence in the muscle's connective tissues contributes significantly to muscle function. Although essentially each opposes the other, together they assure that contractions will be smooth and that the muscle will not be injured by a sudden strong change in either stretch or contraction.

Magnitude of Contractile Tension

The tension developed by a contracting muscle is influenced by a number of factors such as the characteristics of the stimulus (normally nerve signals transmitted to the muscle), the length of the muscle both at the time of stimulation and during the contraction, and the speed at which the muscle is required or made to contract.

The Stimulus

Most of what has been learned about muscle has been derived from studies using stimulation by electrical pulses. Although it is an artificial stimulus, electricity has distinct advantages for experimental purposes because it can be precisely controlled. The intensity, form (time course of rise to and duration of peak intensity), and frequency of pulses can be arbitrarily selected and varied as desired. Measurable responses of the muscle can be correlated with the quantitated stimulus characteristics. Interestingly, new therapeutic locomotor aids are available using controlled electrical stimuli applied to the motor nerve or muscle when regular nerve function is disrupted.

A curarized muscle may be stimulated directly by pulses applied to the muscle tissue or indirectly by pulses applied to its motor nerve fibers. The response of the whole muscle, of a single motor unit, or of one muscle fiber may be studied under controlled conditions.

The Single Pulse

Response of Muscle to a Single Pulse. If a single electrical pulse of adequate intensity is applied directly to a muscle fiber, the fiber will respond in an **all-or-none** fashion. Increasing the intensity of the pulse will not increase the magnitude of the fiber's response. It is important to mention here that the all-or-none response of the muscle fiber is determined by the all-or-none character of its excitation and not by any all-or-none limitations inherent in the contractile mechanism itself.

When a single adequate pulse is applied to a whole muscle, the muscle will respond with a quick contraction, followed immediately by relaxation. Such a response is called a **twitch.** Its magnitude will vary with the number of muscle fibers which respond to the stimulus and this will vary directly with the intensity of the pulse up to a finite maximal intensity.

The twitch is an indication of force development by the muscle. After a short **latent period** tension becomes evident and rises in a hyperbolic manner to a peak (the **contraction period**). It then declines over a slightly longer time course to zero (the **relaxation period**) (Fig. 4.1(a)).

The time course of the development of overt tension in the twitch is influenced by the interaction of the contractile components of the muscle fibrils with the elastic components of the muscle. Figure 4.1(b) illustrates the sequence of events and their influence on the shape of the twitch curve.

1. The active state, which indicates an increased resistance to stretch and a rise in heat production as compared to resting

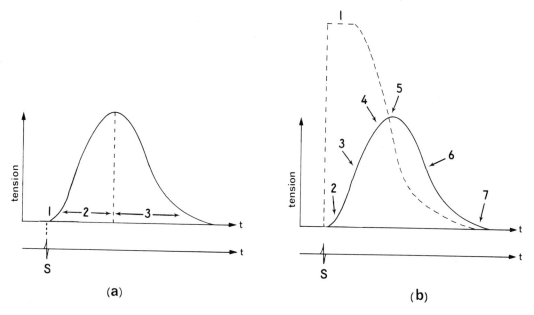

Figure 4.1. (a), tension development in a muscle twitch as a function of time. S, stimulus; 1, latent period; 2, contraction period; 3, relaxation period. (b), relationship of active state and tension development. Active state (broken line) is superimposed on the twitch tension curve (solid line). 1, peak of active state; 2, elastic components begin to exert tension; 3, tension rises rapidly; 4, as active state begins to decline, tension continues to rise but at a decreasing rate; 5, at the peak of the twitch curve, contractile and elastic tensions are in equilibrium; 6, as the active state continues to decay, recoil of the elastic components stretches out the contractile elements and tension falls; 7, active state decay is complete before tension has returned to zero. See text for further discussion.

muscle, is evident even before tension appears. It reaches full intensity abruptly, is maintained for about half of the contraction period, and then progressively declines during the rest of the contraction period. It is reflected diagrammatically in Figure 4.1(b).

2. The contractile components begin to undergo activation during the latter half of the latent period. As they shorten, the elastic components of the muscle are stretched and begin to exert passive elastic tension. Elastic tension is low at first. During this time the contractile elements are able to shorten rapidly.

3. When the active state is at full intensity, about halfway through the contractile period, the elastic tension is rising rapidly.

4. As the active state begins to decline in the latter half of the contractile period, its intensity is still sufficient to continue to stretch the elastic components, and tension continues to mount but at a decreasing rate. The twitch curve begins to round off.

5. At the peak of the twitch curve, tension in the contractile and elastic elements is in equilibrium.

6. Beyond the peak, as the active state continues its decay, developed tension falls below elastic tension and the elastic components recoil, stretching out the contractile components. Overall tension falls.

7. Decay of the active state is completed before tension returns to zero. The fact that tension outlasts the active state is partially explained on the assumption that

the breaking of cross-bridges within muscle elements requires more time than their formation. Therefore, the recoil of the elastic components lengthens the contractile material less rapidly than the rate of decay of the active state (Hanson and Lowy (1960); Walker and Schrodt (1967)).

The time course and intensity of the active state are studied by the techniques of **quick stretch** and **quick release.** Because of the elastic components and the viscosity of muscle tissue, the externally measured force exerted in a twitch is less than the full capability of the contractile material, that is, less than the intensity of the active state. The viscoelastic effect may be counteracted and the full tension characteristics of the contractile elements registered by employing quick stretch or quick release. If, coincident with stimulation, the muscle is given a short, quick stretch which pulls out the elastic elements just slightly beyond what their effective excursion would be, the muscle is relieved of the necessity of stretching out the elastic components and its full tension is revealed. By this means the onset, rise time, and duration of the peak intensity of the active state can be determined.

The time course of the decay of the active state is studied by the method of quick release, in which the fiber is stimulated to contract isometrically until its full active state has been developed. Then it is suddenly released to a slightly shorter length. Tension falls immediately but is quickly redeveloped, at a rate exceeding that in a normal twitch. The peak level, however, is lower. By varying the time of release and plotting redeveloped tension against time, a curve reflecting the decline of the active state is obtained (Fig. 4.1(b)).

Characteristics of the Single Pulse and Their Influence on the Muscle Twitch. An adequate stimulus may be defined as any environmental change, external or internal, which arouses in the contractile material an active state of sufficient magnitude to produce measurable tension. Whether natural or artificial the environmental change must meet certain minimal requirements with regard to its basic char-

acteristics: the magnitude or intensity of the change; its abruptness or rate of rise; and the duration of its application. Within physiological limits, increase above minimum in any of these will induce an increased response in the muscle.*

1. *Intensity.* A single electrical pulse must have a certain minimal intensity to be effective. The level of the minimum is an inverse measure of the irritability of the tissue; the smaller the minimal intensity, the greater the irritability. The minimal effective intensity is designated the **threshold** or **liminal stimulus.** These terms refer to the weakest stimulus which will evoke a barely perceptible response. **Subthreshold** and **subliminal** refer to a stimulus of inadequate intensity. As the intensity of the single pulse is increased above minimal, contractile tension in the muscle increases progressively as a result of the activation of more and more muscle fibers. Finally, an intensity is reached which evokes the maximal response of which the muscle is capable. Presumably all fibers are then active. Further increase in intensity will not be accompanied by further increase in contraction. The weakest stimulus intensity which will evoke maximal contraction of a muscle is called the **maximal stimulus.**

2. *Abruptness or Rate of Rise.* A weak but adequate pulse with a rapid rate of rise from zero to its preset intensity will evoke a stronger contraction than will a pulse of the same intensity with a slower rise. A minimal rate is required even for an intense stimulus. If intensity rises too gradually, there will be no response at all; the stimulus is then ineffectual. For any stimulus of adequate intensity, the more abruptly it is applied the greater will be the response it evokes, within the limits of the muscle's capacity. The greater the intensity the less rapidly it need rise to produce a given level of response.

A common experience illustrates the principle. If the hand is plunged abruptly

---

* In a single fiber, only an increase in the frequency of stimuli will produce an increase in its all-or-none response.

into hot water of about 110°F, the response (sensation of heat) resulting from the abruptness of the change in skin temperature from about 93 to near 110°F will be greater than if the change is made gradually by first immersing the hand in water at skin temperature and then slowly raising the temperature to 110°F. If the rate of temperature change is too slow, the change will be imperceptible.

3. *Duration.* For a stimulus of adequate intensity and rise rate, the duration of its peak intensity will influence its effectiveness. Within limits, the longer its duration the greater will be the muscle's response. Exclusive limits are found at both extremes: the duration can be so short that no response will occur in spite of the fact that the same intensity and abruptness would be sufficient with longer duration, or the duration can be so long that the response decreases until it ceases altogether. The latter is a common experience in the laboratory when direct current is used to stimulate tissue. The muscle responds at the closing of the circuit but ceases to respond as current flow continues at the constant (peak) level. The duration of the peak intensity has exceeded the response capabilities of the tissue.

The relationship of intensity and duration of single current pulses in the production of a barely perceptible contraction is presented in the intensity-duration curve shown in Figure 4.2. Note that both the upper and lower ends of the curve become straight lines, one vertical and the other horizontal, neither meeting the axes. The upper end indicates that even a very strong stimulus must be applied for at least a minimal duration to be effective. The lower end shows that below a certain minimal intensity a stimulus will not induce a response regardless of its duration. Between these limits, the greater the intensity, the less duration is required to produce a response.

For a stimulus of constant intensity and rate of rise, the longer its duration the greater will be the response up to a finite limit (Fig. 4.3(a)). The duration required for a given stimulus to evoke a perceptible

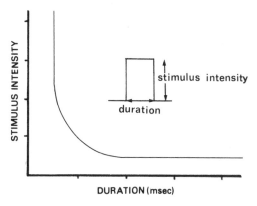

Figure 4.2. Intensity-duration curve. The upper limb of the curve indicates that a very strong stimulus must be applied for at least a minimal duration in order to be effective. The lower limb shows that below a certain minimal intensity a stimulus will not induce a response regardless of its duration. Between two limbs intensity and duration are inversely related.

response is its **excitation time** and is, within the limits discussed above, inversely related to the intensity.

If a stimulus of constant intensity and duration is applied at various rates of rise, effectiveness will be directly related to the rate. The more abruptly the stimulus is applied the greater will be the muscle's response. As the rate decreases, the response will diminish until ultimately, regardless of intensity, the stimulus becomes ineffectual (Fig. 4.3(b)).

The decreased effectiveness of a constant stimulus intensity at long duration and/or low rate of rise is designated **adaptation** or **accommodation.** Many tissues besides muscle adapt to a gradual or persistent stimulus. The physiological changes which are induced by the stimulus are apparently reversed at a rate which is faster than their development under the existing conditions. In the case of muscle tissue, excitatory processes may be inadequate to activate the tissue or, if activated, the magnitude or persistence of the active state may be insufficient to stretch out the elastic components enough to produce overt tension.

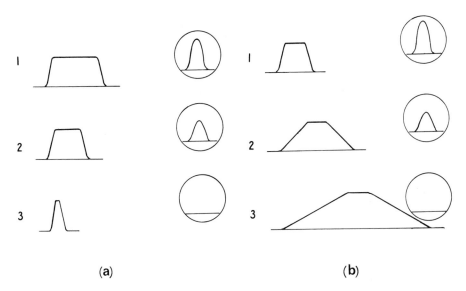

Figure 4.3. Comparison of single pulses with regard to rate of rise and duration of peak intensity. Hypothetical responses are shown in circle insets. (Note: the time scale of the pulses is greatly exaggerated as compared with that of the responses.) (a), duration of pulse. Three pulses of identical intensity and rate of rise but with different durations are shown: 1, moderate duration: probably the most effective; 2, short duration: less effective; 3, duration too short: ineffectual. (b), rate of rise. Three pulses of identical intensity and duration but with different rise times are shown: 1, the most rapid rise: the most effective stimulus; 2, less rapid rise: less effective; 3, least rapid: least effective (response much reduced or absent).

The rate of rise and duration of electric pulses may be varied as required for the principle under study. For most studies of concern to us, a pulse of rapid rise and short duration is used, with variations in its intensity appropriate to experimental objectives.

Repetitive Stimulation

Response to Repetitive Pulses. If an adequate stimulus is applied to a muscle fiber repeatedly at a rate rapid enough so that each succeeding stimulus reactivates the contractile elements before the previous tension has completely subsided, successive responses summate, each building upon the previous until a maximal level is achieved. If stimulation is continued, the contraction peak is maintained at this level. Such a response is known as **tetanus** or **tetanic contraction.** Ultimately, fatigue will cause the peak level to decline pro-

gressively. When stimulation ceases, contraction terminates and the fiber relaxes, tension subsiding quickly to zero. If, however, the repetitive stimulation is too prolonged, **contracture** will result and relaxation will be very much slowed as compared with normal. Unlike rigor, contracture is reversible.

Effect of Frequency of Pulses upon Response. The frequency of stimulation, usually expressed as cycles per second (some investigators employ the recently adopted physical unit, **hertz**), determines both the shape and the magnitude of a tetanic contraction traced on a myograph by an excised muscle. When pulses are delivered with a period which places successive stimuli during the relaxation phase of the preceding response, the contraction approaches a tremor and a scalloped tracing results. This is **incomplete tetanus.** With a period which is short enough to

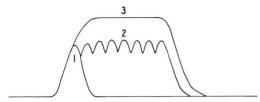

Figure 4.4. Response to repetitive stimulation. Curve 1, single twitch in response to a single stimulus; curve 2, incomplete tetanus in response to low frequency repetition of the stimulus; curve 3, complete tetanus in response to higher frequency repetition of the stimulus. Same stimulus magnitude used in all three.

restimulate during the contraction phase, the tracing is smooth. This is **complete tetanus** (Fig. 4.4). Within physiological limits, the shorter the period (i.e., the greater the frequency) the smoother the curve and the greater the tension development will be. If, however, the period is shortened beyond a certain point, the refractory period will be encountered. The **absolute refractory period** is a short space of time immediately following stimulation during which the muscle cannot be reexcited regardless of stimulus intensity. This is followed by a longer period, the **relative refractory period,** during which irritability is gradually regained and the tissue will respond to a stimulus which is appropriately greater than threshold. The earlier the pulse falls in the relative refractory period the greater its intensity must be to be effective. Although both portions of the refractory period last for a finite time, in muscle both have been completed before tension begins.

Tetanus-Twitch Ratio. The tension developed in response to repetitive pulses is greater than that evoked by a single pulse of the same magnitude. The **tetanus-twitch ratio** varies with different muscles and may be as great as 5 (rat gastrocnemius). To explain the greater tension developed in a tetanic contraction, it has been postulated that in a twitch the short duration of the active state allows too few bridge movements to permit the contractile material to shorten enough to fully stretch out

the elastic components before the active state begins to subside. Hence the full capacity for tension production cannot be realized. Repetitive stimulation, however, by maintaining the active state, permits continuation of bridge activity. The stretching of the elastic components is completed and full tension is developed.

Post-Tetanic Potentiation in Muscle. In many muscles, especially when curarized, if twitch responses to single pulses are recorded before and immediately after a period of tetanic stimulation, the post-tetanic twitch shows an increase in magnitude and a steeper rise of tension than the pre-tetanic control. This phenomenon is known as **post-tetanic potentiation.** The effect occurs whether the muscle is stimulated directly or indirectly by its motor nerve.

Potentiation is maximal shortly after the repetitive stimulation and then decays exponentially at a rate which is dependent on both the frequency of pulses and the number delivered in the train. Short trains produce potentiation without any alteration of the twitch duration, but longer trains result in lengthening of the contraction time and of the half-relaxation time (the time required for tension to drop to 50% of its peak value).

If, as has been suggested (Close and Hoh, 1968), the mechanism of post-tetanic potentiation is located within the muscle fibers, it may involve prolongation of the active state with a resulting increase in the number of fully activated myofibrils in the fiber, or it may be due to an increased liberation of some activator substance, perhaps Ca^{++}, which induces an increase in the number of bridges formed between actin and myosin and in the rate of their cycling.

Conclusion

To be adequate, stimulation must consist of an appropriate combination of intensity, rate of rise, duration, and frequency to excite muscle fibers and to activate their contractile material sufficiently to produce measurable tension. Within the limits discussed, the adequacy of the stimulus is determined by the inter-action of these mutually interdependent characteristics.

Nervous System Stimuli

In the living body an adequate environmental change results in the generation and conduction of a train of action potentials in motor neurons which becomes responsible for excitation of the muscle fibers. The frequency of impulses reflects the effectiveness of the stimulation and determines the magnitude of the muscle tension developed. Because the magnitude and form of the nerve impulses are constant for any given set of body conditions, the frequency of these impulses is the most significant characteristic in determining the muscle's response.

The magnitude and duration of the impulses in any neuron are essentially constant (see Chapter 8), but frequencies vary, sometimes over a wide range. Therefore, the latter is the characteristic which influences the response of muscle fibers and is of concern to us in studying human movement.

The frequency of impulses in human motor neurons generally ranges from 20 to 40/sec. At such stimulation rates, the normal response of the muscle fiber is an incomplete tetanus. Muscle contractions, however, appear smooth because excitation by the various motor neurons is not synchronous and hence motor units respond out of phase. The relaxation in one motor unit is offset by contraction in another. Any oscillations of tension which might occur are further smoothed out by the transmission of tension to the lever through a common tendon. The tension developed by the muscle depends upon the number of motor units activated and the frequency of activation of the muscle fibers composing each unit.

During maximal exertion the frequency of motor impulses may be great enough to produce complete tetanic contractions. If so, synchrony of responses in motor units develops and tremor results, a common experience in all-out effort.

Electromyography (EMG)

Electromyography is an invaluable method for recording the presence of activity in muscles. Like nerve cells, muscle cells at rest are able to maintain an electrical potential difference (voltage) between the inside and outside of the cell, i.e., across the muscle cell membrane. When a muscle cell is stimulated by the arrival of a nerve impulse at the motor end-plate, which acts to transfer the stimulation to the muscle, the latter's membrane potential is reduced and the muscle cell becomes depolarized as migration of ionic constituents across the membrane occurs. While the membrane electrical potential difference undergoes a change which is propagated along the length of the fiber, there is a concomitant migration of ions in the extracellular space. The consequent electrical potentials (action potentials) produced in the extracellular regions are the ones picked up by electromyographic (EMG) electrodes connected to amplifier and display or output system. The EMG voltages usually lie in the range between 10 μV and 2 mV peak to peak and the frequency range is from about 25 to 20,000 Hz. The accuracy with which action potentials can be recorded depends on the electrical impedance of the recording electrodes and the characteristics of the amplifier including input impedance, frequency range, and common mode rejection capability (i.e., the ability to suppress signals of the same polarity which might derive from electrical interference or other biological activity such as the presence of an electrocardiographic (EKG) signal). Precautions often have to be taken to avoid motion artifacts, i.e., signals introduced due to movements of the electrodes, the rubbing together of electrical leads, and the like. Fortunately for most practical purposes, such artifacts encountered in kinesiological studies on humans are low frequency signals (less than 10 Hz in normal walking) and can be easily removed by electronically filtering the combined signals as they are acquired. Surface electrodes or intramuscular electrodes can be used. Surface electrodes are usually attached to the skin above the muscle segment under study. They are of value when there is interest in the activity of an entire muscle group. Silver-silver chloride (Ag-AgCl) electrodes are invariably used on the skin surface. Commercially available electrodes range in diameter from 2 to 10 mm for the active part of the electrode which is usually assembled with a plastic surround to enable adhesion to the skin. These electrodes are in order when gross muscular activity is of concern. However, if a detailed study of the activities of individual muscles in a group is required, intramuscular fine wire electrodes will be necessitated; they are more selective from a spatial standpoint. Often with adjacent muscles, there may be "cross-talk" between their EMG signals and electrode arrangements then have to be carefully considered. In most practical circumstances, bipolar electrodes are used, there being an additional ground connection being made between the subject and the electronic amplifier.

Coaxial needle electrodes can be used in studying isometric muscular contractions. They make it possible to study individual muscle action potentials or EMG activity in small zones. The basic construction consists of an insulated platinum wire (one pole) inside a stainless steel cannula (the other pole) of the electrode system. These electrodes being painful and uncomfortable with movement are therefore displaced by fine wire intramuscular electrodes. The latter are made of thin flexible insulated metal wire with the insulation removed for a few mm at the tips. Twenty-five micron diameter Karma wire with a polyurethane insulating coat has been most widely used. The wires are usually introduced in pairs by injecting them into the muscle with a hypodermic needle (no. 27). First, the wire ends are bent over the rim of the hollow needle to form a barb. This hooks the electrode wires to the muscle tissue when the needle is withdrawn from the muscle. This type of electrode set usually causes little or no discomfort after insertions and the electrodes usually remain in about the same fixed location. When minor dislocations of electrode positions occur, quantitative EMG studies can be influenced and be in error. However, timing or phasic information can still be derived. To minimize motion artifacts, preamplifier stages should be kept close to electrode sites, thus keeping the electrode leads short. Recordings for viewing EMG are most often made using ultraviolet (UV) light recorders. Various UV oscillographs with multichannel capabilities are available on the market. For some purposes, it is often convenient to process the EMG signals before studying them. Often the envelope of the rectified† original signal is derived electronically. This has the advantage that it can be recorded using a recorder with a low bandwidth, i.e., dc to 50 or 100 Hz. Other processing methods are to integrate the EMG signal or to count zero crossings. Additional information on electromyographic technique is presented in Chapter 6.

Muscle Length

The most obvious property of muscle is its capacity to develop tension against resistance. The length of the muscle at the time of activation markedly affects its ability to develop tension and to perform external work. Muscle tension may be measured in terms of the greatest load which can just be lifted or as the maximal tension read-out on a strain gauge or tensiometer.

When a muscle contracts, the contractile material itself shortens, but whether the whole muscle shortens or not depends on the relation of the internal force developed by the muscle to the external force exerted by the resistance or load. The terms "force" and "tension" are often used erroneously as synonyms. Tension is a scalar quantity having magnitude only, while force is a vector quantity having both magnitude and direction. The term **tension** is

---

† A rectified signal may have either the negative or positive portion of the signal removed (half-wave rectifier) or one of the portions would be switched to make the entire signal either positive or negative (full-wave rectifier).

used in this discussion to refer to the magnitude of the pull of the muscle as it would be registered on a strain gauge arranged in line with the muscle axis. **Internal force** is used to refer to the tension magnitude acting in the direction of the action line of the muscles under given conditions, and external force refers to the resistance opposing the muscle.

Types of Muscle Contraction

There are three identifiable types of muscle contraction designated according to the length change, if any, induced by the relationship of internal and external forces. The three types are isometric, isotonic, and eccentric contraction.

Isometric Contraction. If the internal force generated by the contractile components does not exceed the external force of the resistance and if no change of muscle length occurs during the contraction, the contraction is **isometric**. The available energy expended by the muscle and the tension produced against the resistance may be considered to be in equilibrium.

No contraction in the body is purely isometric because at the fibril level the contractile components do shorten. Their shortening, as discussed above under active state, is offset by stretching of the elastic components. By current usage, an isometric contraction is one in which the *external* length of the muscle remains unchanged. (Isometric contraction is sometimes called static contraction.)

Isotonic Contraction. If the constant internal force produced by the muscle exceeds the external force of the resistance and the muscle shortens, producing movement, the contraction is **isotonic.** (This is sometimes called a *concentric* or a shortening contraction.) Energy utilization is greater than that required to produce tension which will balance the load, and the extra energy is used to shorten the muscle. During isotonic contraction work is done by the muscle on the load. This is regarded as *positive work*. Recalling Equation 3.14 work W is the product of the force F and

the distance x it is moved through

$$W = Fx$$

A muscle can develop greater tension in isometric than in isotonic contraction because none of the available energy is expended in shortening. In isotonic contraction the greatest load that the muscle can lift is about 80% of its maximal isometric tension.

Eccentric Contraction. If to an already shortened muscle an external force greater than the internal force is added and the muscle is allowed to lengthen while continuing to maintain tension, the contraction is called **eccentric**. (The term lengthening contraction is sometimes used.) The energy expended by the muscle is less than the tension exerted on the load, but the muscle acts as a brake controlling the movement of the load. In eccentric contraction a muscle can sustain greater tension than it can develop in isometric contraction at any given equivalent static length. During an eccentric contraction work is done by the load on the muscle. This is logically taken as negative work. Naturally, negative work is measured in the same units as positive work.

In lifting and then lowering a given weight W through a vertical distance h the amounts of positive and negative work for the musculature involved (say for biceps) would be identically equal to Wh. While chemical energy is expended by the muscle in both instances, the energy cost of the negative work is considerably less than for the positive work. The difference is indicated by a lower oxygen uptake during the negative work, being about one-tenth as much in human subjects. Other estimates have placed the cost at one-third to one-thirteenth of that required for the equivalent amount of positive work.

Eccentric contractions are very common. Every movement in the direction of gravity is controlled by an eccentric contraction. Examples include sitting, squatting or lying down, bending forward or sideward, going down stairs, stooping, placing any object down onto a surface, etc. In eccentric contractions the active muscles are those which are the antagonists of the same movement when it is made against gravity. Sitting or squatting is controlled by leg extensors, not flexors; lying down, by hip flexors, not extensors; lowering a load, by shoulder flexors, not extensors. Electromyograms (which reflect the electrical activity of muscle and hence provide insights regarding muscle contractions) show not only that anatomically antagonistic muscles are actively controlling the eccentric movement but also that the electrical activity in these muscles is less than when the same muscles are contracting isotonically to do the same amount of positive work with the same load over the same distance and at the same speed.

Concentric Contraction. In contradistinction with eccentric contractions, concentric contractions occur when the muscle is shortened while maintaining tension. The work of Singh and Karpovich (1966) demonstrates that the force developed by elbow flexor and extensor muscles in eccentric contraction exceeds that in both isotonic and isometric contraction at most muscle lengths. Using an instrument designed by Singh, they measured the manifestations of muscle force through the entire range of motion at the elbow joint, simultaneously recording the angle through which the forearm was moving. With the elbow flexors, eccentric force was consistently the greatest of the three over the entire range of motion and isotonic was least (Fig. 4.5(a)). With the elbow extensors, although eccentric force was lowest of the three at the start (shortened position, elbow at 140°), it had exceeded isotonic force when the angle reached 120°, and by 100° and lower it had surpassed isometric force as well (Fig. 4.5(b)).

Daily activities involve a continual shifting from one to another type of contraction and of combinations of the three types, as required. During movements the changing lengths of lever arms and of angles of pull, for both muscle and load, introduce complexities which require complicated processes of neuromuscular integration to properly adjust the number and activity of motor units to the task.

Relationship of Muscle Tension to Length

The **initial length** of a muscle, i.e., its length at the time of stimulation, influences the magnitude of its contractile response to a given stimulus. A stretched muscle contracts more forcefully than when it is unstretched at the time of activation. This is true whether the contraction is isometric, isotonic, or eccentric. Within physiological limits, the greater the initial length, the greater will be the muscle's tension capability. Parallel-fibered muscles exert maximal total tension at lengths only slightly greater than rest length. Muscles with other fiber arrangements have maxima at somewhat greater relative stretch. In general, optimal length is close to the muscle's maximal body length, i.e., the greatest length that the muscle can attain in the normal living body. This is about 1.2 to 1.3 times the muscle's rest length. Tension capability is less at shorter and longer lengths. Therefore, a muscle can exert the greatest tension or sustain the heaviest load when the body position is such as to bring it to its optimal length. In isotonic contractions the increased tension and longer length permit greater shortening; hence, more work can be done or, alternatively, the same work can be done at lower energy cost. The diminished energy cost of eccentric contraction is in part due to this **stretch response,**‡ but other factors are also involved, as evidenced by the capacity to produce greater tension than with either isometric or isotonic contractions at most equivalent lengths.

The relationship of tension to muscle length may be presented graphically in the form of a **tension-length curve** in which tensions in an isolated muscle are plotted against a series of muscle lengths from less

---

‡ The stretch response should not be confused with the stretch reflex. The latter is a response mediated by the nervous system, while the former is a property of muscle tissue independent of nerve.

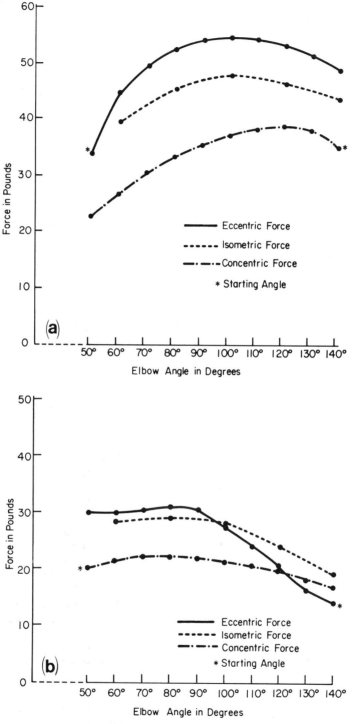

Figure 4.5. Concentric, isometric, and eccentric tension curves. Curves of maximal concentric, isometric, and eccentric tension of (a) forearm flexor muscles and (b) forearm extensor muscles. (From Singh, M., and Karpovich, P.V., 1966. Isotonic and isometric forces of forearm flexors and extensors. *J. Appl. Physiol.* 21: 1435.)

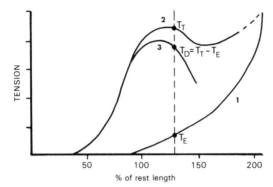

Figure 4.6. Tension-length curves for isolated muscle. Curve 1, passive elastic tension T_E in a muscle passively stretched to increasing lengths; curve 2, total tension T_T exerted by muscle contracting actively from increasingly greater initial lengths; curve 3, developed tension calculated by subtracting elastic tension values on curve 1 from total tension values at equivalent lengths on curve 2, i.e., $T_D = T_T - T_E$.

than to greater than the resting length (Fig. 4.6). Both the passive elastic tension (curve 1) exerted by the elastic components in the passively stretched muscle and the total tension (curve 2) exerted by the actively contracting muscle are plotted. Since total tension T_T represents the sum of elastic tension T_E plus the developed tension T_D of the contractile elements, the latter may be found by subtraction and is represented by the difference $T_T - T_E$ between the curves for total and elastic tensions. Values for developed tension are shown as curve 3. Note the following facts regarding developed tension.

1. At less than 50% of rest length the muscle cannot develop contractile tension.

2. At normal (intact) rest length the muscle is already in slight passive elastic tension. At this length the muscle produces its greatest *developed* tension ($T_D = T_T - T_E$).

3. When contraction is initiated at a length longer than rest length, although *total* tension is greater than at rest length, *developed* tension has already diminished and declines progressively at all greater lengths.

4. At extreme lengths (far right end of the curves) total tension would ultimately become equal to elastic tension, developed tension being zero.

Maximal contractile tension is assumed to be developed when sarcomere lengths are such that maximal single overlap of actin and myosin filaments exists. At greater lengths the number of cross-links diminishes as overlap decreases, and at shorter lengths double overlap results in reduced tension as a result of the antagonistic action of bridges. Gordon *et al.* (1966) investigated the tension-length relationship in frog skeletal muscle fibers at various sarcomere lengths. Their results are plotted in Figure 4.7. In these fibers the mean sarcomere length was 2.5 μ and the mean filament lengths were 1.5 μ for myosin and 1.0 μ for actin. Maximal tension was developed at sarcomere lengths of 2.0 to 2.25 μ. At greater lengths tension decreased linearly, becoming zero at about 3.65 μ. At shorter lengths tension declined gradually with decreasing length until about 1.7 μ and then dropped abruptly to 0 at about 1.27 μ.

Drawing upon these data and the electron microscope evidence of filament relationships in contracted muscle, we may postulate the stages of the sliding filament process most probably associated with significant lengths on the tension-length curve presented above. Figure 4.8, (a) through (d), presents four of these stages. The work of Gordon *et al.* thus provides further evidence favoring the sliding filament theory and supporting the concept of a quantitative relationship between tension and the number of bridges linking actin and myosin filaments.

Knowledge of length-tension relationships is obviously of importance to orthopaedists who undertake repairs of the musculoskeletal system.

Speed of Contraction

Most isolated nonloaded muscles normally shorten by about 50% or less of their rest length. The absolute amount by which any muscle can shorten depends upon the length and arrangement of its fibers, the

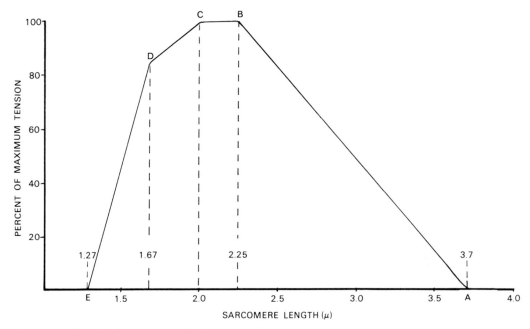

Figure 4.7. Tension-length curve for frog muscle at various sarcomere lengths. The letters on the tension curve and the broken vertical lines relate tension to significant sarcomere lengths. Note that tension is 0 both at the shortened length of 1.27 μ and at the extended length of about 3.7 μ, is maximal over lengths 2.00 μ and 2.25 μ, and declines rapidly below 1.67 μ and above 2.25 μ. (After Gordon, A. M., Huxley, A. F., and Julian, F. J., 1966. Variation in isometric tension with sarcomere length in vertebrate muscle fibers. *J. Physiol. (London) 184*: Fig. 12, p. 185.) See Figure 4.8 for diagrams of probable filament relations in myofibrils at some of the lengths.

greatest shortening occurring in the long parallel-fibered muscles such as the biceps and sartorius. In intact muscle, shortening is further limited by the structure of joints, the resistance of antagonists, and any load which opposes the muscle.

Intrinsic Speed of Shortening

The intrinsic shortening speed of a muscle reflects the rate of shortening at the sarcomere level. It is limited by the rate at which bridges can attach, move, and detach and by the rates of the chemical reactions involved. With muscle attachments severed, shortening speed of the contractile material is maximal but no tension is developed. A muscle can produce tension only when shortening against resistance, and the amount of tension developed is equal to the load.

When shortening against resistance, speed varies inversely with the load. Therefore in isotonic contraction the less the resistance the more nearly maximal is the rate of shortening. This may be explained as follows: the active state arises abruptly upon stimulation and persists for a relatively fixed period of time; the less resistance which is met by the contractile material the more readily the bridges function and the greater the distance of shortening accomplished during the persistence of the active state.

When a muscle is required to shorten more rapidly against the same load, less tension is produced than when it shortens more slowly. This may be due to the fact that fewer links are formed between actin and myosin in the shorter time available and that the bridges which do form are detached more quickly. Consequently at

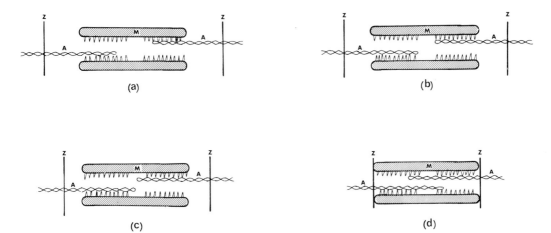

Figure 4.8. Schematic drawings of filament relationships at various stages of the sliding filament process associated with significant lengths on the tension-length curve presented in Figure 4.7. M, myosin; A, actin; Z, Z disc. (a), sarcomere length 2.5 μ. Actin filaments are partially overlapping the myosin filaments in the A band. Not all bridges can attach. Tension capability about 85% maximum. (b), sarcomere length 2.25 μ. Actin filaments are in maximal single overlap with the bridge-containing regions of the myosin filaments. Maximal tension capability. (c), sarcomere length 2.0 μ. Actin filaments have reached the center of the A band. Still maximal tension capability. (d), sarcomere length 1.5 μ. Z discs have collided with the ends of the myosin filaments. Ends of the actin filaments have passed into the bridge area of the opposite half of the sarcomere. Forty-five percent maximal tension capability.

higher speeds fewer bridges will be attached at any given moment and less tension is produced.

Hill (1965) has pointed out that the load determines the rate of the chemical reactions associated with contraction and that the magnitude of the velocity depends on the difference between the actual load being lifted and the maximal magnitude of force of which the muscle is capable. Bárány's work (1967) supports the correlation of ATPase activity and speed of contraction in 14 different muscles of mammals, lower vertebrates, and invertebrates.

In isometric contraction, different levels of tension are achieved at the same rate. Since no shortening is involved beyond that needed to stretch out the elastic components, the rate of tension development is constant, determined by the active state. Hence the time to reach any given tension will be proportional to the tension: lower tensions will be achieved sooner than higher tensions.

Force-Velocity Relation

Speed is a scalar quantity, lacking the component of direction, while velocity is a vector quantity having both magnitude and direction. Therefore, the term *speed* has been used in the discussion of the rate of intrinsic shortening of the contractile material and the rate of tension development within the muscle, for direction was not significant. The term *velocity* is used to discuss the rate of muscle shortening against external resistance, i.e., the rate of movement, for in such considerations direction is an influential factor.

The velocity at which a muscle shortens is influenced by the force that it must produce to move the load. In isotonic contraction the relationship is evidenced by the decrease in velocity as the load is increased (Fig. 4.9, solid line curve). Shortening velocity is maximal with zero load and reflects the intrinsic shortening speed of the contractile material. Velocity

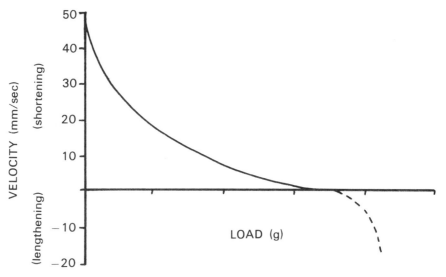

Figure 4.9. Relationship of velocity of shortening to tension in isotonic contraction (——). As the load is increased, velocity of shortening decreases, reaching 0 with a load just too great for the muscle to lift. In eccentric contraction, shortening velocities become negative and tension increases with increased speed of lengthening (- - -).

reaches zero with a load just too great for the muscle to lift; contraction is then isometric and maximal force can be produced.§ When more muscle fibers are activated than are needed to overcome the load, the excess force is converted into increasing velocity and therefore greater distance of movement. A commonly experienced example is the exaggerated movement which occurs when one lifts a light object anticipated to be much heavier.

In eccentric contraction, values for shortening velocity become negative and the muscle's ability to sustain tension increases with increased speed of lengthening, but not to the extent which might be expected from extrapolation of the shortening curve (see broken line in Fig. 4.9, extending the curve from the hyperbola of the force-velocity curve below the abscissa into the area of lengthening velocity).

§ As previously stated, greater forces can be *sustained* in eccentric contraction than can be *produced* in either isometric or isotonic contraction.

In isotonic contractions the length and tension of the elastic components do not change once they are sufficiently stretched to permit the load to be raised. Therefore, isotonic twitch myograms reflect the velocity as well as the extent of shortening. When an excised frog gastrocnemius muscle records twitch responses with different loads, the lighter the load the higher is the twitch curve and the steeper its rising slope (Fig. 4.10). In other words, the lighter the load the greater the amount of shortening per unit of time. With a light load (Fig. 4.10(a)) the muscle's maximal velocity as measured over the steepest part of the contraction is 17 mm in 0.01 second or 1.7 m/sec. With a moderate load (Fig. 4.10(b)) velocity is 12 mm in 0.01 second or 1.2 m/sec, and with a heavy load (Fig. 4.10(c)) velocity is 5 mm in 0.01 second or 0.5 m/sec.

Optimal Velocity

The force-velocity relationship may also be stated conversely in terms of the influence of velocity upon force: a rapidly

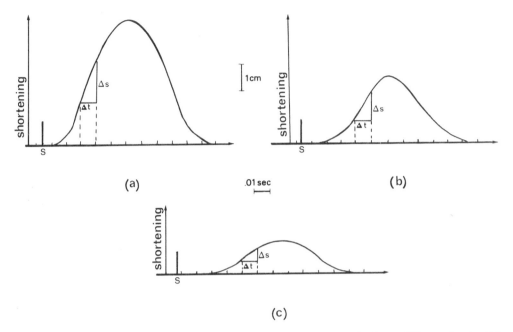

Figure 4.10. Shortening velocity of excised frog gastrocnemius with three different loads. Stimulus, S; Δt, 0.01 second taken at steepest part of contraction; Δs, amount of shortening in millimeters during that time unit. (a), with a light load, maximal rate of shortening (as taken over the steepest part of the contraction) is 17 mm/0.01 sec or 1.7 m/sec. (b), with a moderate load, velocity is 12 mm/0.01 sec or 1.2 m/sec. (c), with a heavy load, velocity is 5 mm/0.01 sec or 0.5 m/sec.

contracting muscle generates less force than does one contracting more slowly. From the standpoint of efficiency, however, energy expenditure is least when work is done at moderate velocity. With any given load, if the velocity of shortening is gradually increased, the work output of the muscle rises at first, reaches a peak, and then declines. Therefore, for any load the optimal velocity lies somewhere intermediate to the slowest and fastest shortening rates. Furthermore, the greater the load the lower is the optimal velocity with which it can be moved.

Velocities below optimum are uneconomical because force must be maintained over a longer time; hence, more energy is expended to achieve the same amount of shortening. Velocities above optimum waste energy because of the need to employ a greater number of muscle fibers to achieve the same force. A plausible explanation is as follows: (1) the tension developed will be proportional to the number of bridges attached at any moment; (2) chemical reaction rates dictate that a finite and constant time is required for a bridge to attach; (3) the greater the speed at which the active sites on the actin filament move past the myosin bridges, the fewer the bridges that can attach and the less will be the tension; (4) as a result, more muscle fibers must be recruited to achieve the necessary force. Optimal speed probably reflects the greatest speed which will still allow a sufficient number of bridges to attach to provide the required tension.

Most individuals will unconsciously perform at optimal velocity if allowed to do so. Optima vary for the same individual with different loads and in different types of activity and among individuals for the same load or activity. In athletic performance efficiency is often sacrificed for other objectives, and rates above optimum are deliberately adopted.

There is evidence that strength increase in a trained muscle is due to an increase in the number of fibrils per fiber. This would mean that there would be more bridges within the fiber which could attach per unit of time, and the force capability at any given velocity would be correspondingly greater. A trained muscle should be able to develop a given force more rapidly. In other words, training should be expected to improve both force and optimal velocity of contraction.

In isolated vertebrate muscle the Q_{10} of velocity[||] is 2.5. Although studies have not been made directly in man, empiric evidence suggests that the velocity of muscle shortening is improved by "warming up." Experiments on the velocity effects of local warming or increased core temperature might indicate that artificial prewarming of athletes to a safe degree would improve their performance records. Physical educators have differing opinions on the value of "warm-ups."

Slow and Fast Muscle Fibers

Although the previous discussion has considered striated muscle in general, there is abundant evidence that there are two types of skeletal muscle, distinguishable by speed of contraction and endurance. Almost 100 years ago Ranvier observed that some muscles of the rabbit were redder in color and that those muscles contracted in a slower and more sustained manner than did the paler muscles of the same animal. Since then the designations of red and white muscles have become synonymous with slow and fast contraction, respectively. In addition to a slower contraction-relaxation cycle, red

---

[||] Q_{10} refers to the extent to which the rate of a chemical reaction is increased by a 10°C rise in temperature. It is usually measured within the range 15–35°C. A Q_{10} of 2 indicates that the rate is doubled. For most biochemical reactions Q_{10} lies between 2.5 and 3.0.

muscles have lower thresholds, tetanize at lower frequencies, fatigue less rapidly, and are more sensitive to stretch than the faster white muscles.

As might be expected, individual muscle fibers reflect these differences in contractile behavior. Investigations by a number of workers have revealed histological and biochemical differences which distinguish the two types of muscle fibers and which correlate with the physiological differences between fast (white) and slow (red) muscles. These are listed in Table 4.1.

Examination of the table suggests obvious relationships between the several categories which are consistent with differences in speed and endurance. The larger size, greater density of fibrils, and lower viscosity of the white fibers should contribute to greater speed. Their abundant sarcoplasmic reticulum and T tubules located at the A-I junctions should, by providing for the transport of glycolytic enzymes and for large scale release of Ca^{++} ions in the vicinity of the cross-bridges, also favor fast response. Conversely, in the red fibers, reduction in the sarcoplasmic reticulum and sparsity of T tubules are appropriate to slower response. The discrete end-plates and multiple folds in the subneural sarcolemma of the fast fibers may be expected to provide for increased activation. Where multiple innervation exists in slow fibers, it is consistent with the small graded action potentials which distinguish them from the fast fibers with their rapidly rising, larger potentials.

The large glycogen stores and high ATPase activity in the white fibers will favor speed of response but quicker fatigue is to be expected because of their lesser blood supply and predominantly glycolytic metabolism. The increased endurance of the red fibers is consistent with their rich blood supply and abundance of mitochondria which support an essentially oxidative metabolism. Furthermore, their small diameters provide a greater surface for exchange of gases, ions, and metabolites than is provided by an equivalent mass of larger white fibers. Rapid K^+ depletion progressively alters the ionic gra-

Table 4.1
Comparison of Fast and Slow Muscle Fibers: Histological, Physiological, and Biochemical

	Fast	Slow
Histological differences		
Size and color	Fibers large and pale	Fibers smaller and redder because of greater myoglobin content
Sarcoplasm	Agranular sarcoplasm	Granular sarcoplasm
Fibrils	Many fibrils	Fewer fibrils
Mitochondria	Large mitochondria but few in number	Numerous small mitochondria
Z discs	Narrow Z discs	Wider Z discs (about 2×)
SR and T system	Sarcoplasmic reticulum abundant and well-developed; T tubules (and triads) at A-I junctions	SR sparse and rudimentary; T system, when present, is found at the Z lines
Innervation	Single innervation by large somatic motor neurons with fast conduction rates	Innervation by small, slow-conducting somatic motor neurons; multiple innervation in some species. Some autonomic neurons
End plates	Discrete (*en plaque*) end-plates with many sarcolemmal folds	Diffuse (*en grappe*) end-plates with few or no junctional folds
Blood supply	Few capillaries except those shared with adjacent slow fibers	Dense capillary supply, located at the interstitial angles between fibers
Physiological differences		
Contraction cycle	Rapid contraction-relaxation cycle	Slower cycle (2–3×); graded contraction in some muscles
Tetanus	Rapid onset of tetanic fusion but only at high frequencies and short-lasting	Slower onset of fusion but at lower frequencies and of longer duration
Potentials	Higher resting potential	Lower resting potential
	Larger end-plate and action potentials	Smaller end-plate and action potentials, latter often graded
Active state	Rapid initial decay of active state	Slower initial decay
Endurance	Rapid fatigue	Greater endurance
Tension capacity	Higher tension which develops rapidly	Lower tension and slower development
Elasticity	Lower coefficient of elasticity	Greater elasticity
Biochemical differences		
Metabolism	Metabolism primarily glycolytic (as indicated by high ATPase activity)	Oxidative metabolism (as indicated by high succinic dehydrogenase activity)
Myoglobin content	Low	High
Glycogen	Large glycogen storage	Variable glycogen storage
Na^+ and K^+	Less Na^+ and more K^+; rapid loss of K^+ during stimulation	More Na^+ and less K^+; rate of K^+ depletion diminishes with continued stimulation
Amino acids	Differences in concentration of various amino acids	

dients and ultimately limits the ability of the fast fibers to perform work. In the slow fibers presumably a steady state is reached in which K^+ gain resulting from recovery processes is in equilibrium with the loss incurred during contraction. As a result, endurance is enhanced. The greater elasticity and slower initial decay of the active state which is characteristic of the red fibers can account for their mechanical fusion at lower frequencies and for their prolonged twitch times.

Most of man's striated muscles contain both types of fibers but in differing proportions which determine the color of each muscle. Some show a characteristic arrangement or zonation of the fiber types within the muscle; in others the two types are randomly distributed. In such muscles as the gastrocnemius, tibialis anterior, and flexor digitorum longus, fast fibers predominate, although slow fibers may also be present. In many mammals the soleus muscle appears to consist entirely of slow fibers. The preponderantly slow-fibered muscles are the antigravity muscles, adapted for continuous body support. Their sensitivity to stretch results in a continuous mild (tonic) activity even at rest. The predominantly fast-fibered muscles are phasic muscles which produce quick postural changes and fine skilled movements. At rest they are electrically silent.

Recent studies indicate that in mammals the slow red fibers should be further subdivided on the basis of differences in enzymatic activity, especially succinic dehydrogenase and ATPase activity, into two subcategories. As a result, muscle fibers may be classified into three types, which have been designated as A, B, and C. Type A fibers represent the classic fast, pale fibers, whereas types B and C represent two types of slow red fibers. Stein and Padykula (1962) found all three types present in the rat gastrocnemius, with type A predominant. In the soleus, A fibers were absent but the muscle contained both types B and C.

Serial sections of human vastus lateralis muscle are reflected in Figure 4.11. Three fiber types are identified.

The significance of slow and fast characteristics of muscle fibers is discussed further in Chapter 10.

Key Observations

The following summarizes key observations relating to muscle contractility:

1. The tension exerted by active muscle is a function of its length and is maximal at about the greatest length that the muscle can assume in the living animal. Tension decreases nearly linearly above and below this length. The shape of the tension-length curve for isometric contraction is similar for all muscles tested.

2. A muscle which is being lengthened while it is contracting can maintain greater tension than it can develop at any given equivalent static length. Therefore, tensions greater than the isometric maximal can be recorded in eccentric contraction.

3. Velocity of shortening of the contractile material decreases with increasing load in a hyperbolic manner. When velocity is expressed as a percentage of maximal velocity at zero force and force is expressed as a percentage of maximal force at zero velocity, the force-velocity curve is essentially the same for all muscles.

4. Quick release produces the same effects in all muscles: tension falls immediately and then re-develops to a value characteristic of the shorter length.

5. The rate of the development of tension depends on the intrinsic shortening velocity of the contractile material, compliance of the elastic components, and the rate of decay of the active state.

The existence of such extensive similarities among muscles provides a measure of confidence in assuming that information on the properties of muscle derived from animal studies may also apply to man. However, reasonable caution and good judgment must be exercised when direct confirmation is lacking.

Muscle Classifications

Various endeavors have been made to classify skeletal muscles in appropriate

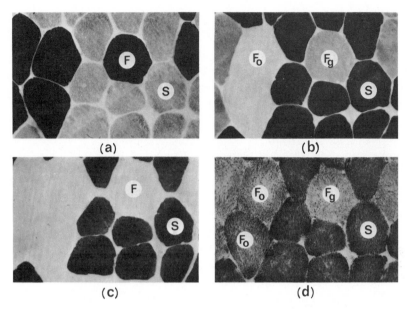

Figure 4.11. Serial sections (using light microscopy) of human vastus lateralis muscle. Magnification approximately ×240. (a), myosin ATPase pH 10.0; demonstrates two fiber types. (b), myosin ATPase pH 4.6; demonstrates three fiber types. (c), myosin ATPase pH 4.3; demonstrates reversal of pH 10.0 reaction. (d), NADH diaphorase (oxidative state); demonstrates three fiber types. F, fast twitch; F_g, fast twitch (glycolytic); F_o, fast twitch (oxidative); S, slow twitch. (Courtesy of G. Elder, McMaster University.)

ways. These include the following categories: (1) red and white; (2) tonic and phasic; (3) expanders and contractors; and (4) spurt and shunt muscles.

Red and White Muscles

As indicated earlier in this chapter, physiologists and histologists have identified both red and white skeletal muscle. The red muscle is slow contracting and the white muscle contracts rapidly.

Tonic and Phasic Muscles

Most anatomists and kinesiologists generally divide muscles on the basis of function as antigravity or postural muscles, and as the more rapidly contracting phasic muscles used in motor skills. More recently the inclination is to speak of the antigravity-postural muscles as tonic on the basis of the continuous low level of contractile activity which is required to maintain a given posture. The muscles of the tonic group contain proportionately more red, slow contracting muscle fibers, while the more rapidly contracting phasic muscles contain a larger proportion of white fibers.

Stockmeyer (1970) points out further characteristics of the tonic muscles: they are mostly penniform with shorter muscle fibers, they lie deeper and more medially, they generally cross only one joint, and they belong to the extensor group functioning as abductors as well as lateral rotators. Phasic muscles are located more superficially and more laterad; they have longer fibers, they may cross more than one joint, and they generally belong to the flexor group whose functions also include adduction and medial rotation. Stockmeyer (1967) also speaks of tonic muscles as stabilizers and of phasic muscles as mobilizers.

Contractors and Expanders

Grant and Smith (1953) point out that skeletal muscles as a whole may be divided into contractors and expanders. Those muscles which pull the body into an approximation of the fetal position, e.g., the flexors, adductors, and medial rotators, are classed as the "contractors." On the other hand, those muscles which expand or open up the body, e.g., the extensors, abductors, and lateral rotators of the limbs, are classed as the "expanders."

Spurt and Shunt

MacConaill and Basmajian (1969) offer a different concept by dividing the muscles on the basis of the relative magnitudes of their stabilizing and rotatory components. Those muscles with attachments farther from the joint axis will have larger stabilizing components and are called shunt muscles, while those which attach closer to the joint axis will have larger rotatory components and are called spurt muscles. This terminology is derived from that employed by 19th century British engineers who used the term shunt to describe a force which prevented a body moving in a curved path from taking a tangential or rectilinear direction; the body was shunted back onto the curved track. Thus, the term shunt as used by MacConaill and Basmajian refers to the centripetal force of the physicist or engineer of today. The 19th century engineers used the term spurt to indicate the force which "provided the

necessary spurt of energy that impelled the body into motion, or if necessary, kept it in motion." This is simply applying a different term to the force defined in Newton's Second Law, which states that force is equivalent to the mass of a body multiplied by its acceleration.

Muscle Function

Motor skill and all forms of movement result from interaction of muscular force, gravity, and any other external forces which impinge on skeletal levers. Muscles rarely act singly; rather, groups of muscles interact in many ways so that the desired movement is accomplished. This interaction may take many different forms so that a muscle may serve in a number of different capacities, depending on the movement. Thus, at different times a muscle may function as a prime mover, an antagonist, or a fixator, or synergically as a helper, a neutralizer, or a stabilizer. A muscle that is the prime mover for one movement will become an antagonist when the movement is reversed, and on other occasions it may perform any one of the other functions listed above, depending upon the circumstances.

Prime Movers and Movers

Whenever a muscle causes movement by shortening, it is functioning as a mover or agonist. If it is believed that the muscle in question makes the major contribution to the movement, that muscle is regarded

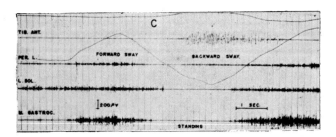

Figure 4.12. Electromyogram of postural sway. Deflection upward of free moving line records forward sway; deflection downward records backward sway. A good example of reflex inhibition via spindle afferents. (From O'Connell, A. L., 1958. Electromyographic study of certain leg muscles during movements of the free foot and during standing. *Amer. J. Phys. Med. 37:* 289.)

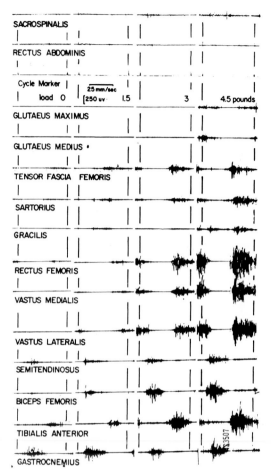

SACROSPINALIS		
RECTUS ABDOMINIS		
Cycle Marker	25 mm/sec	
load O	250 uv 1.5	3 4.5 pounds
GLUTAEUS MAXIMUS		
GLUTAEUS MEDIUS •		
TENSOR FASCIA FEMORIS		
SARTORIUS		
GRACILIS		
RECTUS FEMORIS		
VASTUS MEDIALIS		
VASTUS LATERALIS		
SEMITENDINOSUS		
BICEPS FEMORIS		
TIBIALIS ANTERIOR		
GASTROCNEMIUS		

Figure 4.13. Electromyograms showing reciprocal inhibition of muscles crossing the hip, knee, and ankle joints while riding a stationary bicycle. Cycle marker (third line) indicates foot at 3 o'clock. (From Houtz, S. J., and Fischer, F. J., 1959. An analysis of muscle action and joint excursion during exercise on a stationary bicycle. *J. Bone Joint Surg.* 41-A: 123.)

as the prime mover. Other muscles crossing the same joint on the same aspect but which are smaller or which are shown electromyographically to make a lesser contribution to the movement under consideration are identified as secondary or assistant movers or agonists.

Antagonists

Muscles whose action is opposite to and so may oppose that of a prime mover

are called antagonists. This does not mean that an antagonist always exerts tension against the prime mover; electromyography has again and again demonstrated an absence of electrical activity in opposing muscles (Fig. 4.12). The term antagonist, however, is a carry-over from the days when it was erroneously believed that whenever a muscle contracted as a prime mover, e.g., as a flexor, the opposing extensors always regulated and controlled the movement by exerting a lesser tension against the prime mover. Electromyographic records of simple, familiar movements or of highly skilled movements performed against a light resistance (such as only the weight of the body segments being moved) clearly demonstrate that reciprocal and complete inhibition of opposing muscles is the rule, i.e., the antagonistic muscles relax (Fig. 4.13). Co-contraction in normal subjects has been demonstrated electromyographically but only in hypertense individuals, when an individual is deliberately performing a "tension" movement, or when tremendous effort is being exerted. Under these latter circumstances the antagonistic activity is in all probability synergic in nature as it serves to stabilize the joints against the power exerted by the prime movers.

In various pathological conditions such as hemiplegia and cerebral palsy, cocontractions are observed. Often therapeutic strategies have to be devised to minimize or even eliminate such co-contractions.

Synergy

Synergic action has been defined as cooperative action of two or more muscles in the production of a desired movement. A synergist, then, can be regarded as a muscle which cooperates with the prime mover so as to enhance the movement. Synergic interaction can take many forms and variations as discussed below.

Conjoint Synergists. Two muscles acting together to produce a movement which neither could produce alone may be classed as conjoint synergists. Dorsiflexion of the foot at the ankle is an example. The

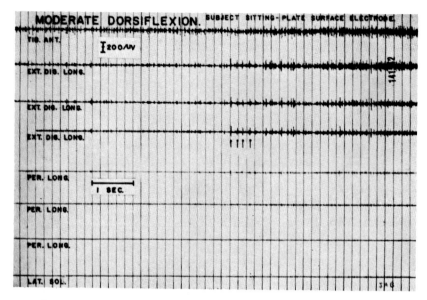

Figure 4.14. Electromyogram of dorsiflexion recorded from 3-mm surface electrodes placed around the lateral aspect of the upper leg. Electrodes approximately 2 cm apart on center. As movement progresses the extensor digitorum longus participates (arrows). Early activity recorded from the extensor digitorum longus is believed to be spread from the tibialis anterior. (From O'Connell, A. L., 1958. Electromyographic study of certain leg muscles during movements of the free foot and during standing. *Amer. J. Phys. Med. 37:* 289.)

movement is produced by the combined action of the tibialis anterior and the extensor digitorum longus (Fig. 4.14). The tibialis anterior alone would produce a combination of dorsiflexion and inversion, while shortening of the extensor digitorum longus alone would produce toe extension, dorsiflexion, and eversion. Acting together, the muscles produce a movement of pure dorsiflexion. Another example occurs in lateral deviation of the hand at the wrist; e.g., ulnar deviation results from the simultaneous action of the flexor carpi ulnaris and the extensor carpi ulnaris.

Neutralizing or Counteracting Synergists. Some muscles cross bi- or multiaxial joints in such a fashion that they are capable of causing more than one action at that joint. The muscles on the medial aspect of the thigh are all prime movers in adduction of the thigh at the hip joint but, if unopposed, the pectineus and the adductors longus and brevis will at the same time flex the thigh and rotate it medially.

Thus, whenever only adduction is desired, other muscles crossing the hip joint must become active to neutralize or counteract the undesired components. Further problems arise whenever muscles cross two or more joints. A muscle producing a desired movement at one joint frequently produces an undesirable effect at another joint (or joints). Therefore, still other muscles must contract synergically to neutralize the undesired actions. The classic example of this situation is that of the finger-flexion wrist-extension interaction. Contraction of the finger flexors to grasp an object also tends to flex the wrist. The unwanted wrist flexion is counteracted by the synergic contraction of the wrist extensors.

Stabilizing Synergists. The *sine qua non* of an effective coordinate movement involves greater stabilization of the more proximal joints so that the distal segments move effectively. The greater the amount of force to be exerted by the open end of a kinematic chain (whether it is the peripheral end of an upper or of a lower extremity), the greater is the amount of stabilizing force needed at the proximal links. Thus, when throwing or striking forcefully or moving a heavy weight, the muscles of the shoulder girdle and shoulder must contract more forcefully to support the scapula and shoulder joint than they do when the hand is making a gesture or lifting food to the mouth. In such instances the opposing muscles interact, contracting simultaneously in such a manner that, while the scapula moves on the thorax during the effort, it still provides a firm base for the movement of the humerus. At the same time the rotator cuff muscles (supraspinatus, subscapularis, infraspinatus, and teres minor) all contribute their opposing tensions to the common end of supporting the humeral head against the glenoid fossa.

Other examples of stabilizing synergy may be mentioned. Executing a Valsalva maneuver (holding the breath while forcefully contracting the abdominal musculature which exerts pressure on the viscera) while running or jumping provides a firm base for the lower extremities to push against so that the body is propelled efficiently. The abdominal muscles are exerting a stabilizing synergy on the trunk. When one foot is supporting the body weight, all of the muscles crossing the ankle joint become active in synergic contractions supporting and stabilizing the ankle joint.

Fixation

When a joint is voluntarily fixed rather than stabilized there is, in addition to immobilization, a rigidity or stiffness resulting from the strong isometric contraction of all muscles crossing that joint. These muscles will forcefully resist all external efforts to move that joint. As fixation can be very tiring, it is seldom used—and rarely useful.

Stabilization vs Fixation

From the above discussion one should recognize the difference between stabilization and fixation of joints. As stated, fixation denotes a rigidity or stiffness in opposition to all movement, whereas stabilization implies only a firmness. Fixation can occur inadvertently when, for example, a student learning a skill misinterprets directions given by an instructor. When a golf instructor stresses the need for a straight left arm (for a righthanded player) during the swing, many beginners will respond with maximal contraction of both the biceps and triceps, with disastrous results. Under these circumstances the long heads of both muscles are pulling on their scapular attachments above and below the glenoid fossa which severely inhibits the free movement of the humerus so necessary to a successful swing. On the other hand, just sufficient tension in the triceps to keep the elbow extended will stabilize that joint and provide a constant lever length or radius for the free and easy swing that will stay "in the groove" and connect forcefully with the ball. Similar instances can arise during the teaching of any skill new to the student.

Economy of movement involves the use of minimal stabilizing synergy and never fixation of joints. Teaching cues and coaching points for motor skills should be geared to this end; e.g., "Keep your arm straight, but not stiff," etc.

Therapists encounter similar kinds of problems, perhaps more magnified, when they endeavor to instill effective therapeutic notions in patients.

MUSCLE ATTACHMENTS AND THEIR EFFECTS ON FUNCTION

The torque about any joint is determined by the muscular forces generated coupled with several anatomical factors which contribute to the length of the moment arm or arms of the musculoskeletal elements under consideration. The contributing factors which should be recognized are: (1) attachment of the muscle to bony prominences; (2) passage of the muscle tendon over bony prominences; and (3) both attachments of the muscle distant to the joint.

Attachment to Bony Prominences

The stress exerted on bones created by the frequent tension exerted by contracting muscles stimulates increased growth of the bone in the area of the attachment. Thus, the very fact that a muscle is anchored to bone causes the protuberances at anchor points. The attachments of the adductors and the two vasti muscles create the linea aspera on the back of the femur; the pull of the glutei increases the size of the greater trochanter; the size of the deltoid tuberosity is directly related to the size of the deltoid muscle attached to it, that of a muscular man being much larger than that of a woman who is comparatively inactive. These prominences and even the bones themselves can change in size, both increasing and decreasing during the lifetime of an individual. Prives (1960) in dealing with the influences of labor and sport on skeletal structure in man reported that X-rays taken of the same individuals over a period of 10 years showed "Variations of bone shape and structure characteristic for the given occupation or sport.... Change of professional trade provokes changes in the structure of bones corresponding to the new type of loading."

In this study laborers and athletes were deliberately changed to a more sedentary way of life while sedentary workers were inducted into labor gangs and their participation in athletics was encouraged. The results which followed indicated that:

1. Physical loading causes the redistribution of certain radioactive isotopes such that they accumulate in the most loaded or stressed parts of the skeleton.
2. Different forms of osseus hypertrophy occur.
3. Physical work favors the growth of the bones in length.
4. Physical work delays the aging of the osseus system which is important in the combat for prolongation of life.

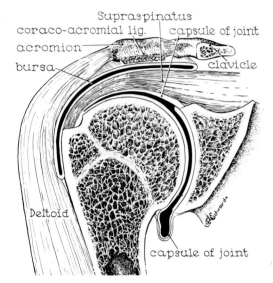

Figure 4.15. Relation of middle deltoid to greater tuberosity of the humerus. (From MacConaill, M. A., and Basmajian, J. V., 1969. *Muscles and Movements.* Baltimore: The Williams & Wilkins Company.)

The increase in the size of the bone occurs, of course, in response to the strain put on the entire musculoskeletal system and in particular to the increased pull exerted by the increased use of muscular force. The osseus hypertrophy of a bony prominence as well as of the bone itself moves the muscular attachment that much farther from the joint axis and so increases the moment arm or arms of the muscle in question. This of course increases the effectiveness of the muscle force as well.

Passage of Tendon or Muscle over Bony Prominences

Passage of a muscle or tendon over a prominence alters the direction of the action line of the muscle, moving it farther from the joint axis, and therefore increasing the moment arm of the muscle. For example, the middle deltoid must pass over the greater tuberosity of the humerus (Fig. 4.15) and tendons of the gastrocnemius and the hamstrings pass over the femoral condyles (Fig. 4.16). The action

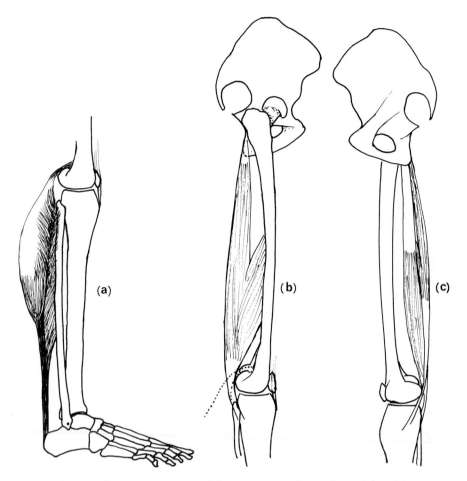

Figure 4.16. Relation of gastrocnemius and hamstrings to femoral condyles. (a), gastrocnemius, lateral view; (b), biceps femoris, lateral view; (c), semitendinosis and semimembranosis, medial view.

line of the quadriceps femoris is deflected away from the knee joint by the patella (Fig. 4.17). Thus, in the instance of a patellectomy, for the same muscular force generated, the reduced moment arm results in the production of less torque about the knee joint.

Both Attachments of the Muscle Situated at a Distance from the Joint

A muscle with both of its attachments at a large distance from a joint axis and without fascia or reticulum to hold the belly or tendon close to the bones would have a large moment arm. However, such an arrangement would be most ungainly and, as restrictions are always in effect so that no "bowstringing" of muscle or tendon can occur, the result is that there are no grossly exaggerated moment arms. Nevertheless, even with these restrictions, when the adult body is in the anatomical position the action lines of several muscles fall from 1½ to 2½ in from the axis or axes of the joint at which they act, as a result of the fact that the closest attachment is 2 to 3 in from the joint center. Probably the

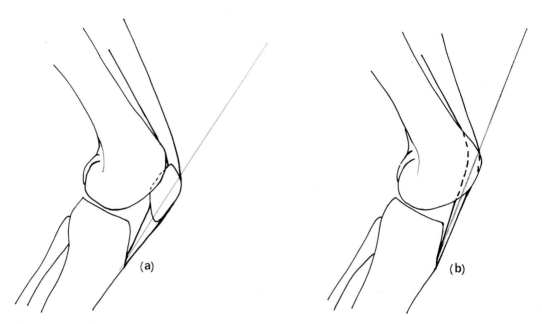

Figure 4.17. Change in angle of pull of quadriceps femoris caused by the patella. (a), patella present; (b), patella absent.

longest muscular moment arms in the body (abdominal muscles excluded) are those of the hamstring muscles at the hip. In the anatomical position their action line passes from 2 to 3 in from the flexion-extension axis through the head of the femur at the hip. The distal attachments of the pectoralis major on the humerus are far enough down the bicipital groove so that the flexion moment arm of the clavicular head can also be as large as 2 in, as may be the adduction moment arm of the sternal head. The action line of the triceps surae (the two heads of the gastrocnemius and the soleus) running upward from the calcaneus can vary from 1½ to 2 or more in. These variations in moment arm lengths result from individual differences in skeletal size and bony development. Thus, the larger boned individual has an immediate advantage over one of slighter build insofar as moment arms and therefore produced torque for the same bulk (cross-section) of muscle involved are concerned.

Muscles with long moment arms, even

when the limbs are in the anatomical position, are for the most part large muscles which have the capability of exerting considerable force (e.g., the glutei, hamstrings, gastrocnemius-soleus, pectoralis major). The fact that their attachments are at some distance from the joint axis means that they must shorten more to move a segment through a given angle than a muscle which attaches closer to the joint (Fig. 4.18). Two facts are obvious from the above. (1) If the two muscles depicted in Figure 4.18 shorten at the same rate, the muscle in (a) will move the segment faster. (2) The moment arm of (b) is longer; therefore, if it exerts the same amount of force per unit shortening as does (a), the moment of force generated by (b) will be greater.

Muscle Forces Acting on Skeletal Levers

When a muscle generates tension it pulls equally against all of its attachments just as a rubber band stretched between two fingers pulls against each. In movement analysis, it may at times prove con-

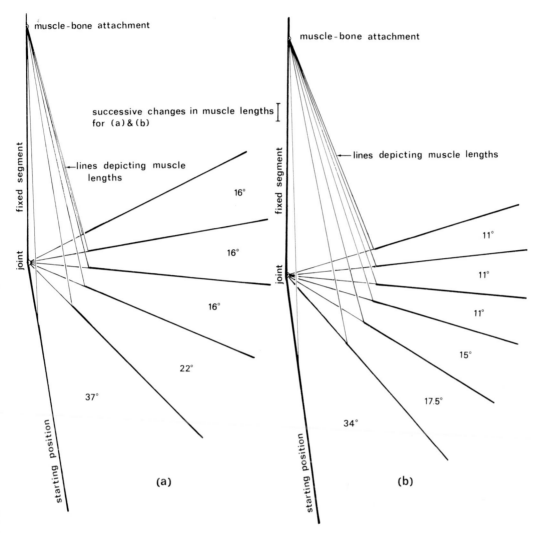

Figure 4.18. Degrees of joint motion caused by equal muscle shortening. (a), muscle attached close to joint axis; (b), muscle attached farther from the same axis. Successive changes in the muscle lengths (shortening) are indicated commencing from the common starting positions shown.

venient to consider the muscular force as being applied only at the attachment of concern.

If a muscle or muscle group is shortening (isotonic contraction) or exerting tension while being lengthened by an outside force (eccentric contraction) then one of the body parts to which that muscle is anchored is being moved on an adjacent part. Also, one of the two segments to

which the muscle is attached is closer to or may even form (as is the case with a hand or foot) the open end of a kinematic chain. In the isotonic contraction the muscle contraction is the moving force. For the eccentric contraction the muscle performs a regulatory service resisting the moving force and so controliing the rate at which the body part is being moved by the outside force. In either case the muscle force

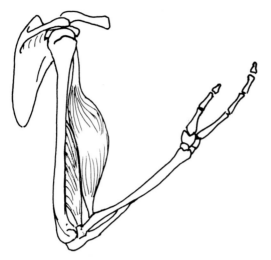

Figure 4.19. Biceps and brachialis are flexing the elbow.

is considered as being applied to the moving segment nearest the *open end of the kinematic chain* where its effect is clearly visible.

In Figure 4.19 the biceps and brachialis are flexing the elbow. This action moves the hand toward the shoulder while the humerus remains motionless and close to the rib cage. There is no movement either of the arm or of the shoulder girdle so the tension exerted by the flexors is effective only at the distal attachment. Under these and similar circumstances we say the effort exerted by these muscles is applied at the distal attachments of the muscles, i.e., via the attachment to the bone of the moving segment.

Imagine a gymnast climbing a rope. To start with, consider his right hand to be supporting his entire weight as he reaches upward to grasp the rope with his left hand. After this reach when the left hand grasps the rope the left elbow flexes and the left arm extends at the shoulder as he pulls himself higher. In this instance the hand is grasping the rope, so it and the forearm remain stationary while the body rises as a result of the combination of

elbow flexion and shoulder extension. Because the body is moving on the motionless forearm, the lever for the elbow flexion is the kinematic chain made up of the body segments proximal to the elbow joint, i.e., the entire body excluding the left forearm and hand. Thus, in order to be effective and to move the body, the force exerted by the elbow flexors (biceps, brachialis, and brachioradialis) must be applied at the proximal attachments of these muscles.

Simultaneously, as mentioned above, there is extension of the left shoulder (which is also accompanied by some adduction). In other words, the angle between the elevated left arm and the trunk decreases as the body is pulled upward by the shortening of the shoulder extensors, including the latissimus dorsi and the sternal head of the pectoralis major. Anatomically these two muscles combine to form a sling which supports the body from the bicipital groove of the humerus; thus, the isotonic contraction of these two muscles moves the body up to the forearm. Again, the effective muscular force is applied at the proximal attachments of these muscles (latissimus dorsi and pectoralis major) on the torso.

The above discussion presents illustrations involving shortening or isotonic contraction, but the same principles apply when considering eccentric or lengthening contraction.

Now imagine that the gymnast is descending the rope and his elbow flexors and shoulder extensors are slowly yielding to the pull of gravity and being lengthened by this outside force. During this descent when the right hand grasps the rope, the right elbow extends slowly while the left shoulder flexes at the same rate. Here the muscles of these joints, the elbow flexors and shoulder extensors, are acting as a resistance to the outside force, and this muscular resistance must be applied to the moving end of the kinematic chain which they control, i.e., at the proximal attachments of the muscles, just as in the case of the isotonic contraction.

Ranges of Muscle Extensibility and Contractility

The maximal degree of angular displacement of a body segment possible at a given joint affects all muscles crossing that joint. It determines both the greatest amount of stretch which such a muscle may be required to undergo and the maximal amount of shortening which it can be called upon to perform *in situ*. The full range of extensibility and contractility of a muscle has been variously termed its "functional excursion" (Brunnstrom (1964)) and its "amplitude" (Scott (1963)).

Any muscle crossing a single joint is normally capable of shortening sufficiently to move the body segment to which it is attached through its maximal angular displacement and, conversely, it is sufficiently extensible to permit a full range of motion in the opposite direction. This is not true of muscles which cross two or more joints.

Mechanics of Multijoint Muscles

A muscle crossing two or more joints has certain characteristics, capabilities, and limitations when compared with those muscles which cross only one joint. When a muscle crosses more than one joint it creates force moments at each of the joints crossed whenever it generates tension. The moments of force it exerts at any given instant depend on two factors: the instantaneous length of the moment arm at each joint and the corresponding amount of force that the muscle is exerting. The joint with the longest moment arm, and hence with the greatest moment of force, is normally the one at which the multijoint muscle will produce or regulate the most action. The hamstring muscles are a case in point: their moment arm at the hip is at least 50% longer than the one at the knee, and electromyography has repeatedly demonstrated that activity such as slow hip flexion as in toe touching is controlled by eccentric contraction of these muscles and that, when the action is reversed, they return the body to the upright posture

Table 4.2
Thigh Muscle Moment Arms

Muscle	Hip	Knee
	cm	cm
Hamstrings	6.7	3.4
Rectus femoris	3.9	4.4

without assistance of the gluteus maximus (Joseph (1960); Basmajian (1974)). According to Elftman (1939) representative moment arms for the anterior and posterior thigh muscles are as indicated in Table 4.2.

If the hamstrings contract with a force of 10 kg, the moment of force at the hip will be 67 kg-cm, and at the knee it will be 34 kg-cm. At the same time, if the rectus femoris contracts with an equal amount of force, its moment at the hip will be only 39 kg-cm and at the knee 44 kg-cm. Such conditions demonstrate how these opposing muscles interact to produce extension at the hip while the knee extension is maintained; each joint responds to the greatest force moment acting on it. If greater force is needed at either location, the single joint muscles (gluteus maximus at the hip and the vasti muscles at the knee) are available. It seems that the hamstrings are more effective as knee flexors than is the gastrocnemius. The latter does not act at the knee joint unless the ankle is kept dorsiflexed. Again the moment arm of the gastrocnemius at the ankle is larger than the one at the knee, and when these moment arms are compared (an approximation of 5 cm at the ankle and 2.5 cm at the knee), the reason for the greater amount of activity at the ankle joint becomes obvious.

Muscle Insufficiency

If a muscle which crosses two or more joints produces simultaneous movement at all of the joints that it crosses, it soon reaches a length at which it can no longer generate a useful amount of force. Under these conditions the muscle is said to be **actively insufficient.** An example of such insufficiency occurs when one tries to achieve full hip extension with maximal knee flexion. The two-joint hamstrings are incapable of shortening sufficiently to produce a complete range of motion at both joints simultaneously. The individual feels considerable discomfort at the back of the thigh as he makes the effort. If he then reaches behind him, grasps his ankle and pulls the leg into full flexion at the knee and the thigh into full extension at the hip, the discomfort disappears from the hamstrings but is transferred to the anterior thigh. The rectus femoris is not long enough to span both joints comfortably under these conditions and displays **passive insufficiency.**

Passive insufficiency of a muscle is indicated whenever a full range of motion at any joint or joints that the muscle crosses is limited by that muscle's length rather than by the arrangements of ligaments or structure of the joint itself. Thus, passive insufficiency of the hamstrings is indicated when one is unable to touch the fingertips to the floor while maintaining full extension of the knees, a situation popularly known as tight hamstrings.

Structure and Chemistry of Muscle

The reader is referred to Chapter 7 where details pertinent to muscle structure and the chemistry of muscle are presented.

CONCLUSION

The material provided in this chapter should facilitate for the reader an understanding of the essential mechanical properties of skeletal muscle.

CHAPTER 5

Musculoskeletal Forces

INTRODUCTION

With the background provided by the foregoing chapters, particularly Chapters 3 and 4, it is now possible to examine a variety of force situations within the living body when it assumes various static positions. Applying the principles of mechanical analysis to living systems takes us directly into the realm of biomechanics. A number of examples are provided dealing mainly with static equilibrium situations. Where dynamic aspects are dealt with, the examples are confined essentially to whole body force analyses. A number of techniques are then presented for determining the center of gravity of body parts and for the whole body.

EXAMPLE 1—Resultant of Forces of Two Gastrocnemius Heads

It is assumed that the forces acting at each of the two heads of the gastrocnemius are represented by vectors $\vec{AB}$ and $\vec{AC}$ in the plane of the paper as shown in Figure 5.1. Given that the forces exerted by the two heads are equal, it is required to determine the magnitude and direction of the resultant force.

By graphically constructing the parallelogram $ABDC$, with the length AB equal to the length AC, vector $\vec{AD}$ is obtained. This represents the resultant force whose magnitude (established by measurement) is 1.82 times AB or AC. Its direction is upward along the line which bisects the angle BAC.

EXAMPLE 2—Resultant of Forces of Two Heads of Pectoralis Major

Figure 5.2 shows a graphical composition of the forces exerted by the clavicular and sternal heads of the pectoralis major. $\vec{AB}$ is the clavicular component of force and $\vec{AC}$ the sternal one. The vector $\vec{AD}$ is the resultant force of the entire muscle acting at point A.

EXAMPLE 3—Three Abductor Forces Acting at the Hip Joint

Consider the three abductors acting about the hip joint: gluteus medius, gluteus minimus, and tensor fascia lata. Each of these three muscles has a different shape and mass and a slightly different angle of pull, although they all attach on or near the greater trochanter of the femur.

117

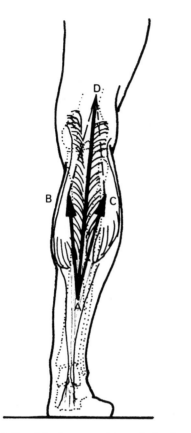

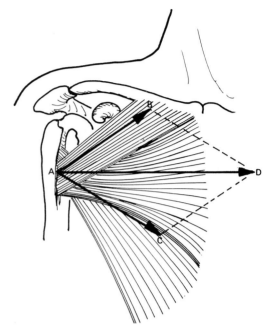

Figure 5.2. Graphical composition of the forces exerted by sternal and clavicular heads of the pectoralis major.

Figure 5.1 Graphical composition of the forces exerted by the two heads of the gastrocnemius.

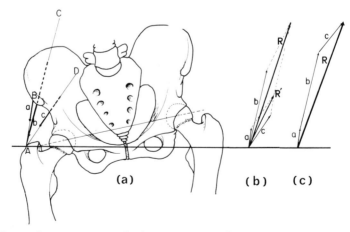

Figure 5.3. Abductor forces acting at the hip joint. (a), pelvis and hip joint with abductor action lines (redrawn after Inman (1949)); (b), graphical solution of resultant by parallelograms; (c), graphical solution of resultant by a polygon of forces.

To evaluate the resultant force in magnitude and direction necessitates a knowledge of the magnitudes and lines of action of the individual muscle forces. Inman (1947) determined the action lines on the pelvis. By postulating that their separate contributions to abduction were proportional to their individual masses, he established the following relative force values:

Tensor fascia lata	1
Gluteus minimus	2
Gluteus medius	4

On this basis and knowing the directions of the action lines, the relative resultant force can be determined. Figure 5.3 indicates the action lines from the trochanter A. That of the tensor fascia lata, AB, is shown as attaching where it does as this point is so close to its junction with the iliotibial band. The action line of the gluteus medius is represented by AC and that of the gluteus minimus by line AD. $\vec{a}$, $\vec{b}$, and $\vec{c}$ are the respective vectors acting on the pelvis. The resultant force of these three muscles may be determined graphically in two different ways: (1) by the parallelogram method illustrated above, and (2) by constructing a polygon of forces.

Solution by Parallelograms of Force. With three forces it is necessary to construct two parallelograms. The three vectors $\vec{a}$, $\vec{b}$, and $\vec{c}$ in Figure 5.3(b) are drawn from a common origin and at the same angles as in Figure 5.3(a). In this case, for reasons of clarity, the first parallelogram is drawn between vectors $\vec{a}$ and $\vec{c}$ with a resultant vector $\vec{R}'$. The second parallelogram is drawn between vectors $\vec{R}'$ and $\vec{b}$ to find the final resultant vector $\vec{R}$.

Solution by polygon of forces. Referring to Figure 5.3(c) it will be seen that the first vector $\vec{a}$ is drawn at the same angle to the horizontal as in Figure 5.3(a). The second vector $\vec{b}$ is added to the end of $\vec{a}$ and the third one $\vec{c}$ is similarly added to $\vec{b}$. The vector that closes the polygon is the resultant, R, of the three forces.

In both solutions, the resultant relative force is found to be just less than 7 units. This value is close to the arithmetic sum

of the relative forces since they all act in approximately the same line.

Solution by Resolution into Horizontal and Vertical Components (Algebraic Approach). Given the proportional forces exerted by each of the muscles in the group, it is convenient to convert the proportions to percentages of the total force exerted by the three muscles. Thus, gluteus medius contributes four-sevenths (57% approximately) of the force, gluteus minimus two-sevenths (29% approximately) and tensor fascia lata one-seventh (14% approximately). The relative amount of force in the resultant can be obtained by using the percentage of the total force available from the group as a whole. For the purpose of this problem we will assume the total abductor force to be 100% and that each muscle is exerting its proportional amount of force:

Table of known data

Muscle	Gluteus medius	Gluteus minimus	Tensor fascia lata
Force	57%	29%	14%
Angle with horizontal	76°	55°	81°
Sine	0.97030	0.81915	0.98769
Cosine	0.24192	0.57358	0.15643

Horizontal component:

$$R_X = \Sigma F_X = 57 \cos 76° + 29 \cos 55°$$
$$+14 \cos 81°$$
$$= 57 \times 0.24192 + 29 \times 0.57358$$
$$+ 14 \times 0.15643$$
$$= 13.79 + 15.63 + 2.19$$
$$= 31.61$$

Vertical component:

$$R_Y = \Sigma F_Y = 57 \sin 76° + 29 \sin 55°$$
$$+ 14 \sin 81°$$
$$= 57 \times 0.97030 + 29 \times 0.81915$$
$$+ 14 \times 0.98769$$
$$= 55.31 + 23.75 + 13.83$$
$$= 92.89$$

The angle θ that the resultant makes with the horizontal is given by

$$\theta = \arctan R_Y/R_X$$

$$= \arctan + 92.89/+31.61$$

$$= +71.2°$$

The resultant force at the greater trochanter is directed upward and to the right, making an angle of 71.2° with the horizontal. Correspondingly, the force acting on the pelvis is directed downward and to the left.

EXAMPLE 4—Calculation of Center of Gravity of Body Segments

The hand of a man weighs 0.9 lb and is 7 in from tip to wrist when extended. His forearm weighs 2.8 lb and is 10 in long from the wrist to the elbow joint. It is desired to determine the location of the center of gravity of the combined forearm and hand.

Referring to the tables in Appendix 4 it is possible to establish the diagram in Figure 5.4. Table A4.3 (Appendix 4) indicates that the center of gravity of the forearm (CG_a) is located 43% of the forearm length measured from the elbow axis; the center of gravity of the hand (CG_h) is at 50.6% of the length of the extended hand from the wrist axis, which places the CG of the hand approximately 10 in + 3.5 in = 13.5 in from the elbow axis.

The total downward force W_T of the hand and the arm is

$$W_T = W_a + W_h$$

where W_a = weight of the arm (forearm) alone and W_h = weight of the hand, i.e.

$$W_T = 2.8 + 0.9 = 3.7 \text{ lb}$$

which acts at CG_T.

Invoking the principle of moments,

$$W_T \times MA_{W_T} = \Sigma M \qquad (5.1)$$

i.e., the moment of force due to the combined weights of the arm and hand must equal the sum of the moments of the individual weights.

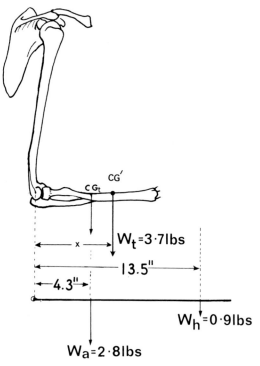

Figure 5.4. Diagram for Example 4.

The left side of Equation 5.1 becomes

3.7 x in-lb since x = MA_{W_T}

The right side is

$$2.8 \times 4.3 + 0.9 \times 13.5 \text{ in-lb}$$

$$3.7 \text{ x} = 12.04 + 12.24 \text{ in-lb}$$

$$= 24.28 \text{ in-lb}$$

$$x = 24.28/3.7 = 6.6 \text{ in}$$

The center of gravity of the combined forearm and hand lies 6.6 in distal to the axis through the elbow.

It should be recognized that the calculation just performed will, at best, produce a good estimate of the location of the center of gravity of the combined body segments. This is so since in vivo measurements on a particular living subject are not easy to perform, and reliance has to be placed on estimates arising from data accumulated by workers such as Dempster (1955) and along the lines reflected in Appendix 4.

EXAMPLE 5—Shift in Center of Gravity with Relative Body Movement

If the hand of the man in Example 4 is flexed to 90° at the wrist, by how much would the position of the combined arm and hand CG shift, and in what direction?

In this case the same mathematical relationships as indicated in Example 4 will obtain, but the CG of the hand will be located at 10 in. (approximately) from the elbow. Let the position of the combined CG relative to the elbow be x'. Then

$$3.7\ x' = 2.8 \times 4.3 + 0.9 \times 10$$
$$= 12.04 + 9 = 21.04 \text{ in-lb}$$

i.e.

$$x = 5.69''$$

The shift in center of gravity is

$$\Delta x = x - x' = 6.60 - 5.69 = 0.91''$$

i.e., there is a shift of 0.91 in toward the elbow.

EXAMPLE 6—Momentum

Consider the problem of the football quarterback carrying the ball being tackled by a lineman. The tackle is approaching the quarterback with a velocity of 20 ft/ sec. The quarterback sidesteps so that the tackle is made at an angle of 60°. The quarterback is still moving at 30 ft/sec and weighs 160 lb as against the 200 lb of the tackle. As the tackle grabs the quarterback they react as a single unit after the tackle is made.

The momentum of the tackle-quarterback system is to be determined, and the resulting velocity of that system found.

Figure 5.5 provides a diagrammatic representation of the situation.

Resolving the momenta into X and Y components before the collision we have:

X components:

$$\Sigma X = m_1 v_1 \cos 0° + m_2 v_2 \cos 60°$$

$$= 200 \text{ lb} \times 20 \text{ ft/sec} \times 1 + 160 \text{ lb}$$

$$\times 30 \text{ ft/sec} \times 0.5$$

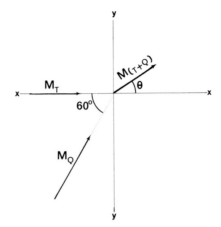

Figure 5.5. Schematic diagram of events before and after a collision on the football field. M_T is the momentum of the tackle weighing 200 lb (m_1) and traveling at 20 ft/sec (v_1); M_Q is the momentum of the quarterback carrying the ball and who weighs 160 lb (m_2) and travels at 30 ft/sec (v_2). $M_{(T+Q)}$ is the momentum of the combined quarterback and tackle system after the collision.

$$= 4000 \text{ lb ft/sec} + 2400 \text{ lb-ft/sec}$$

$$= 6400 \text{ lb-ft/sec}$$

Y components:

$$\Sigma Y = m_1 v_1 \sin 0° + m_2 v_2 \sin 60°$$

$$= 160 \text{ lb} \times 30 \text{ ft/sec} \times 0.8660$$

$$= 4156.8 \text{ lb-ft/sec}$$

The resulting momentum R follows readily from the application of Pythagoras' Theorem

$$R^2 = X^2 + Y^2$$

$$= 6400^2 + 4156.8^2$$

i.e.

$$R = 7631.4 \text{ lb-ft/sec}$$

The direction of R is determined trigonometrically

$$\theta = \arctan \frac{Y}{X} = \arctan \frac{4156.8}{6400}$$

$$= \arctan 0.64937$$

$$= 33°$$

The magnitude of the resulting or final velocity, v_f of the system can be easily obtained by dividing the final momentum by the resultant mass $(m_1 + m_2)$

$$m_1 + m_2 = 200 \text{ lb} + 160 \text{ lb}$$

$$= 360 \text{ lb}$$

$$v_f = \frac{R \underline{|\theta}}{m_1 + m_2}$$

$$= \frac{7631.4}{360} \underline{|33°}$$

$$= 21.2 \text{ ft/sec} \underline{|33°}$$

Immediately after colliding, the tackle and quarterback would move onward with a velocity of 21.2 ft/sec and at an angle of 33° with the direction in which the tackle was traveling, if there were no other forces acting on them. However, gravity and friction with the ground acting on the tackle's body combine to pull the quarterback down and they both come to a halt very shortly.

EXAMPLE 7—Impulsive force

If it is assumed that a similar collision occurred on ice and between hockey players, and if the two men slid as one mass almost without friction, they would travel at an initial velocity $v_i = 21.2$ ft/sec until they slammed into the side wall, where their combined final velocity v_f would become 0 ft/sec. If as they hit their bodies were compressed approximately 2 in (0.167 ft) by the force of impact, we can determine the force with which they hit the boards. In order to determine the amount of this force, the impulse equation (Equation 3.4) is used:

$$F \cdot \Delta t = mv_2 - mv_1$$

$$= mv_f - mv_i$$

The length of time of the impact is also unknown but can be determined from the available data. The amount of compression (0.167 ft) is the distance that the bodies travel during the impact. As their velocity changes from 21.2 to 0 ft/sec during this time, the average velocity v_{av} is used in the final determination by substitution in the formula $v = s/t$ or $t = s/v$.

$$v_{av} = \frac{v_f + v_i}{2}$$

$$= \frac{0 + 21.2 \text{ ft/sec}}{2}$$

$$= 10.6 \text{ ft/sec}$$

By substitution

$$\Delta t = \frac{0.167 \text{ ft}}{10.6 \text{ ft/sec}}$$

$$= 0.0157 \text{ or } 0.016 \text{ sec}$$

$$F = (mv_f - mv_i)/\Delta t$$

$$= \frac{360 \times 0 - 360 \times 21.2 \text{ lb-ft/sec}}{0.016 \text{ sec}}$$

$$= \frac{-7632.0 \text{ lb-ft/sec}}{0.016 \text{ sec}}$$

$$= -47.7 \times 10^4 \text{ lb-ft/sec}^2$$

F is an impulsive force of 47.7×10^4 lb-ft/sec² imparted by the wall to the impacting bodies. To convert to force in pounds weight it is necessary to divide by $g = 32.2$ ft/sec², i.e., $F = -1.48 \times 10^4$ lb.

EXAMPLE 8—Effect on Impulsive Force of Increasing Time of Impact

The effect of increasing the time of impact can be graphically illustrated if a comparison is made of the force generated when a 100-lb boy drops 3 ft from a high bar and lands (1) "hard" with little or no give in his joints so that the total give of his body (shoes, feet, spine, etc.) is approximately 1 in; and (2) when he lands softly on the balls of his feet and his joints all give so that the total drop of his center of gravity *after* his feet touch the floor is approximately 10 in.

The time for a 3-ft drop under the influence of gravity follows from

$$s = \tfrac{1}{2} gt^2$$

i.e.

$$t_{\text{drop}} = \sqrt{2s/g} = \sqrt{2 \times 3/32.2}$$

$$= 0.43 \text{ sec}$$

The velocity at the end of the drop

$$v_f = v_i + gt = 0 + 32.2 \times 0.43$$

$$= 13.8 \text{ ft/sec}$$

The average velocity during impact is

$$v_{av} = \frac{13.8}{2} = 6.9 \text{ ft/sec}$$

"Hard" Impact. Since the distance during impact is 1 in (0.083 ft), time of impact

$$\Delta t_i = \frac{0.083 \text{ ft}}{6.9 \text{ ft/sec}}$$

$$= 0.012 \text{ sec}$$

and as

$$F = \frac{mv_f - mv_i}{\Delta t}$$

$$= \frac{100 \text{ lb} \times 0 - 100 \text{ lb} \times 13.8 \text{ ft/sec}}{0.012 \text{ sec}}$$

$$= 11.5 \times 10^4 \text{ lb-ft/sec}^2 = 3570 \text{ lb}$$

"Soft" Impact. In the second instance, the soft landing, the time of drop, and the velocities are the same, but the distance has increased 10 times, so now

$$\Delta t_i = \frac{0.83 \text{ ft}}{6.9 \text{ ft/sec}}$$

$$= 0.12 \text{ sec}$$

Hence

$$F = \frac{1380 \text{ lb-ft/sec}}{0.12 \text{ sec}}$$

$$= 11.5 \times 10^3 \text{ lb-ft/sec}^2$$

$$= 357 \text{ lb}$$

On the other hand, if the boy uses a greater amount of eccentric, lengthening contraction sufficient to slow his joint actions further, the time would be increased beyond this point and the force on his feet would be decreased. Suppose that the time

lapse from toe touch to deepest bend was 0.36 sec; then

$$F = \frac{1380 \text{ lb-ft/sec}^2}{0.36 \text{ sec}}$$

$$= 3833.3 \text{ lb-ft/sec}^2 = 119.04 \text{ lb}$$

This is considerably less force than in either of the two previous examples; i.e., 3.3% of that of the hard landing and only 33% of that of the first soft landing. These calculations serve to illustrate some of the typical forces to which the human body can be subjected without sustaining serious injury, and they provide objective evidence for using controlled "give" when one is landing from a height.

It should be noted that as the time for impact Δt increases, the resulting force decreases. Since the force cannot be less than 100 lb (the weight of the boy), the impulse equation must be used with caution as the time period Δt grows larger. (In essence, the longer the time, the less the impact, i.e., the force F undergoes variations from instant to instant during impact).

MOMENT ARMS IN THE BODY

In the more complex levers characteristic of the musculoskeletal system, the moment arm (MA) of a muscle is the perpendicular distance from the line of force or action line of the muscle to the axis of the joint involved (Fig. 5.6). This distance from the line of force, the MA, must be used in all calculations of muscle force, rather than the length of the lever arm which is the distance from the point of attachment of the muscle from the joint axis.

Because a muscle always has one attachment close to the joint at which it acts, a muscle lever arm and an even shorter muscular MA are typical of all joints and body levers. When movement is produced by *shortening*, this situation precludes the presence of any second class levers in the musculoskeletal system because the length

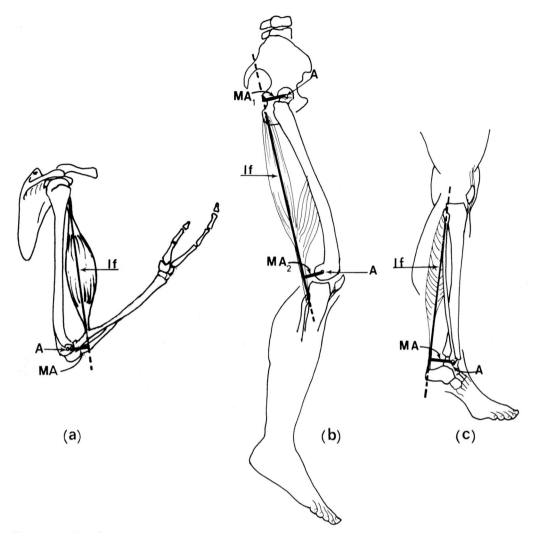

Figure 5.6. Muscle moment arms. (a), biceps brachii (elbow joint); (b), biceps femoris (knee and hip joints); (c), soleus (ankle joint). *If*, line of force (action line) of the muscle; *A*, joint axis; *MA*, moment arm of the muscle.

of resistance moment arm of the movement is always greater than that of the effort moment arm.

On Classes of Levers in the Body

In normal quiet standing the center of gravity is anterior to the promontory of the sacrum, and the line of gravity falls anterior to the ankle joint passing through the metatarsals (Fig. 5.7). Electromyograms of quiet standing show constant slight ac-

tivity in at least one soleus muscle under these conditions (Joseph (1960); O'Connell (1958)). This tension is counteracting the downward-forward moment caused by gravity and so is preventing the body from falling forward to the ground.

Quiet standing then is a static, balanced situation in which muscular effort balances the pull of gravity and posture is maintained. Also, these forces (soleus tension and gravity) operate a first class lever

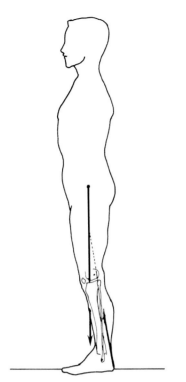

Figure 5.7. Line of gravity passing through the dorsum of the foot.

system, the axis through the fulcrum in the ankle joint lying between the soleus acting behind and gravity acting in front of the ankle axis.

To rise on the toes the center of gravity must be shifted forward until the line of gravity falls over or even momentarily slightly in front of the toes themselves. At the same time the heels come off the floor as metatarsophalangeal extension takes place. This action may simply be caused by the forward sway (i.e., gravity) or possibly by the contraction of the long and short toe extensors pulling on their proximal attachments, thus causing the toe extension while the toes are still weight bearing. (Whether the toe extension is caused or contributed to by one, both, or neither could be determined electromyographically by an interested investigator.) The forward sway immediately elicits increased activity in the calf muscles (Bas-

majian (1974); Joseph (1960); O'Connell (1958)) (see Fig. 4.12), which pull back on the tibia and lower femur to move the body backward at the talotibial joint so that the line of gravity will continue to pass through the decreased base (the heads of the metatarsals and the toes) and, under the circumstances, produce the extreme plantar flexion characteristic of the posture when a subject has risen on the toes.

There are thus two possibilities for the effort of toe extension, gravity, and the toe extensors. In the first instance (Fig. 5.8(a)) we have a first class lever system whether the axis is considered as being at either joint; as both the ankle and metatarsophalangeal joints are between effort (gravity) and resistance (tension in the plantar- and toe-flexing muscles). In the second instance (Fig. 5.8(b)), the axis is at the metatarsophalangeal joints and the effort is contraction of the toe extensors which have a moment arm shorter than that of the resistance due to gravity, so the leverage is third class.

On the other hand, some kinesiologists refer to the brachioradialis acting on the forearm to flex it at the elbow joint as the effort in a second class lever supporting a load halfway between the radial attachment and the elbow (Fig. 5.9). To do this they have made use of the concept of an effort arm (all parts of the lever between the axis and the point where the effort is applied), EA in Figure 5.9(a). They then postulate a weight hung midway on the forearm, R in the same figure. Superficially this seems to be an example of a second class lever, with the resistance between the effort and the axis and the effort arm longer than the resistance arm. However, one cannot use the length of such an effort arm to determine the amount of force that the brachioradialis must exert to support the load hanging from the forearm. Rather, the moment arms indicated in Figure 5.9(b) must be invoked in calculations. Thus, the equivalence with a third class lever as shown in that figure.

Interestingly enough, if we consider the brachioradialis or any elbow flexor as a *resistance* supporting the forearm

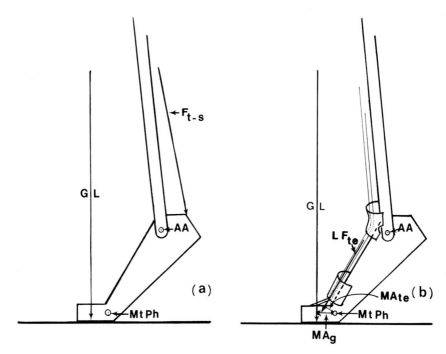

Figure 5.8. Standing on the toes. (a), force diagram as a first class lever (gravity effort). AA, ankle axis; MtPh, metatarsophalangeal axis; F_{t-s}, line of force of triceps surae (gastrocnemius-soleus-plantaris); GL, gravity line. This is a first class lever system as the joint axis is between the line of gravity (the force causing the forward motion of the body which moves the center of gravity over and briefly anterior to the toes) and the resisting force (the tension exerted by the plantar flexors). (b), force diagram as a third class lever (toe extensors are effort). AA, ankle axis; MtPh, metatarsophalangeal axis; LF_{te}, line of force of the toe extensors; MA_{te}, moment arm of toe extensors; GL, gravity line; MA_g, moment arm of gravity. MA_g is longer than MA_{te} and both lie on the same side of the metatarsophalangeal axis.

against gravity while the individual is lowering a weighty object (Fig. 5.10), this lever system becomes second class. The muscle moment arms at the elbow are always shorter than the distance from the joint axis to the point where any load can be applied and so, when the weight of the load is the force causing the movement, the system becomes a second class lever. This will be true in any situation where muscle shortening is the effort of a *third* class lever. It is obvious that, as soon as this muscular tension becomes a resistance to a reversal of joint action caused by an outside force, the lever system is reversed also and becomes a second class lever. Under these circumstances the muscle is

contracting eccentrically and performing negative rather than positive work.

Advantages and Disadvantages of Short Muscle Moment Arms and Longer Lever Arms

Any lever system with a short effort moment arm requires a force greater than the load if the lever is to move that load. This is one of the seeming disadvantages of the human body, in which all muscular moment arms are short in proportion to the levers they move. However, there is in fact an advantage to this situation: a very small movement of the short end of the lever is magnified in direct proportion to

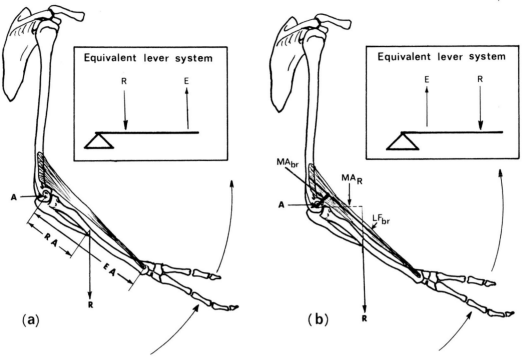

Figure 5.9. Branchioradialis as the effort in elbow flexion: a second or third class lever? (a), erroneously postulated second class lever. A, axis containing fulcrum of elbow joint; R, line of force of resistance (load); RA, resistance arm; EA, effort arm. (b), third class lever. A, axis of elbow joint; R, line of force of resistance (load); MA, moment arm of the resistance; LF_{hr}, line of force of brachioradialis; MA_{br}, moment arm of brachioradialis.

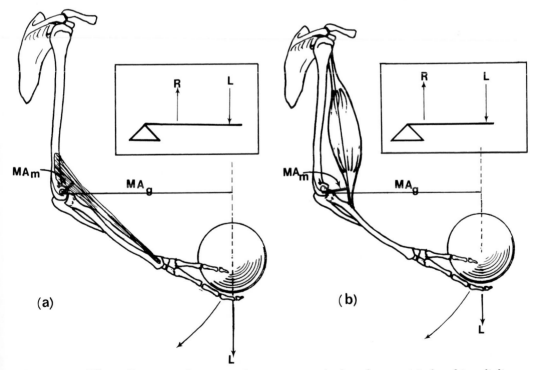

Figure 5.10. Elbow flexors acting as resistance: second class levers. (a), brachioradialis as resistance; (b), biceps as resistance. L, load or gravity as the moving force or effort; MA_g, moment arm of gravity; MA_m, moment arm of resisting muscle. The insets show equivalent lever systems of the second class.

127

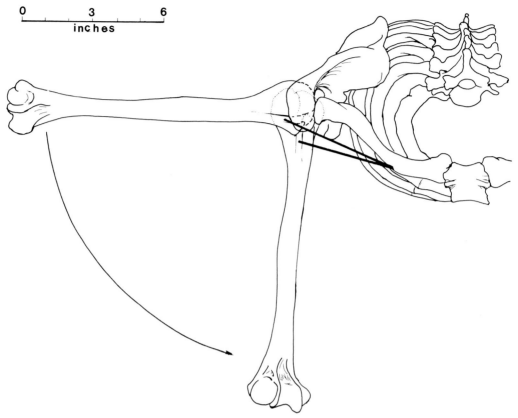

Figure 5.11. Effect of small amount of muscle shortening on range of movement. Clavicular head of pectoralis major has shortened 0.8 in and the humerus has moved through an 83° arc.

the length of the lever being moved (Fig. 5.11). The pectoralis major has shortened only 0.8 in and has moved the upper extremity through an angle of 83°. Since in the same amount of time the distal end of the lever moves a much larger distance than does the point of muscular attachment, the distal end is moving proportionately faster. This is also illustrated in Figure 5.12. The entire arm, forearm, hand, and tennis racket are moving as a unit through the same angular displacement in the same amount of time, but the distance, i.e., the length of the arc, traversed by the elbow joint is obviously shorter than that of the hand, and both are shorter than that of the racket. Under these circumstances the racket head, as it is being moved through the same angle as the elbow has

traveled, has moved curvilinearly almost 3½ times as far in the same amount of time and so is moving at 3½ times the speed of the elbow.

This lengthening of a body lever by an implement in order to gain speed is made use of in any sport which uses a club, bat, or racket, etc. to project an object. The longer the lever the greater will be the speed of the impact surface of the implement, but size, length, weight, and shape are limited not only by the rules of the particular sport involved but by anatomical considerations as well. A tennis racket suitable for Bjorn Borg or Jimmy Connors would be a handicap to most women and certainly to a child or most adolescents. While the overall length of a tennis racket cannot be increased, the weight and grip

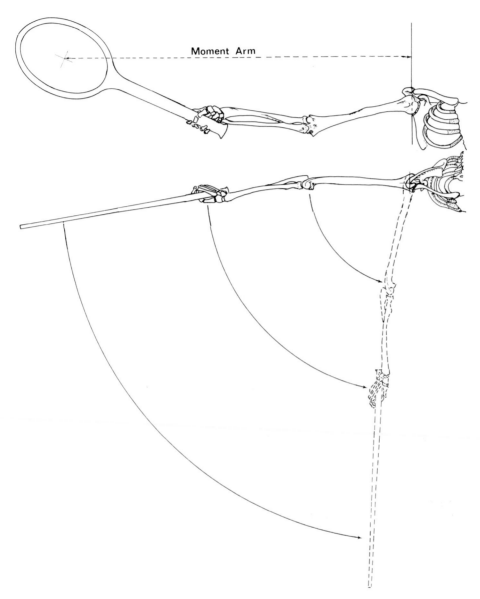

Figure 5.12. Forward swing of a tennis forehand drive.

size are varied to suit a variety of players. The same is generally true of any other striking instrument from baseball bats to polo mallets; in the latter instance, not only the arm length of the player but the size of the polo pony must be considered.

A child that tries to hit a tennis ball or softball with a racket or bat that is too heavy for him will either try to hit the ball while pulling the implement closer to him so he can swing it rapidly with greater ease, or he will compensate for insufficient muscle force by a slower swing, using his limited strength for work rather than speed. In the first instance he has shortened the overall length of his lever so that less effort is needed for the work of moving the load and more of the muscle force is available for speed of movement (Halverson (1966)).

This illustrates the fact that to conserve force for extra speed we can shorten the moment arms of our body levers by flexing distal joints and so decreasing the force necessary at the proximal joint to move the limb. Anyone while running makes use of this principle by flexing the swing leg sharply at the knee as it is brought forward. This moves the center of gravity of the lower extremity closer to the hip axis and so decreases the moment arm of the lower extremity, thereby decreasing the amount of force (and the moment) necessary to move the thigh, leg, and foot forward. This extreme knee flexion is particularly obvious in sprint running (and in Standardbred trotting horses). Because of this sharp knee flexion the hip flexors can shorten rapidly and the speed of the run is maintained.

Elementary body mechanics points out that loads should be carried as close to the body as possible, thus decreasing the moment arm of the load in relation to the joints involved. This of course decreases the strain put on lifting muscles, joint capsules, and ligaments.

Although identifying the various classes of levers as they occur in the body gives insight into the forces necessary to move them, such identification is not vital to movement analysis per se. What is vital is to remember that the body is a system of levers and so obeys the laws of levers and mechanics. It has effort arms and resistance arms, moment arms of the effort (MA_E) and moment arms of the resistance (MA_R). Muscles may be moving forces in a motor skill or they may resist the action of an outside force such as gravity by a lengthening tension controlling and slowing the movement. On the other hand, if a very fast movement is desired, muscles can work with gravity to accelerate the movement, as when an axe is swung down hard to split a log.

Skilled motor activity comes from making optimal use of levers, decreasing or increasing their length as occasion demands, and timing the muscular control to act on them with only an optimal amount of force.

EXAMPLE 9—Muscle and Joint Reaction Forces

It is desired to calculate the total muscle force (TMF) of the biceps and the force on the distal end of the humerus when the outstretched hand of the man in Example 4 is supporting a 10-lb shot as shown in Figure 5.13.

The combined weight of the forearm and hand is 3.7 lb and the center of gravity is 6.6 in from the elbow axis (see Example 4 above); the line of gravity of the shot falls 13.4 in from the elbow axis. (Note that all forces, including the action line of the biceps, are vertical and perpendicular to the long axis of the forearm.)

The moment arm of the biceps is not known, nor can it be accurately measured in vivo. It can be estimated on a living subject who flexes his elbow to the desired angle, by measuring the perpendicular distance from the biceps tendon to the humeral epicondyle; it can be measured with a fair degree of accuracy in a cadaver or on a skeleton.

In this instance the moment arm of the biceps about the elbow will be taken as 2.2 in.

The line diagram in Figure 5.13(b) is a simplified version of Figure 5.13(a), and it does not show the reaction force at the elbow joint. Apart from this, there are two downward forces and a third upward one, all parallel and each with its own moment arm. As this is an equilibrium situation, the sum of the moments is equal to 0:

$$\Sigma M = 0 \text{ or } \Sigma T = 0$$

The weight of the body segments and of the 10-lb shot in the hand tend to rotate the forearm clockwise (CW) about the elbow joint and so are considered positive, while the biceps force tends to rotate the

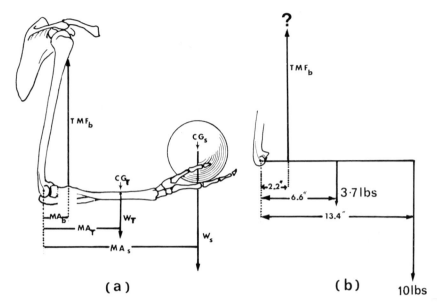

(a) **(b)** 10lbs

Figure 5.13. (a), drawing; (b), force diagram. MA_b, moment arm, biceps; MA_{fh}, moment arm, forearm plus hand; MA_s, moment arm of the shot; CG_T, center of gravity of forearm and hand; CG_s, center of gravity of the shot.

forearm in the opposite or counterclockwise (CCW) direction and so is considered negative:

$$\Sigma M = (W_{fh} \times MA_T) + (W_s \times MA_s)$$

$$- (TMF_b - MA_b) = 0$$

$$= (3.7 \text{ lb} \times 6.6 \text{ in}) + (10 \text{ lb} \times 13.4 \text{ in})$$

$$- (TMF_b \times 2.2 \text{ in}) = 0$$

$$TMF_b \times 2.2 \text{ in} = 24.42 \text{ lb-in} + 134 \text{ lb-in}$$

$$TMF_b = 158.42 \text{ lb-in}/2.2 \text{ in}$$

$$= 72 \text{ lb}$$

It is noteworthy that the biceps must exert about 5½ times as much force as the weight of the load that it supports.

This same biceps force also pulls the forearm against the articular surfaces of the humerus so that there is pressure against them. Let H equal the unknown force of the humerus:

$$\Sigma F = 0$$

$$\Sigma F = + 3.7 \text{ lb} + 10 \text{ lb} - 72 \text{ lb} - H = 0$$

$$H = 13.7 - 72$$

$$= 58.3 \text{ lb}$$

an upward force. There is a vertical force on the distal end of the humerus of 58.3 lb as a result of the biceps supporting the 10-lb shot in the hand. This force is distributed on the articular surfaces of the humerus and the mean pressure is the force divided by the surface area on which the force bears.

EXAMPLE 10—Resolution of Total Muscle Force of Biceps into Rotatory and Secondary Components

Assuming a constant TMF of 100 lb, this force is to be resolved into two components for each of several angles at the elbow, commencing with total extension and proceeding progressively until full flexion. The two mutually perpendicular components to be identified are the rotatory component which acts at right angles to the long axis of the bone that it moves

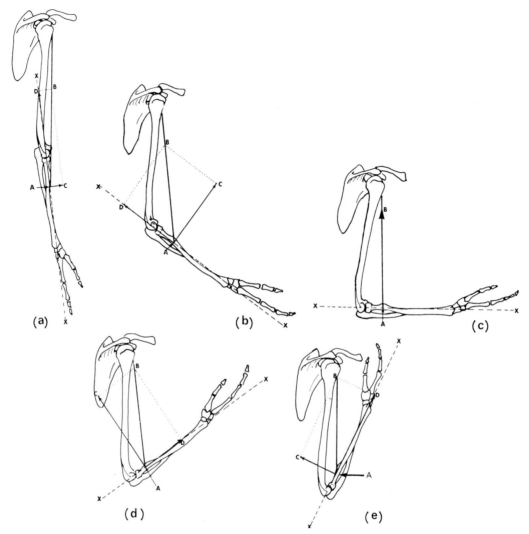

Figure 5.14. Resolution of total muscle force of biceps (TMF$_b$) into rotatory and secondary components. $\bar{A}B$ representing TMF$_b$ is maintained at 100 lb. X-X, long axis of forearm; AC, rotatory component; AD, secondary component. Note the change in direction of the secondary component from that in (a) and (b) in going past (c) as in (d) and (e). See text.

(in this case the forearm); and the secondary component which acts along the long axis. The nature of the latter component is to be discussed.

Figure 5.14 shows the requisite resolution into the two components for a series of different angles. In each case, the action line of the rotatory component originates at point A, the intersection of the component originates at point A, the intersection of the biceps action line with the long axis of the forearm, x-x, and is perpendicular to this axis. The action line of the secondary component is coincident with the long axis, and its direction can easily be determined by constructing the parallelogram of forces around a given TMF of the biceps. Vector $\vec{AB}$ is drawn along the biceps action line and is 100 units long.

In Figure 5.14(a) the forearm is extended and the action line of the biceps makes an angle of approximately 8° with the long axis of the forearm. With a TMF of 100 lb the rotatory component, vector $\vec{AC}$, is only 15 lb, while the component represented by vector $\vec{AD}$ is 98 lb. As the elbow is flexed the angle of pull of the biceps action line increases, as does the rotatory component. Maximal rotatory force is achieved at (c) when the action line of the biceps is at 90° with the forearm axis and all of the 100 lb of contractile force are available for supporting or moving the forearm. With continuing elbow flexion the rotatory component decreases as the secondary component increases but its direction is now away from the joint. In (d) the muscle is pulling the forearm away from the joint with a 92-lb force (vector $\vec{AD}$), and the rotatory force is 41 lb (vector $\vec{AC}$).

It should be noted that when the elbow is in extension beyond the value reflected in Figure 5.14(c), i.e., Figure 5.14, (a) and (b), the secondary force AD is directed toward the joint and is increased with more extension. Being directed toward the joint, this force is of a stabilizing-compressing nature, tending to hold the joint together. As the joint goes into flexion beyond the position shown in Figure 5.14, i.e., Figure 5.14, (d) and (e), the direction of AD is now away from the joint; the secondary force has become one of dislocation-decompression, increasing with greater flexion.

LOCATING THE CENTER OF GRAVITY

Much time and energy have been expended in efforts to pinpoint the location of the center of gravity in the human body. The earliest recorded effort is that made by Borelli in 1679, who balanced the body over a prismatic wedge in three different planes (Dawson (1935)). There the matter rested until the 19th century when a number of different studies were made, notably by Mosso (1884) and by Braune and Fischer in 1889 (Duggar (1966)). Much work has been done in this field since then, and by the end of the first third of this century the method now commonly used in kinesiology laboratories was coming into general use.

With Gravity Board and Scale

The method now used in college kinesiology laboratories was, according to Cooper and Glasgow (1968), devised in 1909 by Lovett and Reynolds (Fig. 5.15). This method is based on the fact that when a body is in equilibrium the sum of the gravitational moments acting on the body is equal to zero. The subject is weighed (weight $= W_t$) and then steps onto the gravity board. The board is tared* by setting the scale to zero so that the weight registered on the scale includes only the gravitational force acting on the body parts.

Taking moments about the line of action of the body force (which line of action must pass through the board), the clockwise moment CWM is the product of the force recorded on the scale W_s and the perpendicular distance from the line of action to the knife edge resting on the scale. If the distance between the knife edges is D, then the counterclockwise moment CCWM about the body is the product of the weight supported by the distal knife edge (total weight minus weight registered on the scale, i.e., $W_t - W_s$) and the perpendicular distance d_2 from the distal knife edge and the plane containing the line of gravity ($d_2 = D - d_1$). By convention CW forces are considered positive and CCW forces are considered negative.

For equilibrium

$$\Sigma M = 0$$

---

* If this is not possible the weight of the board registered on the scale must be subtracted from the weight registered when the subject steps on the board.

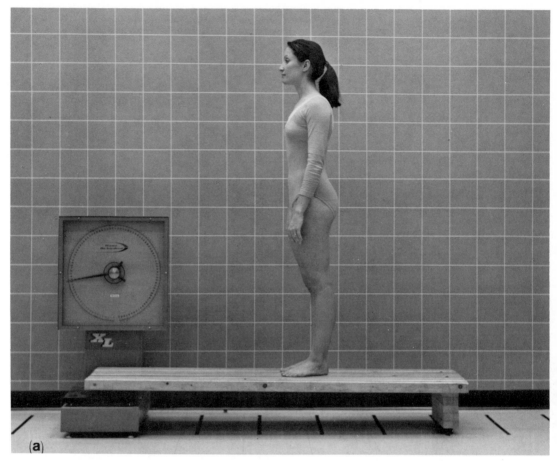

(a)

Figure 5.15. Equipment for determining primary planes of the center of gravity. Note knife edges supporting the gravity board. It is of interest to note the increased reading on the scale for (b) as compared with (a) when the subject rises on her toes, thus moving her center of gravity ahead of the position it occupied in (a).

i.e.

$$\Sigma M = CWM - CCWM = 0$$

$$CWM = W_s d_1$$

and

$$CCWM = (W_t - W_s)d_2$$

$$(W_t - W_s)d_2 - W_s(D - d_2) = 0$$

$$(W_t - W_s)d_2 = W_s(D - d_2)$$

$$W_t d_2 - W_s d_2 = W_s D - W_s d_2$$

$$W_t d_2 = W_s D$$

$$d_2 = W_s D/W_t$$

The line or plane of gravity is d_2 units from the *knife edge on the floor*. If the subject stands in his footprints drawn on a piece of paper which is placed with one edge at a given marker, say 100 cm from the distal knife edge of the board, the plane of gravity can be easily recorded on the footprints.

This method is useful in obtaining the location of the line of gravity in relation to the base of support at a specific instant, i.e., the time when the scale was read. When a subject stands on a gravity board, the pointer on a dial, or the beam of the balance, of a supporting scale is never

(b)

Figure 5.15 (b)

completely still as there is a continuous slight swaying of the body on the heads of the tali, as was first reported by Helle-brandt *et al.* (1937).

When an attempt is made to locate the primary transverse plane by this method (the transverse plane passing through the center of gravity), only an approximation can be made, for several reasons. When the subject is supine there is a change in the distribution not only of the blood throughout the circulatory system but of the abdominal viscera as well. Also, Mosso (1884) and Cotton (1932) discovered that the locus of the center of gravity of a subject in a recumbent position shifted slightly with each deep inspiration (Dug-

gar (1966)). This effect is undoubtedly a result of the abdominal viscera being forced caudad by the contraction of the diaphragm. Such a deep inspiration might well give a center of gravity a locus closer to that when the body is erect.

The use of the gravity board requires that the subject assume a static position, and some fairly elaborate apparatus has been set up to determine the locus of the line of gravity simultaneously in two planes. One technique now becoming pop-ular in kinesiology laboratories makes use of three scales, with the gravity board sup-ported by three adjustable pointed bolts, one resting on each scale (Fig. 5.16). When the board has been leveled it is ready for

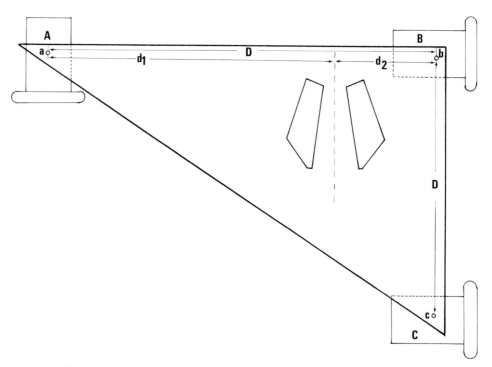

Figure 5.16. Schematic diagram of type of gravity board currently in use. A, B, and C, scales supporting the board; a, b, and c, pointed bolts supporting the board on the corresponding scale; D, the distance between bolts a and b; D', the distance between bolts b and c. While the diagram is annotated for determining the location of the dotted line via d_1 and d_2 thus defining the sagittal plane, it should be evident that the location of a similar (but perpendicular) line described by d_1' and d_2', where $D_1' = d_1' + d_2'$, and indicating the location of the frontal plane, can be specified.

use. Rasch and Burke (1967) use a large rectangular board so that they can place a subject in various side-lying positions corresponding to different diving, or gymnastic conformations. Waterland and Shambes (1970) use a similar arrangement but with a smaller board, and they recommend one in the shape of a right triangle. Two of the scales are so placed that their dials face in the same direction and at 90° to the face of the third scale. The subject is photographed simultaneously from the side and front by synchronized cameras, one picture including the face of scale A and the other the faces of scales B and C (Fig. 5.16).

Calculations for locating the primary planes which pass through the center of gravity are based on the same principles presented earlier, and procedures are similar.

Taking moments about the dotted line (in Fig. 5.16) which represents one plane of gravitational forces on the body:

Clockwise moments: $CWM = W_A d_1$

where W_A is the weight registered by scale A and d is the perpendicular distance from the plane of gravity to bolt a. Counterclockwise moments: $CCWM = (W_B + W_C)d_2$ where W_B and W_C are the respective indications of scales B and C, and d_2 is the perpendicular distance from the line connecting bolts b and c to the plane of gravity.

$$\Sigma M = CWM - CCWM = 0$$

leads to

$$W_A d_1 - (W_B + W_C)d_2 = 0$$

Recognizing that $d_1 + d_2 = D$, $d_1 = D - d_2$ and substituting in the moment equation gives

$$W_A D - W_A d_2 - W_3 d_2 - W_C d_2 = 0$$

i.e.

$$W_A D - (W_A + W_B + W_C)d_2 = 0$$

Since the weight of the subject

$$W_t = W_A + W_B + W_C$$

it follows that

$$W_A D - W_t d_2 = 0$$

thus

$$d_2 = W_A D/W_t$$

To find the plane of the center of gravity parallel to the line between bolts a and b, the same procedure is followed, i.e.

$$\Sigma M = W_C d_1' - (W_A + W_B)d_2' = 0$$

where $d_1' + d_2' = D'$ and d_1' is the perpendicular distance from C to the line representing the plane of gravity perpendicular to line BC,
i.e., $d_2' = W_C D'/W_t$.

Swearingen (1949) used equipment consisting of five platforms mounted one above the other. The top platform supported the subject in an adjustable seat, the second and third platforms were horizontally adjustable, and the fourth platform was separated from the fifth or, base, by a ball and socket joint in the center and electrical contact points at each corner. Measurement of the CG [center of gravity] location for the upright seated posture, for example, was made by balancing the system with the subject tipped horizontally, then rebalancing with the seated subject tipped approximately 20 degrees from the horizontal. Two planes passing through the CG were thereby established, and the exact location could be measured with respect to any suitable reference point (Duggar (1966)).

By Segmental Method

All of these methods are moderately accurate subject to the limitations mentioned earlier, as well as to the accuracy of the scales used in the study. However, none of these methods can be used in dynamic situations. A solution to this problem was suggested by Dawson as early as 1935. He suggested that the subject be "appropriately fotograft & the foto transfered to coordinate paper. The subj. is then weighed & measured & compared with the subj. described in the tables (Braune and Fischer's measures, Table A4.1, Appendix 4). One of the 4 cadavers is selected as being most like the subj. in bild & the data from this cadaver ar therefore used in the following calculations. The wts. of the parts of the subj. (wt.) are computed on the assumption that the weight of the part of the subj. =

(wt. part cadaver × wt. whole subj.)/wt. of whole cadaver

"Having determined the wt. of the various parts of the subj. & noted upon the coordinate paper the position of the cc. g. [centers of gravity] of the parts, it is a simple mathematical calculation to determin the position of the c. g. of the body as a whole. By the use of a system of coordinates the calculations wer stil further simplified so that it became necessary merely to insert the values ascertained in an appropriate equation which could be solved."†

This approach has become greatly simplified with the use of time-motion study film projectors and microfilm readers. Also, it is no longer necessary to rely solely on Braune and Fischer's data.

---

† The above quotation is an accurate reproduction from The Physiology of Physical Education by Percy Dawson, M.D. Dr. Dawson was a member of a group known as the Simplified Spelling Board and insisted on this format for what was, at the time, an excellent physiology text.

Dempster (1955) collected similar material on eight cadavers and went even further. After determining segmental weights and centers of gravity he measured the specific gravity of each segment. All of these data are presented in Appendix 4, Tables A4.1 through A4.4. Dempster's cadavers averaged approximately the same height as those used by Braune and Fischer but averaged about 8 to 9 lb lighter in weight and were considerably older, having died from natural causes. Braune and Fischer's cadavers died of acute strangulation and were obtained from the hangman!

Dempster also measured the volumetric displacement of limb segments of 38 young men, 6 rotund, 11 muscular, 10 thin, and 11 of medium physique (see below for description of this technique). Table A4.5, Appendix 4, presents fractions of body weight calculated from these data by taking the product of each segmental displacement (weighed to the nearest gram) and the specific gravity of the segment. The average was then taken for each segment in each group. The availability of these data makes it possible to calculate the moment of force exerted by the weight of each segment of the body. As the center of gravity (CG) is located at the point where the sum of all moments is equal to zero, it becomes a simple matter to locate this center by the mathematical composition of parallel forces.

When using any table of fractions of body weight, it is not necessary to calculate the moment of each segment: all that is necessary is to take the product of the fraction of body weight given in the table and the moment arm of that segment. The sum of these figures presents the length of the moment arm (MA) of the CG from the selected axis.

It should be remembered that all of these data are based entirely on measurements made from men and so do not yield as accurate results when applied to women. The authors have found that there may be a discrepancy of several centimeters between the location of the primary transverse plane (CG) as calculated by the segmental method when compared with

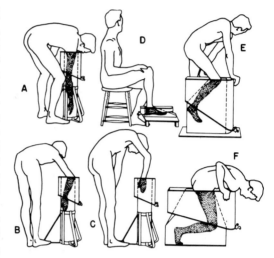

Figure 5.17. Immersion tanks used by Dempster (1955).

that determined from the supine position on the gravity board.

Immersion Technique for Determining Segmental Weights. Greater accuracy can be obtained by the limb immersion technique. Figure 5.17 illustrates Dempster's immersion tanks which were made of stainless steel. Versions less expensive but just as accurate can be made of plastics by any good plastics workshop (Fig. 5.18). Table A4.3, Appendix 4, presents Dempster's landmarks and directions for the immersion technique. These joint centers, detailed in Table A4.3, can be marked on the subject with a skin pencil or felt point pen so that when the limb is immersed it can be lowered only as far as the desired mark. The displaced water is then drained from the jacket into a tared vessel and weighed to the nearest gram or ounce. This weight of displaced water is then multiplied by the specific gravity of the segment measured (Table A4.4, Appendix 4) to obtain the segmental weight. There will, of course, still be some slight inaccuracy as the specific gravities were obtained from elderly male cadavers, but this is not considered significant at this time. Hopefully some day similar specific gravity measures for women will become available. However, comparison of the CG location for

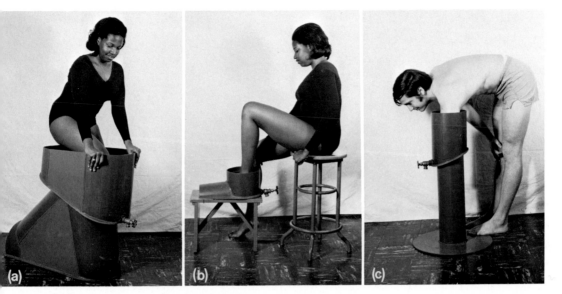

Figure 5.18. Immersion tanks made of plastic.

women (as well as men) supine on the gravity board with that determined by the above technique on the same subject and in a similar posture shows a closer agreement than with the other methods.

Mathematical Determination of Segmental Weights. A reasonably accurate mathematical derivation of segmental weight has been devised by Barter (1957) the United States Air Force. He used data from both Braune and Fischer's and Dempster's reports on cadaver measures and developed the series of regression equations presented in Table A4.6, Appendix 4. This material is also available in Duggar (1966), who states that "The standard deviations of the residuals do not incorporate the uncertainty of the regression value itself based on only 12 cadavers, so that the total error of prediction will be greater for body weights differing from the average cadaver weight, 59.4 kg. Extrapolation of these regression equations to include persons weighing more than the heaviest cadaver, 75.3 kg, is unreliable."

Complete accuracy in locating the CG is not possible at this time. "... The most exacting measurements cannot locate the CG of a particular man to within 0.3 cm"

(Duggar (1966)). However, a fair estimate of the accuracy of the segmental calculations can always be made by comparing the results of such calculations made from a standing posture with those obtained with the subject supine on the gravity board. "Calculations based on appropriate segment data and applied to particular subjects whose (link) length and posture can be measured should be accurate to within 1 cm" (Duggar (1966)).

EXAMPLE 11—Location of Line of Gravity by Segmental Method

Figure 5.19 shows tracings made from film strips synchronized when using two mutually perpendicular cameras. The particular pose is a segment of a gymnastic stunt, the scale, and known in ballet as an arabesque.

It is desired to determine the positioning along the X-axis (in Fig. 5.19(a)), of the gravity line for the portion of the body which is supported by the right hip joint, using the segmental method alluded to earlier.

To begin, Figure 5.20, a link diagram of

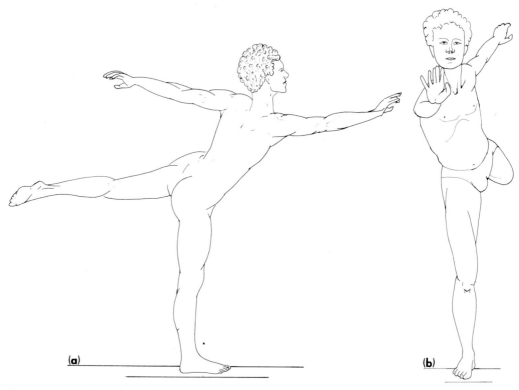

Figure 5.19. Tracings from film. (a), from the side; (b), from the front.

Figure 5.19(a), is constructed to scale using the performer's measurements. Next, the horizonal or X-moment arms in the sagittal plane for each segment are measured and the moments about the origin (intersection between X and Z axes located at the right hip) are computed. Those moments occurring to the right of the supporting hip axis will cause a clockwise (CW) rotation of the trunk on the right femur and are by convention considered as positive. Those moments occurring to the left, on the other hand, will cause counterclockwise (CCW) moments and are considered as negative; i.e.,

$$\Sigma M = (W_{ut} \times MA_{ut}) + (W_{lt} \times MA_{lt})$$
$$+ (W_h \times MA_h) - (W_{lth} \times MA_{lth})$$
$$- (W_{ll} \times MA_{ll}) - (W_{lf} \times MA_{lf})$$

$$+ (W_{ra} \times MA_{ra}) + (W_{rfa} \times MA_{rfa})$$
$$+ (W_{rha} \times MA_{rha}) + (W_{la} \times MA_{la})$$
$$- (W_{lfa} \times MA_{lfa}) - (W_{lha} \times MA_{lha})$$

where W is weight of the segment, MA is the moment arm of the segment, and the lower case subscripts identify the segment: ut, upper trunk; lt, lower trunk; h, head; lth, left thigh; ll, left leg; lf, left foot; ra, right arm; rfa, right forearm; rha, right hand; la, left arm; lfa, left forearm; lha, left hand.

Table 5.1 summarizes the data and facilitates the required calculations.

$$\Sigma M = 522.609 - 473.798 = 48.811 \text{ in-lb}$$

In any parallel force system the product of the resultant and its moment arm is equal to the sum of the moments; i.e., $R \times$

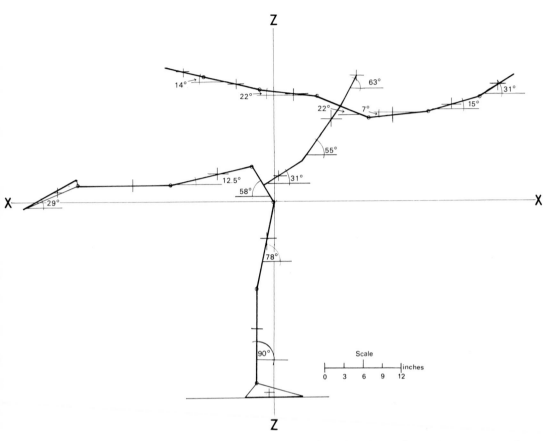

Figure 5.20. Link diagram drawn from Figure 5.19 to scale, using the performer's measurements.

$MA_R = \Sigma M$. From Table 5.1, R is approximately 111 lb; then

$$111 \text{ lb} \times MA_R = 48.811 \text{ in-lb}$$

$$MA_R = 0.44 \text{ in}$$

Thus, the gravity line passes through a point 0.44 in anterior to the right hip (origin).

Should it be necessary to locate the center of gravity of the portion of the body supported about the right hip joint, the same process would be followed, calculating the Z or vertical moments and then the Z position of the center of gravity. At this juncture, we would, of course, have information about the center of gravity only in

the sagittal plane. Frontal plane information can be derived from the tracing in Figure 5.19(b) and by carefully describing mathematically body components in that plane; the Y component can be calculated in a similar vein. For repetitive calculations involving a variety of body poses, computer-aided approaches are indicated (see Chapter 6).

EXAMPLE 12—Shift in Line of Gravity with Position

Figure 5.21 shows an altered scale position in link diagram form. Note that both arms are flexed at the shoulder in this

Table 5.1
Segmental Data: X Moments (see Fig. 5.20)

Segment	Weight	Plus Moment Arms	Plus Force Moments	Minus Moment Arms	Minus Force Moments
	lb	*in*	*in-lb*	*in*	*in-lb*
Upper trunk	23.7	7.3	173.01		
Lower trunk	29.8	0.66	19.668		
Head	11.1	11.0	122.10		
L thigh	20.05			100	200.50
L leg	7.1			24.4	17.24
L foot	1.9			35.0	66.50
R arm	5.25	18.5	97.125		
R forearm	2.55	27.78	70.8135		
R hand	0.85	36.0	30.60		
L arm	5.25	1.77	9.2925		
L forearm	2.55			7.66	19.533
L hand	0.85			16.5	14.025
Total	110.95		522.609		473.798

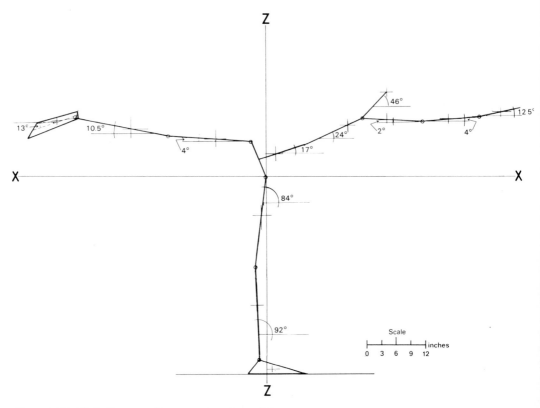

Figure 5.21. Link diagram of a second scale position.

Table 5.2
Segmental Data: X Moments (see Fig. 5.21)

Segment	Weight	Plus Moment Arms	Plus Force Moments	Minus Moment Arms	Minus Force Moments
		in	in-lb	in	in-lb
Upper trunk	23.7	13.5	319.950		
Lower trunk	29.8	2.25	67.050		
Head	11.1	24.25	269.175		
L thigh	20.05			12	240.60
L leg	7.1			27.5	195.25
L foot	1.9			42.5	80.75
Both arms	10.5	31.75	333.375		
Both forearms	5.1	37.0	188.70		
Both hands	1.7	49.0	83.3		
Total	110.95		1261.55		516.60

case. It is desired to determine the shift in the gravity line from the situation shown in Figure 5.20.

Table 5.2 shows the tabulated calculations in a manner similar to Table 5.1.

$$\Sigma M = 744.95 \text{ in-lb}$$

i.e.

$$R \times MA_R = \Sigma M$$

$$MA_R = (\Sigma M)/R$$

$$= 744.95/110.95$$

$$= 6.71$$

Thus, the line of gravity is now 6.71 in anterior to the hip axis. There has, therefore, been a comparative anterior shift relative to the conditions depicted in Figure 5.20 of 6.71 − 0.44 = 6.27 in.

EXAMPLE 13—Estimation of Total Muscle Force Exerted by Hamstrings

For the subject depicted in Figures 5.19 and 5.20 it is desired to determine the muscle force exerted by the hamstrings which must play a major role in maintaining the subject's pose. Next, the effect on the hamstring force required for the modified position shown in Figure 5.21 is to be determined.

It should be realized that it is impos-sible to measure moment arms of muscles or to indicate a muscular line of pull with complete accuracy in vivo. If a skeleton or a cadaver could be positioned exactly as the subject, measures of moment arms and angles could be made and used. The next best method is to draw the skeletal structures within the body outline as accurately as possible and to work from these. This has been done in Figures 5.22(a) and 5.23(a) which provide a means whereby appropriate approximations and estimates can be made. The essence of the approach to be made is in the proper construction of a force diagram which represents the situation of concern.

To properly construct the force diagram we must know the horizontal distance of the line of pull of the hamstrings from the hip axis as well as the similar distance to the line of gravity. A horizontal line is drawn through the hip axis of Figures 5.22(a) and 5.23(a) and another through the hamstrings from the femoral condyle (which alters the line of pull somewhat from attachment to attachment) to the attachment on the ischial tuberosity. The intersection of this line with the horizontal gives the distance from the hip axis as well as the angle the line of pull makes with the horizontal. As we have already located the line of gravity (the X or sagittal component of the center of gravity) in each

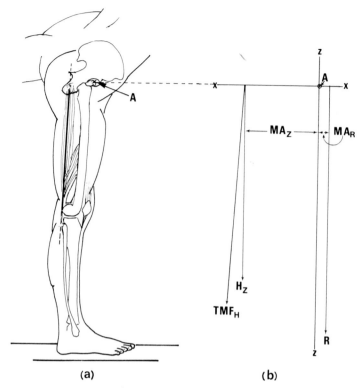

Figure 5.22. Supporting limb. (a), details pertinent to Figure 5.19 with bony structures and hamstring action line. (b), force diagram for determining hamstring TMF. x-x, horizontal coordinate through hip axis A; z-z, vertical coordinate through hip axis; MA_R, moment arm of the resultant weight of the supported body segments, 0.44 in; R, line of force (action line) of the resultant weight; TMF_H, line of force of hamstrings; MA_Z, moment arm of Z component of hamstring force; H_Z, Z component of hamstring force.

case of concern (Examples 11 and 12), we can proceed with the force diagrams as shown in Figures 5.22(b) and 5.23(b).

For Figure 5.22 the line of gravity falls 0.44 in anterior to the hip axis. This is shown in Figure 5.22(b) as the line R located the perpendicular distance MA_R from the 5.38 hip joint axis A. By measurements made upon Figure 5.22(a) we find that the line of pull of the hamstrings crosses the horizontal 3.4 in posteriad and forms an angle of 81° with the horizontal.

Thus, TMF_H, which represents the total muscle force of the hamstrings can be located as shown, its distance from A on the X-axis being MA_Z.

The gravitational forces on all of the body segments supported by the right hip joint exert a clockwise moment about that joint while the supporting muscles (hamstrings) exert a counterclockwise moment.

Since this is an equilibrium situation where $\Sigma M = 0$, we can solve for the unknown force (TMF_H). The vertical component of TMF_H is depicted in Figure 5.22 by H_Z. For equilibrium then, it follows that

$$R \cdot MA_R - H_Z MA_Z = 0$$

i.e.

$$H_Z = R \cdot \frac{MA_R}{MA_Z}$$

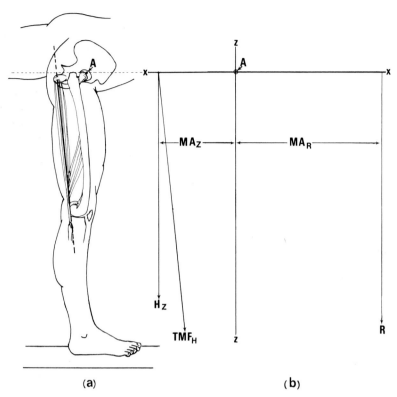

(a)   (b)

Figure 5.23. Supporting limb. (a), details pertinent to Figure 5.21 with bony structures and hamstring action line. (b), force diagram for determining hamstring TMF, X, horizontal coordinate through hip axis A; Z, vertical coordinate through hip axis; MA_R, moment arm of the resultant weight of the supported body segments, 6.71 in; R, line of force (action line) of resultant weight; TMF_H, action line of hamstrings; MA_Z, moment arm of Z component of hamstring TMF; H_Z, Z component of TMF_Z.

substituting the values of these elements

$$H_Z = 110.95 \times 0.44/3.40$$

$$= 14.36 \text{ lb}$$

From trigonometric considerations, it follows that

$$TMF_H = H_Z/\sin 81°$$

$$= 14.36/0.9877$$

$$= 14.54 \text{ lb}$$

Proceeding in a similar fashion for the situation shown in Figure 5.23, the line of gravity falls 6.7 in anterior to the hip axis, and by measurement upon Figure 5.23(a)

the hamstring line of force crosses the horizontal through the hip axis 3.6 in posteriad and forms an angle of 84° with it.

Thus

$$H_Z = 110.95 \times 6.71/3.6$$

$$= 206.80 \text{ lb}$$

and

$$TMF_H = H_Z/\sin 84°$$

$$= 206.8/.9945$$

$$= 207.94 \text{ lb}$$

It is interesting to note that this is about 14 times the muscular effort required in the first position shown in Figure 5.22.

It should be realized, however, that these figures are derived by making a number of assumptions, foremost of which is that the gluteus maximus is not participating; second, that the moment arm of the biceps femoris as measured from the drawing is also that of the semimembranosus and semitendinosus muscles (i.e., the entire hamstring group); and third, that the drawings used were at least moderately accurate for their purpose. We have also reduced a problem to one in which only two dimensions are considered. For more exactness and specificity it would be necessary to calculate all of the force components along the mutually perpendicular three axes X, Y, and Z.

EXAMPLE 14—Estimation of Muscle Force Exerted by the Abductors: Gluteus Medius, Gluteus Minimus, and Tensor Fascia Lata

It is required to analyze and determine the muscle forces exerted by the abductor group acting about the hip joint.

There are three abducting muscles: gluteus medius, gluteus minimus, and tensor fascia lata. Their lines of pull all converge at or near the greater trochanter (Fig. 5.3). Rather than selecting an arbitrary resultant action line, we can compose these three lines of force into a single resultant, basing our solution on the work done by Inman (1947) (see Example 3). Figure 5.24 is also estimated from Inman's (1947) description, and the angles components, a, b and c form with the horizontal are measured. With these items in mind, we can build a vector diagram of the three muscles and, by following any one of the three different methods discussed in Example 3 determine the resultant line of abduction force.

The algebraic approach for determining this force seems to be the most appropriate. The first step of this solution involves resolution of each force vector into its horizontal or Y component (which lies

in the frontal plane) and its vertical or Z component. The Y component is always equal to the cosine of the angle that the vector makes with the horizontal multiplied by its magnitude, while the Z component is equal to the sine of the same angle times its magnitude. As the Y and Z ordinates are drawn through the point of convergence of the three vectors (Fig. 5.24), we see that two of the vectors fall in the first quadrant and so will have positive Y and Z values. The third vector, however, falls in the second quadrant and so will have a negative Y value while the Z value remains positive.

Starting with the horizontal or Y component:

$$R_Y = \sum f_Y = \sum f_Y \cos \theta = 14 \cos 55°$$

$$+ 29 \cos 73° - 57 \cos 91°$$

$$= 14 \cdot 0.57358 + 29 \cdot 0.29237$$

$$- 57 \cdot 0.01745$$

$$= 8.03 + 8.48 - 0.99$$

$$= 15.52$$

Similarly:

$$R_Z = \sum f_Z = \sum f \sin \theta$$

$$= 14 \sin 55° + 29 \sin 73° + 57 \sin 89°$$

$$= 14 \cdot 0.81915 + 29 \cdot 0.95630$$

$$+ 57 \cdot 0.99985$$

$$= 11.47 + 27.72 + 56.99$$

$$= 96.19$$

θ, the angle that the abducting resultant makes with the horizontal is

$$\theta = \arctan R_Z/R_Y$$

$$= \arctan 96.19/15.52$$

$$= 80.83°$$

Assuming that each of the three abductor muscles contributes proportionately to maintaining the pose, we can now build our force diagram and calculate the total muscle force (TMF) exerted by the abductors. As before the weight of the body supported by the right hip is about

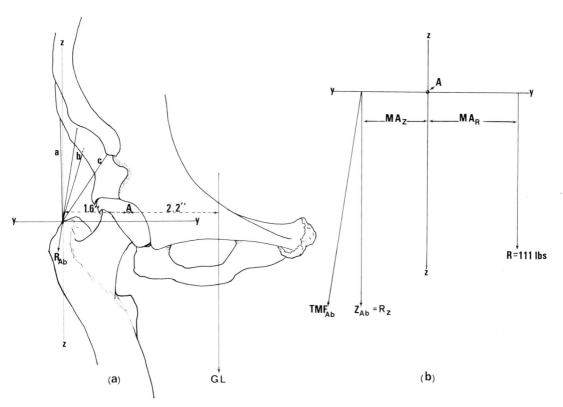

Figure 5.24. Abductor forces acting at the right hip. (a), enlargment of the right pelvis and proximal femur from Figure 5.19.6(b). a, action line of the gluteus medius; b, action line of the gluteus minimus; c, action line of the tensor fascia lata. Y and Z coordinates drawn through the point of convergence of the three action lines. (b), force diagram of the forces acting at the hip in the frontal plane, Y and Z coordinates through the hip axis A, R, the action line of the resultant weight of the lever supported by the hip; MA_R, moment arm of the resultant; TMF$_{ab}$, total muscle force exerted by the three abductors; Z_{ab}, the vertical or Z component of the abductor force; MA_z, moment arm of the Z component.

111 lb. Examination of Figure 5.19(b) shows the pelvis with an upward tilt at the right supporting hip. Here it is not necessary to sum the moments of each segment as we know that the gravity line must fall within the base of support, the right foot. Experience has shown us that if a subject takes this pose on a gravity board to determine the primary sagittal plane, the scale needle (or beam) is constantly vibrating, indicating lateral sways of the body over the foot. Thus, we can safely assume a vertical line passing through the foot as a line of gravity (Fig. 5.25). On Figure 5.24(a)

the horizontal distance mediad from the hip axis is 2.2 in, while that from the abductor resultant is 1.6 in laterad from the axis. With this information we can draw the force diagram in Figure 5.24(b) and proceed with the problem.

For equilibrium about the joint $\sum M = 0$ and therefore the vertical muscle component

$$R_Z = 111 \times 2.2/1.6 = 152.63 \text{ lb}$$

and the total muscle force of the abductors

$$\text{TMF}_{ab} = R_Z/\sin 80.53°$$

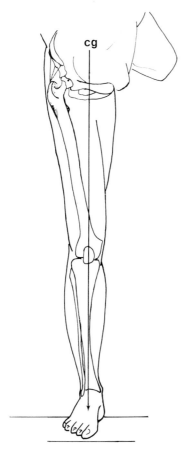

cg

Figure 5.25.Supporting limb, (see Fig. 5.19(b)) showing right pelvis and femur.

i.e.

$$TMF_{ab} = 152.63/0.9864$$
$$= 154.74 \text{ lb}$$

EXAMPLE 15—Compression Forces at the Acetabulum

It is required to determine the compression forces acting on the acetabulum for the static subject depicted in Figures 5.19 and 5.20. In effect, the amount of force with which the head of the femur is pressing against the joint is to be calculated. Such force is a result of the combination of body weight and muscle actions. The muscular actions to be considered are the hamstring forces and the abductor forces.

At the outset, the resultant of the hamstrings force and the effective body weight borne at the hip joint will be determined.

From Figure 5.22(b) and the pertinent calculations of TMF_H depicted in Example 13 it is evident that the resultant force acting at the hip joint, A, must be the vector sum of TMF_H and R, the resultant weight.

Resolving the forces into horizontal and vertical components, there results:

Horizontal components:

$$- 14.54 \cos 81° = -2.2746$$

Vertical components:

$$= - 110.95 - 14.54 \sin 81°$$
$$= - 110.95 - 14.36$$
$$= - 125.31$$

Resultant force

$$= \sqrt{(125.31)^2 + (2.2746)^2} \left| \arctan \frac{-125.31}{-2.2746} \right.$$
$$= 125.33 \text{ lb.} \qquad \left| 180° + 88.96° \right.$$

i.e., the resultant of the hamstrings and effective body weight is 125.33 lb directed at $180° + 88.96°$ relative to the horizontal axis. The reaction force due to this at the acetabulum is directed in the opposite direction (i.e., $180°$ away). Thus, the acetabular force is 125.33 lb at $88.96°$ to the horizontal (X) axis in the sagittal (X − Z) plane.

TMF_{ab} was found to be 154.74 lb at $180° + 80.53°$ in the $Y − Z$ plane; the acetabular component will be 154.74 $\lfloor 80.53°$.

If it is argued that this force occurs at the same time as the previously determined acetabular force, the resultant force acting on the acetabulum will be the vector sum of these two forces in three-dimensional space.

Resolving into X, Y, and Z components as follows:

X: $125.33 \cos 88.96 = 2.27 \text{ lb}$
Y: $154.74 \cos 80.53 = 25.46 \text{ lb}$

Z: 125.33 sin 88.96 + 154.74 sin 80.53

= 125.31 + 152.63 = 277.94 lb.

The resultant force $R^2 = X^2 + Y^2 + Z^2$

i.e.

$R^2 = 2.27^2 \ 25.40^2 + 277.94^2$

= 5.1529 + 648.2116 + 77250.6436

= 77904.0081

R = 279.11 lb

The resultant direction in the $X - Y$ plane is arctan 25.46/2.27 with respect to the X-axis, i.e., 84.91°.

The resultant direction in the $Y - Z$ plane with respect to the Y axis is arctan 277.94/25.46, i.e., 84.77°.

CONCLUSION

At this juncture the student should be familiar with a number of applied mechanics techniques as they relate to determining musculoskeletal forces, essentially under static conditions. Having dealt with pertinent aspects of static biomechanics, we are now in a position to analyze dynamic biomechanical conditions and to examine the various tools required for this. Chapter 6 takes us into these realms.

Analysis of Human Movements

INTRODUCTION

In virtually all of the biomechanical examples presented in the material of preceding chapters, we have been concerned with descriptions of forces and their related reactions associated with static situations. The body or its segments have been taken to be in static equilibrium. In proceeding toward a full analysis of human movement this background is essential, but the next step is to examine the geometric and temporal qualities of human motion, which is to say that we are concerned with kinematics. Kinematics per se is a descriptive endeavor made through the specification of displacements, velocities, and accelerations of a body or body parts with cognizance being given essentially to the identification of the forces involved with the production of the observed phenomena. A complete biomechanical analysis of a movement would incorporate both kinematic and kinetic information. Kinetics per se is the science that deals with the motion of masses in relation to the forces acting on them. Thus, in kinetics the precise determination of

forces acting from instant to instant is of importance.

ON DESCRIPTIVE TECHNIQUE

The ex-football player or football fan will follow the sequence of events after the ball has been snapped and, as the action unfolds, he will know almost immediately what play is being made long before it is completed. If asked, he could diagram the play, indicating the line shifts and moves of each player in technical football-ese such that any coach could present the play to his squad and they could reproduce it. However, this onlooker cannot do the same for a single skill, such as a punt or a tackle, in technical anatomical terms such that the kinesiologist ignorant of football could reconstruct the exact movements occurring at each of the performer's joints and the forces that produced them.

Being able to describe a skilled performance, even an unskilled one, requires a precise vocabulary as well as a complete understanding of the skill. To say that the ballet dancer performed an entrechat, a

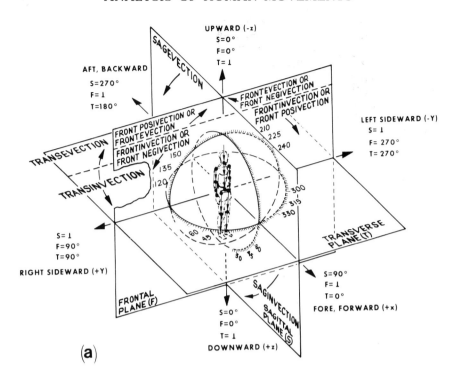

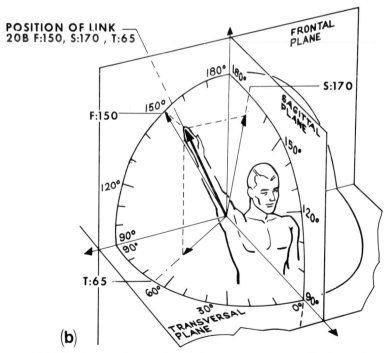

Figure 6.1. Mobility terminology. (a), tri-planar angular coordinate system. (b), example of shorthand notation using terminology presented in (a). (From Roebuck, J. A., Jr., 1966. Kinesiology in Engineering. Paper presented at the Kinesiology Council, Convention of the American Association for Health, Physical Education and Recreation, March 1966.)

fencer made a riposte in tierce, or that a gymnast performed a dislocate on the rings would have instant meaning only to the initiate of the discipline. The initiate would immediately visualize the entire series of bodily movements that were executed by the performer, but such generic terminology is insufficient for the kinesiologist, who should use those words or phrases most widely understood and applicable.

Adjectives are needed in order to describe anything, and the adjectives and verbs used in kinematics of movement analysis are terms used to describe joint movements and/or resulting body positions. Probably the most widely accepted terminology is that used in anatomy and medicine. The American Academy of Orthopaedic Surgeons (1965) has published a booklet which they hope will lead to a standardized terminology in the area. A recent publication by Esch and Lepley (1974) is of value. There have been efforts directed at creating movement notations to facilitate movement description on the one hand, and on the other, scripting of movements to be reproduced (Eshkol and Wachmann (1958); Birdwhistell (1970)).

Kinesiologists and other students of human movement are often dissatisfied with the use of standard anatomical terms inasmuch as they lack precision. This is particularly true of the engineers designing space suits who have built their own vocabulary of descriptive terms (Fig. 6.1). This latter terminology, combined with suitable angular measures, is as precise as the angles and directions measured. Anatomists and kinesiologists may in time come to adopt this or a similar system of notation. However, in the meantime it is still possible, using the more familiar anatomical terminology as presented in Chapter 1, to give an adequate description of any sequence of movements.

KINEMATIC ANALYSIS

Movement analysis requires recordings of body movements and many tech-nological developments have contributed to making effective studies in the field. The most popular technique has been to acquire motion picture sequences which can be analyzed frame by frame.

Any kinematic analysis should start with a description of the bodily movements which take place during the performance of the activity and should include the following data:

1. The name of the movement and the time or frame number at which the movement starts and finishes.

2. The joint(s) at which movement occurs.

3. The lever, i.e., the segment or segments making up the kinematic chain being moved as a result of the joint actions.

4. The force(s) producing the movement, muscular shortening (i.e., isotonic or concentric contraction), gravity, or some other imposed force.

5. Where this force is applied.

6. The force resisting the movement, if any: gravity or eccentric muscular contraction (muscle being lengthened while exerting tension).

7. Where this force is applied.

8. Stabilized and/or relaxed joints in the lever.

9. Forces which stabilize these joints.

10. Stabilized and/or relaxed joints outside the lever.

11. Forces which stabilize these joints.

Many students have been taught to think of the skeleton as a system of levers acted on by muscles, and that each body segment is represented by its core of bone and so forms a single lever. This is a very useful concept, but it might tend to circumscribe one's thinking when analyzing even a simple motor skill. The student is soon involved with any number of moving segments; one segment moving on another, which in turn is moving on a third, etc. He is expected to determine the levers involved in each joint action and to determine the types of internal and external forces involved in moving them. Under these circumstances the concept of the body as a system of links forming one or more kinematic chains becomes very use-

Figure 6.2. Football punt.

kinematic chain made up of these same three links. Under these conditions he will not be tempted to say that the skeletal lever for hip flexion is only the femur or thigh. Similarly if he considers the knee extension, the chain is made up of the leg and foot links and is moved by shortening of the quadriceps femoris, so he knows that the segments forming these two links also form the "lever" in question.

Complete movement analyses of this type are normally made from motion picture sequences such as those shown in Figures 6.3 and 6.5. Multiple, synchronized views might be necessitated and three views, front or back, side, and from overhead are often sufficient. It is often important for the body segments of concern to be in view in two planes at all times. The films can be studied by running them through a time-motion study projector, a viewer-editor, or a microfilm reader such as a Recordak.

EXAMPLE 1—Kinematic Analysis of the Scale Shown in Figure 6.3

From the starting position in frame 1 to the pose held in frame 32 (a scale), the performer goes through what appears at first glance to be a rather simple series of movements. Just how simple these movements are is illustrated by the following analysis.

Movement at the right hip is apparent at frame 4 where extension begins; by frame 14 it becomes apparent that lateral rotation is also taking place concurrently. In each case the action is taking place at the right hip joint, and the lever includes all of the parts being moved: the thigh, leg, and foot. The effort, or moving force, for hip extension is shortening contraction of the hamstrings and gluteus maximus. As the thigh, leg, and foot form the kinematic chain being moved, the effort as always is applied to the moving part, i.e., at the distal attachment of the muscles. The lateral rotation is most probably brought about by the tension exerted by the gluteus maximus which is already contracting as it

ful as it is evident that no proximal link of a chain can be moved without causing movement in one or more adjacent links as well.

Consider the actions occurring during a football punt, those of the kicking leg in particular as it starts the forward swing to contact the ball (Fig. 6.2). There are three different actions involved: (1) flexion of the hip, (2) extension of the knee, and (3) plantar flexion of the ankle. In order to identify the skeletal levers involved in these actions, the student must determine the body segments that are being moved by the action given. Flexion at the hip joint moves the femur on the pelvis, but the thigh is not the only segment moved by this action. The contracting muscles flexing the hip must move the weight of all of the segments attached to the distal end of the femur as well. If the student thinks in terms of an open kinematic chain made up of three links, thigh, leg, and foot, being moved by the shortening of the hip flexors, he should have no trouble in understanding that the "lever" in this case is the

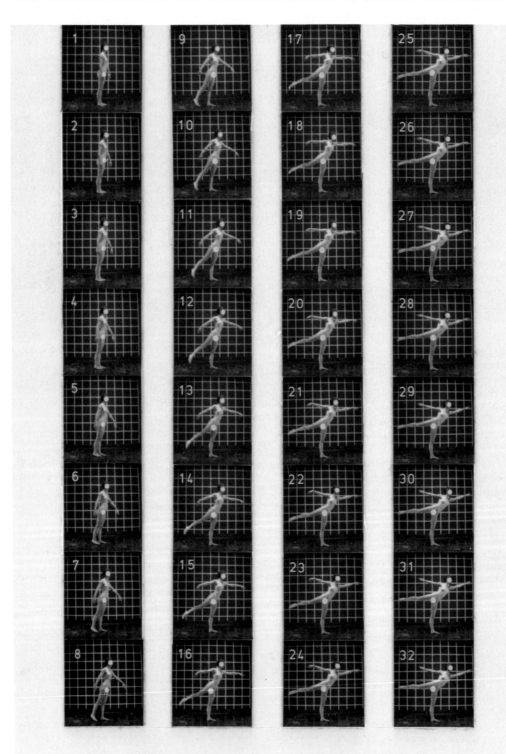

Figure 6.3. Consecutive film strips from motion pictures of the scale. See text for discussion.

pulls upward and backward on the femur. (It is also possible that the six small, deep lateral rotators are contributing to this rotation, but this has not been confirmed electromyographically as has the activity of the gluteus maximus* and hamstrings.) The resistance for both of these movements is the weight of the lever (the kinematic chain formed by the thigh, leg, and foot), i.e., gravity, applied at the center of gravity of the lever, which is approximately at the axis of the knee joint.

During this same time, from frames 4 to 6, the right foot is slightly dorsiflexed, and from frames 7 to 20 the foot is plantar flexing and full plantar flexion is maintained to the end, frame 32. Dorsiflexion is caused by a shortening (isotonic/concentric) contraction, chiefly by the tibialis anterior, but in all probability there is some contractile activity in the other muscles which cross the dorsum of the ankle. As the foot is the moving segment, the effort is applied to the distal attachment of the contracting muscles (on the foot itself); the resistance is the weight of the foot (gravity) and is applied at the center of gravity of the foot. The plantar flexion starting at frame 7 is a result of shortening tension in the posterior leg muscles, chiefly the gastrocnemius and soleus and the weight of the lever, at least through frame 13, when the foot becomes perpendicular to the floor. For this reason there is no resistance to the movement up to this point. (In this type of analysis the resistance offered by ligaments and other connective tissues is not taken into consideration.) After frame 13, as the foot moves beyond the line of its own gravitational force through its center of gravity, gravity becomes a resistance and is, of course, applied at the center of gravity of the foot.

Now consider the actions occurring in the left or weight-bearing extremity. Close examination of the angle formed by the left leg and foot reveals that there is a slight decrease in that angle through frame 12; the body is swaying forward at the

ankle, and the movement is dorsiflexion. The foot is fixed on the floor and does not move; it is the leg, thigh, trunk and the trunk's other appendages that sway forward. Therefore, the lever is the entire body cephalad to the left ankle. The effort is *not* concentric contraction of the tibialis anterior in this case. The reader should look carefully at the first frame. The performer is standing in good posture with a slight forward lean, which means that the gravity line falls anterior to the ankle axis. Under these circumstances the posterior muscle of the calf, the soleus, is tonically active in maintaining the upright posture (Fig. 6.4). Even a slight decrease in this tonic activity permits the body to sway forward in response to the pull of gravity (gravity as the moving force), and this sway immediately evokes (via the stretch reflex) phasic activity in the gastrocnemius (Fig. 4.12). Note that the distal attachments of these muscles are on the foot which is stationary and weight-supporting; therefore, this muscular resistance, applied to the moving part or parts, is effective at the proximal attachments of the muscle involved.

Later, between frames 13 and 17, the angle between the left leg and foot increases slightly so that the movement is plantar flexion. In this case the lever is unchanged from that above, i.e., the body cephalad to the ankle, but the calf muscles are now exerting tension while shortening as they pull the body back over the foot by applying their force at the proximal attachments. The resistance under these circumstances is the weight of the lever, and this force is of course applied at the center of gravity of the lever, i.e., the body cephalad to the left ankle.

Flexion of the trunk on the left hip starts at frame 11 and continues to frame 30. The left lower extremity is for analytical purposes considered motionless as the angle between the trunk and left thigh decreases (hip flexion). Thus, the lever, the kinematic chain made up of all parts moved by this action, includes not only the head, neck, torso, and both upper extremities but the right lower extremity as well. Again, this lever is more easily de-

---

* While the gluteus maximus is not active in the early stages of hip extension, it is extremely active in the latter stages.

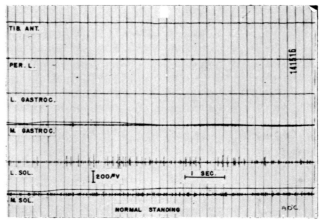

Figure 6.4. Electromyogram of quiet standing. Tonic activity in soleus maintains center of gravity over base of support. (From O'Connell, A. L., 1958. Electromyographic study of certain leg muscles during movements of the free foot and during standing. *Am. J. Phys. Med. 37:* 289.)

scribed as the "body cephalad to the left hip." By inspection it seems possible that at frame 11 (the start of hip flexion) the center of gravity of this lever lies close or possibly posterior to the flexion-extension axis of the left hip. If this is true, it may be assumed that there is a momentary contraction of the hip flexors sufficient to move the center of gravity anterior to the flexion-extension axis of the left hip. At this moment the hip flexors (chiefly iliopsoas) exerting muscular tension while shortening (isotonic/concentric contraction) are the effort and, as it is the body cephalad to the joint that is being moved, this tension or effort is applied at the proximal attachment of the muscles. Gravity of course is the resisting force, which is applied at the center of gravity of the lever.

Once the line of gravity of the above lever passes anterior to the flexion-extension axis of the left hip, gravity becomes the moving force and, as always, is applied at the center of gravity of the lever. At the same time the hip extensors (the hamstrings and gluteus maximus) now become the resistance supporting the weight of the lever against the pull of gravity, and this resistance, effective in controlling the movement of the lever, is applied at the proximal attachment of the muscle. The

performer thus attains the graceful balance required of the skill.

Static Analysis

The kinematic analysis to this point indicates the forces which are *moving* the body segments; it has not accounted for the muscular forces which stabilize the various joints. Further study of Figure 6.3 will show that no movement occurs at either knee joint, and yet they both remain extended against the tension of the hamstrings, which with the gluteus maximus are extending the right thigh at the hip, while those of the left side of the body are resisting the pull of gravity on the trunk by controlling the flexion at the left hip. The extension at both knees must be maintained by isometric tension in the quadriceps femoris. This knee stabilization serves more than an esthetic purpose as it provides a firm base for the hamstring tension to pull against so that these muscles can act effectively as extensors at both hips.[†] Because of this and similar situations, a kinematic analysis should include

---

† This quadriceps femoris activity can also be considered synergic in nature as it counteracts the undesired knee-flexing action of the hamstrings.

a description of stabilized joints both inside and outside of the identified levers.

The reader should also be aware that a moving joint may also be stabilized. In the "scale" the entire body weight is balanced on the left talus and, while some dorsiflexion and plantar flexion take place as described earlier, there is undoubtedly activity in the muscles whose tendons cross the lateral and medial aspects of this joint as well. Electromyographic recordings taken from these muscles under like circumstances have repeatedly demonstrated this condition. Also, muscles crossing some joints may remain relaxed for considerable periods of time. For example, the right wrist is fully relaxed until frame 24, and wrist extension can be seen starting with frame 25. On the other hand, the left wrist is only relaxed until frame 11 and the extension starts with frame 12. When the extension reaches the point where the hands are in line with the forearm, the hands are supported in this position throughout the remainder of the movement.

On Acquiring More Detail

The foregoing movement analysis thus describes the movement or skill, the joint actions and/or the condition (relaxed or stabilized), as well as naming the forces concerned with each case.

Considerably more detail of the movement can be derived by recognizing that the grid in the background of each frame in Figure 6.3 serves as a reference for scaling distance information and the camera frame speed provides a time reference. Appropriate optical criteria have to be met to ensure the precision of scaling distances, i.e., the camera should be located at a very much greater distance from the background than is the performing subject; also, the camera should be carefully set up so that the film plane is parallel with the plane being observed. If the camera speed is 50 frames/sec each frame provides a sample of movement every ¹⁄₅₀ second. (The shutter speed is usually less than the camera speed; higher shutter speeds facilitate

"freezing" of the action, but necessitate greater film sensitivity and/or illumination.) When timing information is of consequence, the film transport speed must be precise. Either a well made camera can be used, or else a suitable clock or timer can be simultaneously photographed by using an additional optical pathway to the camera.

Extracting information systematically frame by frame can lead to the production of data formats which may be more amenable to detailed analyses. For example, the X-Y coordinates of various anatomical landmarks can be plotted; joint angle variations as a function of time can be derived. From movement-time characteristics, velocities and hence accelerations of body segments can be derived. Coupling the masses of the moving members with their velocities permits the calculation of the momenta of these members. When accelerations of body segments are allied with their respective masses, the corresponding segment forces follow.

It will be evident that there is much tedium involved in extracting and processing the data acquired in the aforementioned manner. Considerable efforts have been made to introduce pertinent technology in studies of human movement so as to minimize or eliminate this tedium. Special data digitizers, sensors, computers, computer programs, and display devices have emerged to facilitate these efforts. While they are alluded to in what follows, basic principles are stressed.

Value of Motion and Static Analyses

The type of analyses presented in Example 1 is invaluable in accounting for the muscular activity recorded on electromyograms (EMGs) synchronized with film. All too often beginning investigators have been content to point out that muscles were contracting during a certain phase of a performance, but they have failed to determine, or even hypothesize, the purpose of that contraction. The beginning investigator, and even the expert, may well find gaps in his analysis when he compares it with the electromyographic activity recorded from the various muscles. He may find that some of this activity may not be accounted for in the preliminary analysis. The investigator then returns to the movement data (film) to determine the situation that accounts for the unidentified muscular activity recorded on the EMG.

Once this part of the analysis has been satisfactorily completed, the investigator has the basis for the selection of one or more critical components for further investigation. These components include the muscular, gravitational, or other forces that may be acting, limb or segmental accelerations, and the contribution of these and of segmental velocities to the skill or activity. When the selection has been made, the appropriate portion of the movement record (film) is re-examined for the necessary data.

In essence, when using filmed data, link diagrams such as those depicted in Figures 5.20 and 5.21 are constructed. The number of frames so diagrammed depends upon the type of investigation selected. Only one such frame from each camera was necessary for the study discussed in Examples 5.11 and 5.12.

If the investigation is concerned with angular acceleration patterns of various body parts during the performance of the skill or of a particular phase of the performance, a series of such diagrams made from a regular sequence of frames would be necessary. The nature of the movement would decide the timing of the frames; the more rapid the movement, the greater should be the number of frames per second to ensure that all of the pertinent movement information is captured.

Figure 6.5, while not so evenly spaced, was derived from such data.

EXAMPLE 2—Derivation of Linear Translational Displacement, Velocity, and Acceleration from Processed Film Data

It is desired to extract the linear translational displacement velocity and acceleration components in the forward (x) and vertical (y) directions of the midpoint of the link connecting the hip and shoulder for the subject depicted in Figure 6.5.

This problem is best tackled using enlarged diagrams so as to minimize errors when extracting data. The relative x- and y-values for the midpoint of the requisite link are plotted as a function of frame number (i.e., time) as in Figure 6.6(a). The starting position of the pertinent point in frame 1 is taken as the x-origin for this plot and the relative scale of 1 unit corresponds to the distance from the position in frame 1 to that in frame 58. The y-origin is taken as the lowermost point reflected in frame 54. Smooth curves are drawn through the acquired data points.

At a first glance, it is seen that the x-coordinate varies practically linearly with time, while the y-value at first decreases, then increases, remains steady and then declines to the lowermost level. Taking the slopes of these two curves as the frame number increases, the velocity graphs of v_x and v_y, shown in Figure 6.6(b), are arrived at. (Velocity units are expressed as distance units/frame.)

From Figure 6.6(b) it is seen that v_x commences at its highest level, dropping to a lower level at about frame 22 when it remains practically constant until around frame 48 after which it declines rapidly to zero. v_y has negative and positive components as time progresses. Initially v_y has a negative value since the link segment center is moving downward with time. The segment center is then momentarily stationary at about frame 5 when the y-movement changes direction at increasing velocity until about frame 12 when the maximum positive velocity is attained. Then the velocity v_y proceeds to decline to zero at which value it persists for about 10 frames. Following on this, the link segment center drops, thus causing a negative velocity profile with the plateau and negative peak, and finally the velocity passes through zero again.

Again, taking the slopes of the curves of v_x and v_y as the frame numbers increase renders an impression of the acceleration

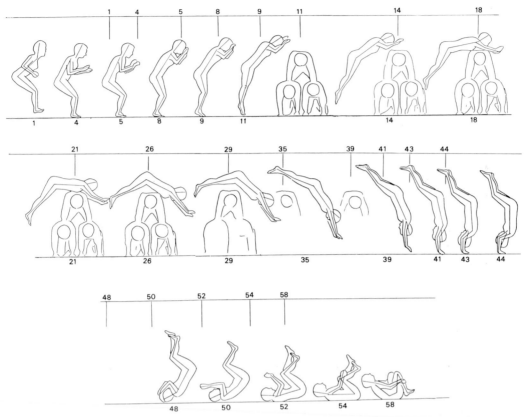

Figure 6.5. Dive for height. Series of tracings made from motion picture film. Numbered intersections of vertical line with upper horizontal mark a fixed point in the background. Number marks location of fixed point with figure bearing same number. (Courtesy of University of Wisconsin Department of Physical Education for Women.)

components a_x and a_y as depicted in Figure 6.6(c).

Essentially, a burst of acceleration a_x is involved around frames 18 to 22. This is most likely derived from the forces from the upward kick provided by hip extension and knee extension while the arms are thrust forward (see Figure 6.5, frames 14–26). A burst of deceleration a_x is produced between frames 48 and 54. A component of "impulsive" force on landing and taken up by the rolling back is evidently acting to brake the forward movement.

a_y reflects an acceleration component acting for the first 12 frames. Lower limb extensions coupled with the initial impulsive reaction force when the feet strike the ground serve to provide the vertical propulsive forces. The subsequent deceleration phase acting until about frame 20 reflects the utilization of the effects of gravity on the motion. From frames 20 to 30 vertical motion of the center of the segment of concern is minimal and there is no net accelerating or decelerating force influencing this. Between frames 30 and 34 a deceleration burst occurs as the body begins to descend to the ground. This is primarily due to gravity. The forward and upward propulsion of the legs which are extending adds an acceleration component in the vertical direction which generates a force to nullify portion of the gravitational pull acting on the body. Thus, the accel-

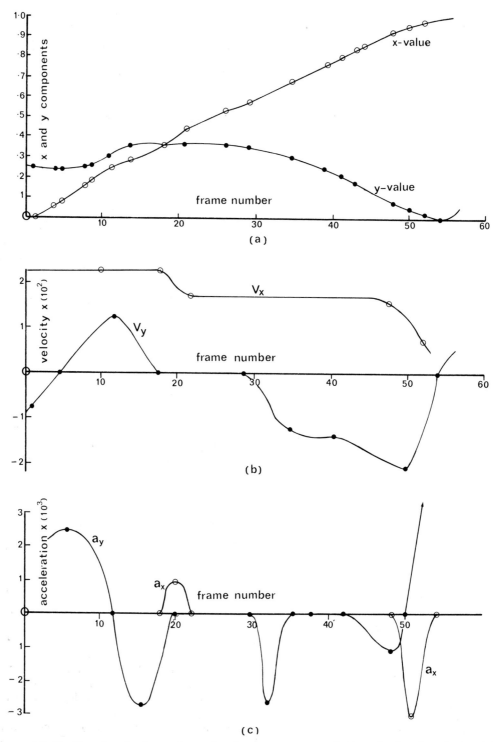

Figure 6.6. Positional, velocity, and acceleration components of "dive for height" depicted in Figure 6.5. (a), x and y positional components vs frame number; (b), x and y velocity components vs frame number, and derived from (a); (c), x and y acceleration components vs frame number, and derived from (b).

eration component a_y is virtually zero from frame 34 to around frame 42 when gravity forces again play a major role but are countered by the reaction of the arms when the hands contact the floor. The upward thrust component and that due to rolling lead to the substantial value of a_y beyond frame 50.

If the normalized distance scale is such that 1 unit corresponds to 12 ft and the frame speed was 24 frames/sec, the scales for the diagrams in Figure 6.6 would have to be modified. Thus, in all cases, the frame numbers would have to be multiplied by 0.042 (1/24) to render time in seconds. The distance scales in Figure 6.6(a) would be multiplied by 12 ft; the velocity scales in Figure 6.6(b) would be multiplied by 12/0.042 = 285.7; and the acceleration scales in Figure 6.6(c) would be multiplied by 285.7/0.042 = 6802.7. (Note that the peak value of the deceleration phase in Fig 6.6(c) would then turn out to be 20.41 ft/sec^2 which is reasonable since it is less than

and close to 32.2 ft/sec^2, the value for gravity alone.)

In general, attempts should be made to use equal spacings between frames of data. This will serve to facilitate subsequent data processing.

Acceleration of Body Parts

Example 2 illustrates the calculation of linear acceleration for a body component. Movement of any sort involves angular acceleration of body segments. The link segments in Figure 6.5 are seen to vary their angular dispositions in the frame sequences. The calculations of angular velocities and accelerations are performed in a fashion similar to that illustrated in Example 2 except that cognizance must be taken of the angular variations of concern.

In the analysis of gymnastic activities when the body is rotating unsupported in air or around a bar, the angles that a segment makes with one of a set of x and y

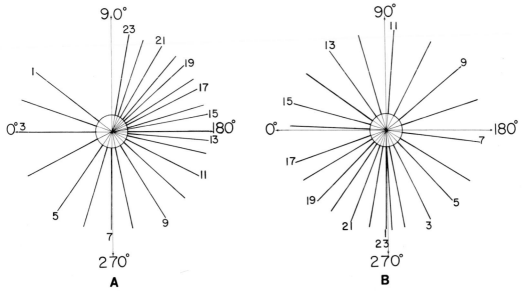

Figure 6.7. Graphs of part of a forward somersault on the trampoline. *A*, of the trunk; *B*, of the lower extremities, x and y axes through hip. Performer was a highly skilled college man. Compare greatest angular displacement of trunk per time, frames 1 through 5, greatest angular displacement of lower extremities per unit time, frames 6 through 11. (From Reuschlein, P. L., 1962. An analysis of the speed of rotation for the forward somersault on the trampoline. Unpublished paper prepared at the University of Wisconsin.)

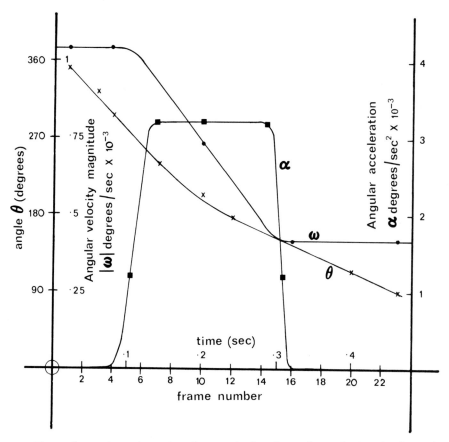

Figure 6.8. Plots of angular orientation θ, magnitude of angular velocity $|\omega|$, and angular acceleration α derived from Figure 6.7(a).

axes are used and the results can be conveniently plotted on a circular graph. Figure 6.7 is made from data collected during a pike somersault performed from a trampoline. In each frame plotted, the x and y axes were drawn through the hip joint and the angles were plotted in their appropriate quadrant. Figure 6.7(A) is the graph of the trunk, while Figure 6.7(B) is that of the lower extremities.

EXAMPLE 3—Angular Velocity and Acceleration Determinations

Using the data presented in Figure 6.7(A) plot the angular velocity and angular acceleration of the trunk for a forward somersault if the camera was operating at 50 frames/sec.

First, a smooth graph of angular position against frame number and time ($t =$ frame no. $\times$ 0.02 second) is plotted as in Figure 6.8. It should be recognized that the angles for frames 1 and 2 can be conveniently plotted by adding 360° to them.

As can be seen, the angular variations commence rapidly, but then slow down. It can be anticipated that there are two distinct velocity phases, an initial rapid one and then a slower one.

The angular velocity, ω, is determined by extracting pertinent slopes from the graph of angle, θ, against time and then plotting the relevant values. For convenience the magnitude of ω, $|\omega|$, is plotted. The two angular velocity phases are borne out. Next, the angular acceleration, α, is derived from the ω-t curve. Its peak is

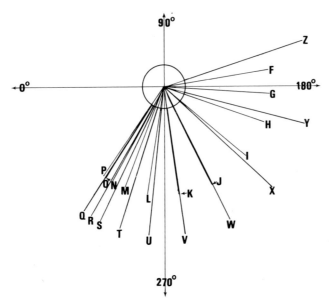

Figure 6.9. Angular displacement of the arms during a glide kip on the uneven parallel bars. x and y axes drawn through wrists. At frame P the performer is in full layout, parallel to the floor and ready to start the kip. The swing, as indicated by arms, has been decelerating from frame J and reverses as kip is performed. Note that in this case the reference position 0° is along the −x axis and clockwise movements are considered to be positive. (From Gowitzke, B. A., 1969. An analysis of levers contributing to the glide kip. Unpublished paper prepared at the University of Wisconsin.)

sustained from frames 7 to 15. α is in fact the manifestation of a decelerating force acting on the trunk.

EXAMPLE 4—Glide Kip on Uneven Parallel Bars

Figure 6.9 depicts the angular disposition of the upper extremities in the sagittal plane of the body during the performance of a glide kip. The origin of x and y coordinates is located at the wrists and the distal end of each line represents the shoulders. The gymnast maintains elbow extension throughout the performance, one of the characteristics of skillful execution. Principal changes in angular velocity are readily seen: the movement is underway at frame F and angular velocity increases during the downward/forward movement of the body to frame I; a short period of almost no acceleration (i.e., main-

taining angular velocity) is seen in the time period between frames H and I and between frames I and J; then, the angular velocity decreases progressively until frame P, the reversal of direction of the body, when velocity is zero; near the point of reversal of direction, velocity is changing quickly and therefore acceleration is approaching its maximum for this skill; as the body now rides backward and upward toward the bar, the arms can be seen to change position rapidly at an increasing rate and are showing their greatest velocity values near the end of the movement as depicted by the changes between frames X and Y and between frames Y and Z.

The time between frames was 0.076 second: measurement of each angle of inclination of the arms provides the information necessary to plot the angular displacement of the arms against time. Such a plot reveals the characteristics of a single oscillating swing which starts at an angle of 171° at frame F, changes direction at an angle of 306° at frame P, and returns to an angle of 162° at frame Z. If the resulting plot is compared with a typical sine wave pattern, it will be seen to approximate a portion of a sine wave for a single oscillating system. The plot will have some irregular shape to it because the human body does not behave quite like a simple pendulum system. That is to say, there are forces acting within the human body in the form of muscle contractions with resulting momenta exchanges which influence the angular displacement of the arms. The first derivative of the angular displacement-time plot by comparison closely resembles that of a cosine wave and will show the velocity changes highlighted in the above discussion. The second derivative of the displacement plot also depicts some similarity with sinusoidal variations with one burst of positive acceleration at the start of the movement for the purpose of getting underway, and another at the end of the movement, for the purpose of decelerating to a stop. (The interested student is invited to construct the various plots alluded to in this discussion.)

ANALYSIS OF SEGMENTAL VELOCITIES

Two examples of analyses of segmental velocities of body parts which contribute to the release velocity of a projectile are presented to familiarize the reader with the techniques involved and the problems associated with them. In any projectile skill, much of the muscular activity and resulting joint actions serve only to provide a base on which the throwing, kicking, or hitting extremity moves to project a given object; the remainder of the muscular force and resulting joint actions contribute directly to the initial velocity of the projected ball or other object. Study of films of various throwing, kicking, and striking activities have led investigators to

a consensus concerning what joint actions move the kinematic chains which contribute directly to the initial velocity of the ball, javelin, etc. In the overhand throwing pattern, for example, the major joint actions involve rotation of the pelvis on one or both hips, trunk rotation around the intervertebral joints, medial rotation of the arm at the shoulder, and wrist flexion. Protraction of the shoulder girdle at the sternoclavicular joint has been largely overlooked by many investigators but can play quite an important role. While the elbow does extend during an overhand throw, this action only serves to increase the radius of the arc through which hip and spinal rotation and shoulder protraction move the hand.

Investigators have long been aware of these actions, but it is only comparatively recently that an ingenious method of determining the contribution made by each of these different actions to the initial velocity of the projectile was devised by Glassow and Roberts in the late 1950's and early 1960's at the University of Wisconsin.

TECHNIQUE

This method requires synchronized films taken from the front or back, from the side and, whenever possible, from overhead (Fig. 6.10). (The overhead sequences increase the accuracy of all measurements of body rotation.)

Taking careful note of the sequences in Figure 6.10, one can see that the subject is wearing fins attached to the hip and shoulders. Figure 6.11 shows a subject equipped with these fins which facilitate the measurement of trunk and pelvic rotations. The fins are made of styrofoam set in aluminum holders and the belts are elastic with Velcro fastenings. The fins are supported at the sacrum and between the shoulders. The holders and belts are designed to maintain the fin perpendicular to the sacrum and the thoracic spine. The maximal length of the fin appears as the performer poses for the scale photographs. These fins may be marked off in alternat-

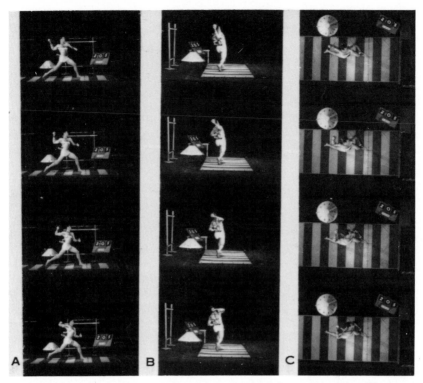

Figure 6.10. Overhand throw. Performer is photographed from (A), side; (B), rear; and (C), overhead cameras. The three films are synchronized with the aid of markings on a revolving cone. (From Blievernicht, D. L., 1967. Courtesy of the University of Wisconsin Kinesiology Laboratory and R. B. Glassow.)

Figure 6.11. Scale photographs of performer wearing styrofoam fins. Side view.

ing black and white stripes or in 6-in or 10-cm sections and, as the body rotates, the image of the fins becomes foreshortened as they leave the picture plane.

The performer in Figure 6.11 also holds a scale (e.g., a yard- or meter-stick) with a predetermined length, 1 ft or 10 or 20 cm clearly marked. Similar photographs are done in each plane of view so as to calibrate the distance scales in each filmed sequence.

From Figure 6.12 it is obvious that the full length of the fin forms the hypotenuse of a right triangle while the foreshortened length forms the adjacent side. As the cosine of an angle is the ratio of the side adjacent to the angle to the hypotenuse, it is possible to determine the angle of the trunk and pelvis from the length of their respective fins. Thus, in views such as

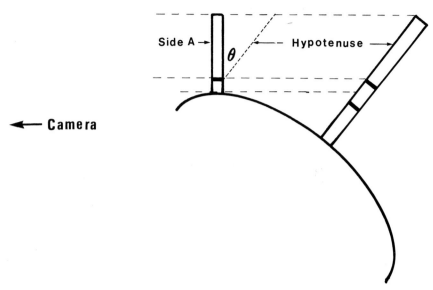

Figure 6.12. Use of fins for measuring trunk and pelvic rotation. As trunk or pelvis rotates from picture plane, fins become foreshortened. This length becomes the adjacent side of a right triangle, while actual length of fin becomes the hypotenuse of this triangle whose angle θ is the amount of rotation of the body part.

Figure 6.11 and for the overhead camera the subject is posed in such a way that the full length of the fins feature in the picture planes.

EXAMPLE 5—Analysis of Velocities of Body Segments Contributing to Overhead Throw

Tracings from several frames of film which recorded an overhead throw are shown in Figure 6.13. Note the inclusion of the scale measures. The time between frames was 0.016 sec. It is required to calculate the linear velocities resulting from all pertinent joint actions and to compare the summation of their contributions at the hand with the initial linear velocity of the ball.

Angular Velocities of the Body Parts

At the instant of release, these velocities may be ascertained by determining their angular displacements during the last brief interval before release. The hand dis-

placement is obtained by measuring the angle formed by the long axis of the forearm with a line drawn from the center of the ball to the wrist axis (Fig. 6.13, (a) and (b); Table 6.1). The angular displacement of the forearm resulting from medial rotation of the arm at the shoulder is obtained by measuring the angle of inclination of the forearm, i.e., the angle which the forearm makes with the horizontal, as shown in Figure 6.13, (a) and (b) (see Table 6.1).

Pelvic and Spinal Rotations

In Figure 6.14(a), the marked area of the fin between the shoulders is 21 units‡ long; in the scale, the corresponding part of the fin is 22 units, so the cosine of the angle that the trunk makes with the picture plane is 21/22 or 0.95454 and the angle is 17°20′. Similarly, in Figure 6.13(b), that area of the fin measured 19.5 units in length so the cosine is 19.5/22 or 0.88636

---

‡ On the original tracing enlargement, 1 ft measured ⁵³⁄₆₀ of 1 in, so these are the "units" referred to.

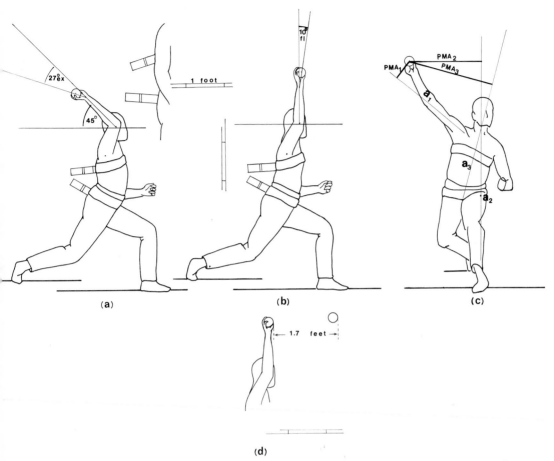

Figure 6.13. Tracings of overhand throw. (a), frame just before release, side view; (b), release frame, side view; (c) release frame, front view. Axes of rotation; a_1, at the shoulders; a_2, at the hip, a_3, of the spine. (d), distance ball has traveled from release frame to frame immediately following.

Table 6.1
Overhand Throw

Segment Joint Action	Angle (a)	Angle (b)	Angular Displacement	Angular Velocity	Radians per sec	PMA*	Linear Velocity	Contribution
				°/sec		ft	ft/sec	%
Pelvic rotation	27°35′	27°35′	0					
Spinal rotation	17°20′	27°35′	10.25°	640.6	11.18	2.34	26.16	24.62
Wrist flexion	27° ext.	10° flex.	37°	2312.5	40.35	0.38	15.33	14.42
Shoulder (medial rotation), angle with horizontal	45°	90°	45°	2812.5	49.08	0.58	28.45	26.77
Sternoclavicular protraction	$d = 2″$; length clav., 7.5″		15.5°	968.7	16.9	2.15	36.34	34.20
			Total linear velocity				106.28 ft/sec	
			Linear velocity calculated from Figure 6.13(d)				106.25 ft/sec	

* Projection moment arm.

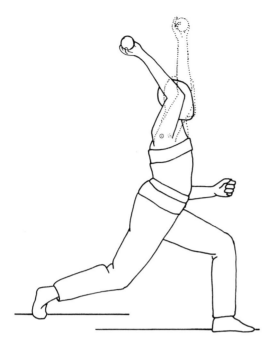

Figure 6.14. Release throw. Dotted line, superimposed over prerelease throw; solid line, shows protraction of the shoulder girdle.

and the angle is 27°35′. The difference is 10°15′ or 10.25°, which is the angular displacement of the trunk caused by spinal rotation just before the release of the ball. There is no change in the length of the pelvic fin from (a) to (b), indicating that the pelvis was held motionless just before the release.

Method of Determining the Amount of Protraction Occurring at the Sternoclavicular Joint

A tracing of Figure 6.13(a) superposed on (b) coincides with (b) to the level of the thoracic belt. (This is always the case when shoulder protraction is taking place, if there is forward motion of the entire body in relation to a fixed point during this pre-release interval or if the entire torso shoulder protraction is not indicated.)

When the long axis of the arm in (a) is compared with that of (b), it is found that

the center of the axilla has moved forward 9 units (Figure 6.14). As 53 of these units equal 1 ft, there are 4.4 units to the inch. Under these circumstances the shoulder girdle has moved forward (protracted) 2 in. This subject placed his arm in the position of Figure 6.13(a), and the distance from his right sternoclavicular joint to a point in line with the center of his right axilla was measured and found to be 7.5 in. An arc with a radius of 7.5 in was drawn and the angle subtended by a 2-in chord was measured and found to be 15.5°.

Calculations of Linear Velocity Resulting from Joint Actions

As all angular displacements from frame to frame took place in 0.016 second the degrees of displacement per second are easily calculated (Table 6.1). The procedure for converting degrees per second of a given joint to the linear velocity allied with the motion at an extremity, follows from Equation 2.12

$$v = \omega r$$

where v is the linear velocity; ω is the angular velocity in radians/sec; and r is the radius arm of concern. Thus,

$$v = \text{degrees/sec} \times \frac{2 \pi r}{360}$$

$$= \frac{\text{degrees/sec}}{57.2957} \cdot r$$

The radial length r in each case is the perpendicular distance from the center of the ball to the axis around which that particular angular displacement occurred. For the sake of convenience this distance will be termed the projection moment arm or PMA. The PMA for wrist flexion is measured from Figure 6.13(b), while the other PMAs are measured from Figure 6.13(c). The PMA for the sternoclavicular joint action is parallel to that for spinal rotation and is taken as 1 in or 0.083 ft shorter than that for spinal rotation (Table 6.1).

All data necessary for the final calculations have now been obtained from the

Table 6.2
Velocity Contributions*

	Range	Angular Velocity	Moment Arm	Linear Velocity
Hip rotation	20.6	824	2.12	30.5
Spinal rotation	9.6	384	2.75	18.4
Shoulder rotation	38.5	1540	0.27	7.3
Wrist flexion	60.0	8571	0.49	73.3
Total linear velocity				129.5

* From Cooper, J. M., and Glassow, R. B., 1976. *Kinesiology*, Ed. 2. St. Louis: The C. V. Mosby Company.

film tracings, and only simple arithmetic is necessary to obtain the actual linear velocity contributed by each joint action (Table 6.1). Figure 6.13(d) is a composite of Figure 6.13(b) and the frame immediately following. The only part of the post-release frame added to Figure 6.13(b) is the ball so that the actual initial velocity can be measured. As the ball has traveled 90 units or 1.7 ft in the elapsed 0.016 second, this initial velocity is 106.25 ft/sec. This value compares extremely well with the sum of the linear velocity contributions. The last column of Table 6.1 shows the percentage of this velocity contributed by each joint action.

Comparison of these data with those from Cooper and Glassow (1976) (Table 6.2) indicate at least two weaknesses in the particular throw reflected in Table 6.1: lack of pelvic rotation and a comparatively small amount of wrist flexion. On the other hand, the Cooper and Glassow data indicate the entire absence of shoulder protraction. This absence seems rather unusual in the light of electromyographic studies which have always indicated that considerable activity in the serratus anterior accompanies any forward motion of the corresponding arm.

EXAMPLE 6—Analysis of Velocities of Body Segments Contributing to Underhand Throw

Tracings from film frames of an underhand softball pitch are shown in Figure 6.15. The time between frames was 0.016 second. It is required to determine the linear velocities resulting from pertinent joint actions and to compare the summation of their contributions at the hand with the initial linear velocity imparted to the ball.

The same procedures are followed as in the overhand throw, but in this instance the characteristic shoulder action is flexion (Fig. 6.16; Table 6.3). There are, however, two problems which did not arise in the analysis of the overhand throw: first, the problem of *effective* spinal rotation when the pelvis is also in motion at the time of release; and second, that of calculating the amount of wrist extension in frame (a).

Effective Rotation of Spine and Pelvis When Pelvis is also in Motion

Whenever the pelvis is rotating on the heads of one or both femurs, the torso is naturally carried through the *same angular displacement*. Under these circumstances any one of the following three situations may be occurring.

1. The angular displacement measured at the thoracic level is **less** than that calculated for the pelvis; therefore, the spinal rotation taking place at the intervertebral joints is in a *direction opposite to* the pelvic rotation. In such an instance the amount of spinal rotation must be subtracted from the total pelvic displacement to arrive at the **effective pelvic rotation.** The spinal rotation, being in the opposite direction, makes no contribution to the speed of the ball.

2. The angular displacement measured from the thoracic fin is the same as

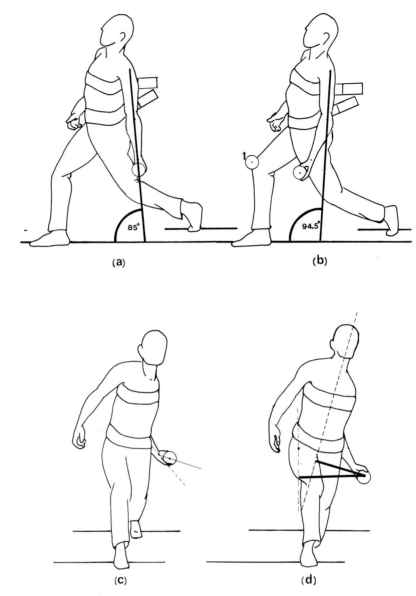

Figure 6.15. Tracings of underhand softball pitch. (a) and (c), prerelease frames, side and front; (b) and (d) release frames, side and front. 1, location of ball next frame after release.

that calculated for pelvis, so there has been no rotational activity at the intervertebral joints.

3. The angular displacement resulting from spinal rotation is **greater** than that measured at the pelvis; therefore spinal rotation has been taking place and is in the same direction as that of the pelvis. In order to arrive at the actual velocity contributed by the spinal rotation, that amount of angular displacement caused by the rotating pelvis must be *subtracted* from that measured at the thoracic level; e.g., in Figure 6.15(a), the angle of the pelvis with

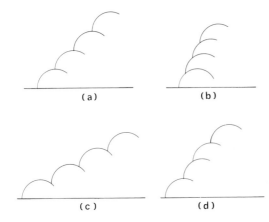

Figure 6.16. Schematic diagram of summation of segmental velocities. (a), maximal velocity of each segment at release/impact; (b), peak segment velocities not reached at release/impact; (c), peak segmental velocities passed at release/impact; (d), one segment out of phase. (After Morehouse, L. E., and Cooper, J. M., 1950. *Kinesiology*. London: Henry Kimpton.)

the picture plane was determined to be 23°37′ and in (b) the angle was 16°36′, so the difference is 6°58′. The thoracic angle in (a) was 28°37′ and in (b) it was 16°36′ so that the total displacement was 12°1′. However, as 6°58′ of this displacement were due to the *pelvic* rotation, only 5°3′ resulted from rotation at the intervertebral joints, and this amount is considered the **effective spinal rotation** (Table 6.3).

Determination of Amount of Wrist Extension just before Release

This problem arises because the arm is laterally rotated and the forearm supinated (Fig. 6.15, (a) and (c)) so that the amount of wrist extension cannot be measured from tracing (a) alone as was possible in the overhand throw. This necessitates a second front view tracing, that of the pre-release frame, Figure 6.15(c). Examination of this tracing shows that there

Table 6.3
Underhand Softball Pitch

Segment	Angle		Angular Displacement	Angular Velocity	Radians per sec	PMA*	Linear velocity	Contribution
	(a)	(b)						
			°/sec			ft	ft/sec	%
Pelvic rotation	16°36′	23°34′	(6°58′) 6.96°	435	7.59	1.58	11.99	16.4
Spinal rotation	16°36′	28°37′	(12°1′) 5.05°†	315.6	5.42	1.24	6.72	9.9
Shoulder flexion, angles of inclination	85°	94.5°	9.5°	593.75	10.36	2.5	25.9	36.79
Wrist flexion	20° flex.	22.2° ext.	42.2°	2637.5	46.03	0.41	18.87	25.6
Sternoclavicular protraction		d = 0.9″, length clav., 7.5″	7°	437.5	7.63	1.16	8.85	12.10

Total linear velocity		72.33 ft/sec
Linear velocity calculated from Fig. 6.15(b)		73.12 ft/sec

* Projection moment arm.
† Effective spinal rotation: 12°1′–6°58′ or 5.05°.

Table 6.4*
Velocity Contributions*

	Range	Angular Velocity	Moment Arm	Linear Velocity
Hip rotation	12	400	1.45	10.12
Spinal rotation	6	200	1.6	5.59
Shoulder flexion	22	733	2.5	31.98
Wrist flexion	30	3750	0.35	22.90
Total linear velocity				70.59

* From Cooper, J. M., and Glassow, R. B., 1976. *Kinesiology*, Ed. 2. St. Louis: The C. V. Mosby Company.

is considerable radial deviation as well, and that the volar surface of the forearm is not fully in the picture plane, i.e., the forearm is facing obliquely forward and to the left of the thrower (i.e., a supinated forearm and lateral rotation at the shoulder). Consequently, if the distances from the wrist axis to the center of the ball in tracings (a) and (c) are taken as side adjacent (frame (a)) and the hypotenuse (frame (c)), it is possible to determine (or at least make a good estimate of) the degrees of wrist extension. In (a) this distance measured 12.5 units and in (c) it measured 13.5 units, so that the cosine is 0.92592 and the angle is 22°12′ or 22.2° (Table 6.3)

From the data in Table 6.3 and from similarly derived data, the investigator can draw conclusions as to the efficacy of the various joint actions in contributing to the throw. Comparison of the data in Table 6.3 with those from Cooper and Glassow (1976) in Table 6.4 indicates that this throw was somewhat faster and that shoulder flexion makes the greatest contribution to the initial velocity of the ball.

Value of Determining Segmental Velocities

From Examples 4 and 5 it will be evident that the analyses undertaken can pinpoint any particular weakness or strength in a given performance.

Movement analyses have demonstrated that the body parts which exert the greatest moment of *resistant* force, either as a result of weight, or length of moment

arm, or both, are those parts which initiate the forward motion of a throw, hit, etc. On the other hand, the part or lever lightest in weight and with the shortest moment arm (least moment of resistant force) is the last segment to be moved. Thus, in a throwing or striking skill the forward rotation of the pelvis on one or both hip joints may be initiated even before the upper extremity completes the backswing phase of the throw or hit. The wrist snap and/or finger extension occur as the ball is released or hit. Ideally, all angular displacements have accelerated to the desired maximal velocity at the instant of release or impact so that optimal results are achieved. When maximal velocity and distance are desired as in a throw from the outfield, a javelin or discus throw, or a golf drive, this early action of the proximal joints results in a "crack the whip" effect which, aided by the more distal joint actions, contributes significantly to the speed at which the object leaves the hand or striking implement.

This summation in optimal sequence has been illustrated by Morehouse and Cooper (1950) and reproduced here in Figure 6.16 where each curve represents a plot of the changing linear velocity contributed by the hand(s) or foot by the joint action it represents. In theory each joint action should impart maximal linear velocity at the instant of release or impact (Fig. 6.16(a)). However, it is quite possible that only one joint action may be out of phase, e.g., the wrist snap in a hitting skill, or that several actions are not timed properly (Fig.

6.16, (b) and (c)). Heretofore there has been no means of determining just how adequate the coordination and/or timing of a skill might be, except by the end result. Now by using the technique of segmental velocities, it is possible to determine the instantaneous linear velocity of the hand(s) or foot contributed by any joint action at any given point on the recorded data. By this means the instant of greatest velocity resulting from a given joint action may be ascertained. Comparison of this velocity with that of the release or impact reveals the degree of coordination and timing of the performance.

ON INSTRUMENTATION

Technological innovation has played a vital role in promoting studies in kinesiology. In what follows, emphasis is laid on instrumentation mainly for studying locomotor function, but the possible ramifications for wider applications should be readily apparent to the serious student.

Capture of Body Motions

The basic ideas concerning the capture and subsequent presentation of kinesiological information are by no means a product of this decade. Our advances hinge upon technological developments and innovation that have occurred particularly over the last century. At the turn of the century the French physiologist Marey studied facets of human locomotion aided by his own advancements in photography. He possessed a locomotion cart used to track an ambulating subject and in which the necessary film processing could subsequently be performed. He produced diagrams reflecting the trajectories of the head, shoulder, hip, knee, and ankle in the sagittal plane. Fundamentally we use the same techniques but have the distinct advantage of a much more sophisticated technological armamentarium to aid and abet us.

To capture the motions of the body as it progresses through space, we obviously must record information about the movements in three-dimensional space, of conveniently selected and marked body sites. Sites usually considered most convenient are the bony prominences close to the joints. The recording system must be able to perceive the marked body locations in three dimensions at all times. This can usually be achieved by the use of several cameras each of which is placed for an appropriate plane of observation. Provided that a particular site can be viewed at all times, views from two different but known planes are sufficient to enable extraction of the required motion information pertinent to it. The practicalities and economics of particular situations will usually dictate the method of data capture. In using photographic methods a marked background or reference plane is required so that views of the moving subject can be related to the environment in which he moves. When using still cameras in fixed positions, simple background marking schemes will enable adequate referencing and scaling of photographed information. In the more general situation where one is interested in capturing information over a wide range of motions and the cameras have to be moved to track the subject, more sophisticated schemes for marking the background might be called for. When processing the captured information, particularly when substantial quantities of data have been collected, special aids are necessitated. These are for minimizing or even eliminating operator participation and fatigue. The use of computers to facilitate processing is increasing, and they can play a special role in providing for numerous possible ways of displaying information to aid in its interpretation. Some examples relevant to the various facets of the foregoing discussion will follow.

A Simple Economic Method for Clinical Use

For relatively unskilled operators collecting information in clinical situations it is important to see almost immediately whether a satisfactory record has been obtained. The use of a camera with a Po-

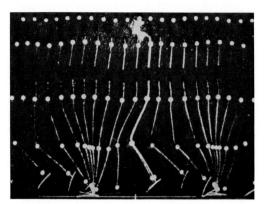

Figure 6.17. Stick diagram acquired from typical "strobe" picture.

laroid back and a stroboscopic flash unit in a darkened room fulfills the requirements (Milner *et al.* (1973a)). The patient might have to be dressed in a black cat suit and reflectors are attached to the various points of the anatomy which it is desired to track. To enable interpretation of the trajectories the points between the reflectors are connected with adhesive reflecting tape. This results in the "stick" diagram of Figure 6.17. Knowledge of the flash rate permits the time dimension to be incorporated in the picture and with a little practice the various patterns of motion are well recognized. Users of such records are soon adept at recognizing the general features of normal walking and can readily distinguish abnormalities. To prepare the patient and take the recording should take no longer than about 15 minutes. With severely handicapped patients such as hip arthritics this time could extend to 30 minutes.

Typically, the subject is made to walk a distance of about 8 m along a straight line. The perpendicular distance from the straight line available for locating the camera, is about 3.5 m. The stroboscopic light source is mounted directly below the camera. A black cloth curtain is suspended to serve as a back-ground. At the edge of the walkpath, white markers are located at fixed intervals to enable scaled measurements to be made from the filmed records.

When records are taken, the room is darkened. Figure 6.17 illustrates a typical "strobe" picture for a normal subject taking a normal walk. The strobe rate is 10 flashes/sec. From the distance scale the heel-strike to heel-strike length is 1.16 m; coupling this with the number of flashes involved renders the average speed of the walk as 3.8 km/hr. The record obtained shows the x − y coordinates of the various spots at each flash. The various trajectories are readily appreciated and, if desired, additional information can be extracted, e.g., the angles of knee flexion from moment to moment.

To facilitate further processing of the filmed data the spot coordinates can be extracted from negative films by means of a device such as a Vanguard Motion Analyzer with punch card capabilities. Setting the x and y cross-hairs over the spots and pressing a button causes the coordinate values to be punched directly on IBM computer cards. Coordinate references are acquired from background markers in each photograph. A series of computer programs enable, among other things, the original stick diagram to be plotted.

Movie Photography and the Use of a Flying Spot Scanner

A system was devised utilizing a movie camera and a computer system in which the major requirement was to preserve the intelligibility of the film for the viewer while still being suitable for automatic data processing (Kasvand and Milner (1972)). Thus, the gray levels in the film had to be retained for the human viewer, while from a computer programming viewpoint a black and white film and a minimum number of marks would have been much more desirable. A suitable compromise was achieved by marking, with black dots, the positions of the limbs which were to be followed. For enhanced contrast the background around the dots was whitened.

To obtain the highest possible resolution and measurement accuracy the camera is moved parallel to the walking subject during photography. Hence, on each frame there must be information regarding the absolute location of each frame and appropriate markings to serve as the coordinate reference for the particular frame. This problem was solved by constructing a special walkway with a gray-coded background.

Of necessity the filming technique must be executed with care, and despite this, films are usually not entirely perfect for processing. This is due largely to inherent factors following from limb movements which can lead to obscuring a marker, rotation resulting in changed marker shapes or the marker becoming invisible.

Filmed information is entered to the computer with the aid of a flying spot scanner under computer control. Essentially, the flying spot scanner is comprised of an oscillographic device which enables a small bright spot to be moved in the plane $(x - y)$ of a screen over which a filmstrip can be located. On the other side of the filmstrip there is a photosensor which measures the quantity of light transmitted through the film. By appropriately controlling the movements of the spot the gray levels (i.e., density) of the film can be examined at any $x - y$ coordinate set. With computer controls on the flying spot scanner the Graycode for each frame could be read and the location and identification of the marker dots on the limbs effected.

Initially, three frames were manually examined via an appropriate display unit, the marker dot positions located and entered to the computer. With this information the dot locations anticipated in each subsequent frame could be determined and thus only restricted portions of the frames could be scanned to save time. Five spots per frame could be tracked in 15 seconds. While the system could operate independently for lengthy periods, it was desirable to monitor the code-reading and dot-extraction results which were continuously displayed.

The programs have several levels of sophistication. If the simplest and fastest procedure fails, the program falls back on more complicated and time-consuming procedures incorporating additional displays for the benefit of the operator. The operator could then decide whether or not to intervene. The trajectory information is then available for further processing.

In order to determine whether the marker dots have been correctly identified, the computer programs are required to perform pattern recognition tasks. Philosophically, this appears to be most useful from the point of view of studies which might be conducted where numerous markers have to be tracked (e.g., traffic studies where many pedestrians are involved). With different shaped markers for the various subjects, provided the markers can be recognized by the computer, a most powerful kinesiological tool would emerge.

Videotaping

A different technological approach to kinesiological data recording is offered by videotaping (Winter et al, (1972b)). A major advantage of this system is the fact that the data collected are immediately available for processing. Film, on the other hand, requires special arrangements for rapid development and errors and omissions are discovered too late for corrective action.

Body markers consisting of halves of ping-pong balls covered with reflective tape and filled with low density polyfoam are attached at anatomical landmarks with double sided adhesive tape. Background markers consisting of larger discs of reflective tape provide a spatial reference for the TV recording system which utilizes 525 lines. The subject is tracked using a television camera and monitor mounted on a cart at about 3 m to the side, The lighting is adjusted to provide a very high contrast image which enables a one-bit conversion of this data into the computer (1 for white, inside a marker, 0 for black, outside a marker.).

The TV recording is replayed on a videotape recorder and converted into digital format via a TV-computer interface. The area to be sampled is determined by positioning a sampling window over the appropriate portion of the image: typically a 96 × 96 matrix sampling every second TV line is used. The videotape is read to the computer at normal speed. A typical run of three to four strides would therefore yield about 300 TV fields and require 5 seconds for conversion.

Computer programs reduce the converted data, cluster the points from each marker, and calculate the absolute coordinates of the geometric centers of the markers. At the same time corrections are made for parallax error.

This system has the advantages of a 24-hour turn-around time and the potential for on-line generation of trajectories. It satisfies two other requirements considered essential for a clinically acceptable system: (1) Minimal encumbrance to normal gait, with easy and quick preparation of subjects who walk on a regular floor surface in a normally lit environment; and (2) sufficient strides can be analyzed to allow stride-to-stride variations to be assessed.

Some Other Methods

Kinesiological data can be acquired and processed in other ways depending upon the requirements of a study and the constraints, financial and otherwise, which might be imposed.

Bullock and Harley (1972) have evaluated stereophotogrammetry as a method of studying dynamic posture. They were concerned with observing large range trunk and pelvic movements in three dimensions. The use of two survey cameras, repeating flash units, an appropriate background and suitable markers upon the subject enables the data to be collected photographically. A stereoplotter with automatic recording capabilities enables extraction of the relevant three-dimensional coordinates by a human operator.

Güth et al. (1973) have developed a method for automatic motion analyses by an on-line computer. The points whose motions are to be recorded are marked by applying small photodiodes. The photodiodes are exposed periodically to a bright V-shaped figure projected by a rotating mirror. The electrical signals resulting from the photodiodes are analyzed by using a computer and the trajectories of the marked spots are thus obtained on-line.

A commercially available system, the Selspot (Selcom (1976)) uses a silicon photosensor capable of transducing the coordinates (in a plane) of light-emitting diodes (LEDS). Multiple locations are tracked by sequentially switching the LED markers. The LEDS are connected through individual cables to the switching circuitry and power supply. A computer system is required for processing beyond the representation of trajectory data and must provide for data storage as well. Woltring (1974) comments on various opto-electronic sensors and concentrates on lateral photodiodes which are used in the Selspot system.

Storing only the coordinates of the markers extracted from a video signal, Cheng (1974) uses a normal television camera as the optical sensor. In this, the x − y data are transferred to a PDP 11/10 computer under program control. It is claimed that two points on the same scanning-line can be detected only if they are sufficiently far apart (⅙ of the picture width).

Andrews and Jarrett (1976) developed a multiple camera system interfaced with a PDP-12 computer. A maximum of five horizontal coordinates can be generated on any one television line.

Bruegger and Milner (1978) have reported on a system which enables virtual on-line scaled displays of tracked body motions. In this they utilize a charge coupled device image sensor having an area of about one's thumbnail and possessed with a 190 × 244 matrix of individual light-sensitive elements. The charge coupled device is scanned electronically under computer control and the light levels falling on it are determined. The subject to be tracked has small bright incandescent lamps placed at anatomical landmarks. By setting the threshold for the detection of light levels to those of the lamps, it becomes possible to reproduce the trajectories of the selected landmarks rapidly. By computer processing of the data numerous display possibil-

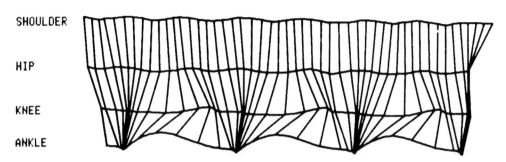

Figure 6.18. Stick diagram for right side of hemiplegic patient. Diagram is generated on a computer graphics terminal within minutes of the subject's walk past the sensing camera system.

ities are afforded. For example, Figure 6.18 shows a stick diagram for the right side of a hemiplegic subject walking from left to right. Shoulder, hip, knee, and ankle joint movements have been tracked for three full strides.

Presentation of Data

Clinical digestibility of the information acquired is very important aside from the scientific aspects of concern in gait studies. Numerous display possibilities exist including:

1. Trajectory plots—x vs y for each body marker (contained in Fig. 6.18).

2. y coordinate for each marker vs time (contained in Fig. 6.18).

3. Velocity (x,y or magnitude) of each marker vs time.

4. Acceleration (x,y or magnitude) of each marker vs time.

5. Limb and joint angles vs time or vs phase of walking cycle.

6. Angular velocity or acceleration vs time or vs phase of walking.

7. Average velocity in the forward direction, stride length, stance/swing phase ratio.

8. Angle-angle diagram, e.g., knee angle vs hip angle at corresponding instants of time.

By way of an example, Figure 6.19 depicts angle plots for a normal subject whose gait was recorded using the stroboscopic method indicated by Milner et al. (1973a). Knee-angle and hip-angle plots are presented both for distance along the walkway and with time (Milner et al. (1973b)). It is possible to differentiate the angle-time or angle-distance records. In the former case, angular velocity would follow. A successive differentiation would lead to angular acceleration. Differentiation with few samples leads to a noisy record and therefore caution must be exercised.

Figure 6.20 shows an angle-angle diagram relating to the hip-knee variations of a normal subject walking at 6.91 km/hr. Essentially, for corresponding strobe flashes (indicated by progressively increasing numbers) θ_k has been plotted against θ_h. Since walking is a cyclic process, a closed loop results. This form of display was first suggested by Grieve (1968) and it has been utilized by Milner et al. (1973b, 1974) in describing pathological gait.

It is interesting to note that the kink

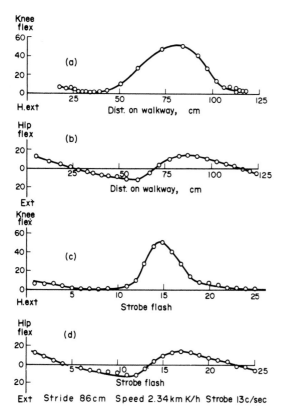

Figure 6.19. Angle plots for a normal subject. (a), knee angle θ_K vs distance X along walkway; (b), hip angle θ_H vs X; (c), θ_K vs time t (expressed as strobe flash number); (d) θ_H vs t (From Milner, M., Dall, D., McConnell, V. A., Brennan, P. K., and Hershler, C. 1973a. Angle diagrams in the assessment of locomotor function. S. Africa Med. J. 47: 951.)

ative angle-angle diagrams in the case of a candidate for total hip joint replacement due to arthritis, It can be seen from Figure 6.21 that preoperatively the right hip joint has a severely limited range of motion and the range of knee movement relative to the left side is diminished. The right hip is always extended relative to the left which covers a wide range of movement. The compensatory actions on the left side include a maintained flexion bias. While the walking speed for the subject is low (around 2 km/hr) the knee flexion-extension "kink" for the left side from strobes 1 to 4 has a substantial amplitude. This is due to the fact that the patient experiences pain when being supported on the right side, and therefore flicks the left leg forward rapidly to take the load on the left side. The postoperative patterns are somewhat similar in shape, but the ranges for the right side do not come up to those for the left. Evidently, the diagrams provide an effective, useful objective record.

which occurs in the angle-angle diagram of Figure 6.20 about heel-strike evolves as one proceeds from low speeds (where it is not apparent) to higher speeds such as that indicated, and for which it is marked. The reason for the flexion-extension sequence of the knee for strobe flashes 13-1-2-3-4-5 while the hip progressively extends is to ensure that the shock of heel-strike (particularly for higher speeds and hence higher kinetic energies) upon the joints in the lower extremity is relieved by being absorbed by the musculature.

Figure 6.21 shows pre- and postoper-

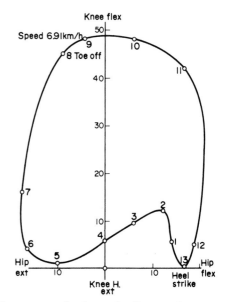

Figure 6.20. Angle-angle diagram for a normal subject walking at a fixed speed: plots of θ_K vs θ_H (From Milner, M., Dall, D., McConnell, V. A., Brennan, P. K., and Hershler, C. 1973a. Angle diagrams in the assessment of locomotor function. S. Africa Med. J. 47: 951.)

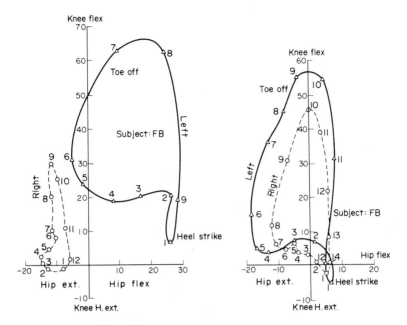

Figure 6.21. Pre- and postoperative angle-angle diagrams in a case of total hip replacement. Leftmost diagram: preoperative condition—Strobe rate = 13/sec. Rightmost diagram: Postoperative condition 12 months—Strobe rate = 10/sec. (From Milner, M., Dall, D., Ruff, A. L., and Brennan, P. K. 1974. Pre- and post-operative angle diagrams in cases of total hip reconstruction. Digest 5th Can. Med. Biol. Eng. Conf. 17: 3a, b.)

On Angle-Angle Diagrams

Angle-angle diagrams in the saggital plane have been obtained on line by Lamoreux (1971) using exoskeletal goniometric devices.

Figure 6.22 shows an exoskeletal electrogoniometer (after the design of Lamoreux and used in the author's laboratory) fitted to a subject. (Taking the output of the electrogoniometer to an electronic differentiator can provide an analog or record of angular velocity. Since differentiators are notorious for emphasizing noise in signals, care must be exercised in their use.)

Figure 6.23 depicts typical angle-angle displays generated in the author's laboratory. Essentially, the electrogoniometer outputs are sampled by a digital computer which also samples footswitches used to determine the contacts of the foot (heel and toe) with the ground. After performing simple calculations on the data and incorporating pertinent entered patient information, the display is produced.

Page and Infante (1973) used small flux gate magnetometers to determine the orientation of human limb segments during walking. These magnetometers are sensitive to the intensity of the earth's magnetic field and by adding suitable electronics were made to produce an output voltage proportional to the angle between the sensing axis and the line of the magnetic field vector. For a constant, known magnetic field these devices can confidently be used as angle sensors. Page and Infante have obtained angle-angle diagrams by using this method.

A commercially available system which utilizes polarized light, appropriate sensors, and special electronics enables remote (a few feet away) goniometric records to be obtained.

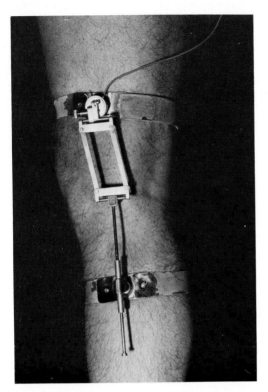

Figure 6.22. Electrogoniometer. The electrogoniometer is fitted to measure knee-joint angle variations in the sagittal plane. The mechanical linkages used ensure that movements only in the plane of concern affect the precision electrical potentiometer which is used to transduce joint angle.

Footswitches

Perhaps the simplest studies of human locomotion can be performed utilizing switches which can signal the contact of the extremities with the ground. If studies are to be undertaken in a fixed place, it is fairly easy to lay down a length of aluminum plate 1.5 mm thick, 45 cm wide and several meters long to serve as one contact of a switch, the other contacts being metallic members attached to the subject's footwear or even to the bare feet. Reasonably economic and useful footwear such as tennis or running shoes can be simply fitted with metallic contacts. Some characteristics of walking which can be delin-eated are the stance and swing phases when heel and toe contacts are used. Increasing the number of contacts enables the sequences of different components of footfall to be studied. Several footswitch responses can be recorded on one recording channel by placing different resistors in series with each footswitch lead, having the current in the footswitches pass through a fixed resistor across which the recorder is connected.

A battery or a suitable d.c. power supply energizes the complete circuit. Utilizing this method, Milner and Quanbury (1970) have studied aspects of control in human walking. In this work, timing relationships as a function of speed and pace period are dealt with for several normal subjects. An interesting observation is that the speed of walking is determined by both frequency of stepping and pace length such that these two components are linearly connected. It would seem that neuro-musculo-skeletal system defects might be characterized by deviations from this particular control law.

Other footswitch designs are possible. One method which has been used takes advantage of the capacitive variations that a subject produces by walking over a specially prepared capacitor (Jones *et al.* (1966). The advantage here is that no electrodes are needed on the feet.

Microswitches have been employed, and a design by Winter *et al.* (1972) where several specially engineered switches are included in a shoe works very successfully to delineate five points of contact underfoot.

A relatively simple two contact switch can be fabricated using 0.1 mm thick brass shim. Three identical pieces about 4 cm square are cut and are separated by 2.5 mm thickness widths of rubber foam placed in the form of stripes on the faces. The whole assembly is finally wrapped with duct tape and the switches can be readily taped to footwear. The author's experience with this design has been most satisfactory.

It is possible also to use specially made footswitches utilizing the elements em-

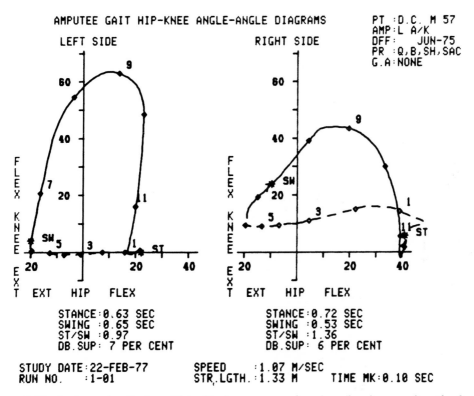

Figure 6.23. Angle-angle display. This display was produced as hard copy directly from a computer controlled unit. The angle-angle data are for a left above-knee amputee. As can be seen, distinctive angle-angle patterns are produced for the amputated and intact sides. The dashed lines represent single support phases; the full lines double support. ST/SW is the stance-swing (period) ratio. From a knowledge of the average walking speed and the heel-strike to heel-strike period, the stride length is calculated. Time marks on the diagrams are at 0.10 sec intervals. ST and SW on the diagrams, respectively, mark the commencement of stance and swing phases.

ployed in modern automobile seatbelt warning devices which detect whether a seat is being sat upon.

Electromyography

Electromyographic (EMG) records in the course of locomotion have been taken by various workers. The signal levels appearing on the skin surface of major skeletal muscles are in the order of 2 mV peak-to-peak with a frequency spectrum in the range about 10 Hz to 2 kHz with most energy concentrated around 100 Hz. Bringing surface electrodes aligned along the muscle length close to each other generally

results in an increase in the observed higher frequency content. Various types of surface electrodes are available. Among them are commercially available Beckman electrodes. Unless one is painstaking about maintaining these electrodes, it is the author's experience that they are prone to the generation of motion artifacts. Disposable type electrocardiographic electrodes can also be used and apart from their large size are most satisfactory

Electrode-electrode impedance can be as low as 1000 ohms. Indwelling electrodes of different designs are possible but perhaps the most useful one was made by Basmajian and Stecko (1962). Essentially

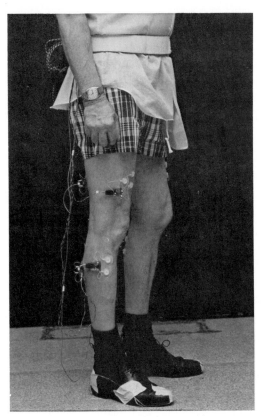

Figure 6.24. Subject instrumented for electromyography. The surface electrode pairs and amplifiers can be seen. Footswitches are taped to the heel and toe of each shoe. Note also the "umbilical cord" which enables data to be directed to a remote station.

this calls for differential amplifiers of high input impedance—around 1 megohm each input leg to ground. A low output impedance amplifier aids in the minimization of electromagnetic noise pick-up which might occur in leads, and it is usually best to locate the amplifiers as close as possible to electrode sites. It is possible to fabricate amplifiers less than matchbox size and such units can be adhered to the subject with suitable adhesive tape. Figure 6.24 shows a subject instrumented for a gait study involving electromyography using surface electrodes.

For convenience all of the connecting leads for amplifiers, their power supplies and also footswitch circuits can be taken to a junction box which can be supported by a belt worn around the waist of the subject. To avoid problems with leads connecting to the recording apparatus it is best to support them overhead in some fashion and different methods of support can be used depending on the available length of walkpaths. One method is to use a cantilever arm hinged to one wall. As the subject walks from one end of the room to the other the arm can be made to move the leads along with him. This should be convenient for short walkpaths about 10 m in length. For longer walkpaths an overhead "curtain-rail" type support is a good solution. The leads can be looped from one mobile support or runner to the next with restraining wires included to avoid stretching the leads. Particular attention has to be paid to providing relief from mechanical stresses at the terminal ends of leads.

Recording of information can be done in a variety of ways. Ultraviolet type recorders have frequency response characteristics depending upon the selection of galvanometers from d.c. up to about 13,000 Hz and simultaneous records of footswitch and EMG data can be effectively made with such recorders equipped with several writing pens. Paper speeds around 100 mm/sec provide good resolution and higher speeds permit a more detailed examination of EMG traces. A new type of oscillographic recorder is now on the market (Honeywell Visicorder model 1858). It

fine insulated wires about 25 μ in diameter are passed down a hypodermic needle (usually no. 27) and bent back at the tip for a few millimetres over the needle body. When the needle is injected into muscle and then withdrawn, the wires remain in place because of the bend in them which serves as a hook. They can then be connected to electronic amplifiers, the most popular connector being a short spiral spring which can grip the fine wire between two adjacent turns after being pulled open to admit the wire and then released. After use, the wire can be easily withdrawn. The electrode-electrode impedance is often around 50 kilohms and

does away with separate writing "pens" and has 18-channel capability.

The inclusion of additional signal processing circuitry will permit the use of more economical recorders better suited to lower frequency signals (in the order of 100 Hz). For instance, by full-wave rectifying the raw EMG signal and passing the product to a low pass filter with a time constant of about 5 msec, it is possible to obtain a display of the envelope of the EMG wave form on such a recorder.

Storage oscilloscopes, when available with multichannel facilities, are most useful in the early phases of work since records can be rapidly reviewed and photographed when desired. Multichannel tape recorders are perhaps the best means of recording information which can be submitted to automatic data processing. With the rapid advances in and availability of computer technology, it is reasonable to assume that most institutions concerned with locomotion studies would have access to some facilities for the automatic processing of data. From the point of view of clinical data handling, it would appear that there are opportunities to take advantage of computational equipment for the processing and display of information. In a study by Milner et al. (1971), EMG signals from indwelling electrodes and footswitch data together with other control signals (including footswitches) to facilitate automatic processing of the data were collected on five channels of a seven-channel FM tape recorder. The tape was input to the analog-digital converters of an SDS 920 computer via signal conditioners. A series of computer programs enabled the automatic processing of the data which were first digitized. Next the footswitch timing characteristics were analyzed and averages obtained. With these analyzed data the EMG data were subdivided and a mean wave form for each muscle at each step was computed, and various data were output in typewritten and graphic form. From the processed data average EMG characteristics were plotted as functions of speed and pace periods; phasic activities for these various conditions were also plotted.

The footswitches were fine wire mesh electrodes taped to the heel and toe areas of the sole of the subject's shoe. The subject walked on an aluminum walkway connected to the system ground. Whenever an electrode made contact with the plate a voltage, unique to that particular electrode, was produced.

The EMG records and the footswitch signals were stored in raw form on five channels of an FM magnetic tape recorder. These data were later played back into a hybrid analog-digital computer for processing and analysis. Before being fed into the computer the EMG signals were preprocessed by absolute average circuits so that only the envelopes of their absolute values remained.

The computer program first analyzed the footswitch signals and plotted out the downtimes of the heel and toe for each step in the walk and also the average downtimes for the particular walk (Fig. 6.25). It also typed out the mean period, and downtime, for both heel and toe, time from heel-down to toe-down in the same step, swing and stance times and the ratios of all the above times to the heel period. Also typed out were the mean deviation, variance, and standard deviation for each time period.

The next step in the computer analysis was to process and analyze the EMG signals. With the information obtained for an average period the EMG data were sequentially divided up into the same number of steps as existed in the actual walk. The resulting segments were then "overlaid" and the mean values were computed together with the maximal and minimal and standard deviations and the sample times when they occurred. The resulting average EMG waveform for one step was then plotted out for two cycles as shown in Figure 6.25 together with the footswitch information.

Ground Reaction Forces

Thus far, mention has been made of footswitch, EMG and trajectory information relating to different parts of the body.

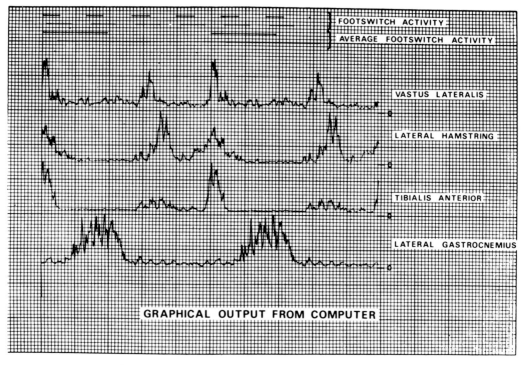

Figure 6.25. Averaged EMG envelopes of the activities of four major leg muscles. Six steps have been averaged. Time is depicted horizontally, EMG amplitude vertically. Each small division corresponds to 0.02 second for the EMG timebase. For the instantaneous footswitch records shown at the top, the scale factor is 0.1 sec/div. (From Milner, M., Basmajian, J. V., and Quanbury, A. O., 1971. Multifactorial analysis of walking by electromyography and computer. *Am. J. Phys. Med.* 50: 235.)

To specify more completely the locomotion process it is necessary to add to the foregoing information the nature of ground reaction forces. Coupling these latter data with anatomical information and trajectory data it is possible to use classical mechanics notions to calculate the forces acting at the various joints of the lower extremities. Information of this kind is obviously also of much value in the design of prosthetic and orthotic devices. Measurements of ground reaction forces of walking subjects have in recent times been made almost entirely with the aid of walkpaths equipped with force plates. Typically, these force plate systems enable records to be made of the magnitudes of the resultant vertical force, torque and horizontal shear forces and the location of the center of pressure of the foot. Numerical data are acquired from the force plate system. Many modern systems are derived from the design of Cunningham and Brown (1952). In essence a stiff flat metallic plate about 45 cm square is supported at the four corners by hollow steel pylons instrumented with several strain gauges. These gauges are responsive to the various strain components produced when the pylons are stressed by a load imposed upon the flat plate which is parallel with, and appears as a part of, the floor. Combining the electrical signals generated by the various strain gauges and the associated electronics results in the generation of the values referred to earlier.

Bresler and Frankel (1950) used the force plate developed by Cunningham and

Brown to measure ground-to-foot forces which enabled them to determine the time variations of the three force components and three moment components transmitted at the ankle, knee, and hip joints during walking on a level surface. Using a similar force-plate Paul (1965, 1966, 1967) obtained similar results but made extensions to obtain resultant joint forces acting at the hip and knee joints. In Paul's work only one step can be recorded at a time. Work by Rydell (1972) aimed at evaluating the results of reconstructive surgery of the lower extremity utilizes two identical 5-m long walking plates to enable 3 consecutive steps for each foot to be recorded. The system yields information regarding the vertical and horizontal forces. Kenwright et al. (1972) utilized Rydell's system in studying 25 normal and at least 20 each of knee, ankle and hip arthrodoses. They made the point that little is known about the effect of such operations on gait mechanics. Baon et al. (1972) have performed studies using the same system to evaluate the progress of 75 patients who had undergone hip prosthetic replacement. A British innovation "The R.A.E. Stanmore Sandal" was developed by the Royal Aircraft Establishment at Farnborough. The entire sole of the sandal is effectively a load measuring transducer which couples electronically with an oscillator housed in the heel. The load variations are telemetered to an external receiving system. This system measures essentially the resultant vertical force component on the foot. It is a step toward providing a system with greater flexibility in use, since the subject need not be confined to a walkpath or unduly constrained in his activities. Its disadvantages are that shear forces are not measured nor is the spatial distribution of measured forces available.

A commercially available force plate system produced by Kistler utilizes piezoelectric transducers. With this system it is possible to measure all of the pertinent foot-floor reactions and as well to determine the instantaneous location of the center of pressure as the foot rolls over the force-plate.

An innovative development by Cook et al. (1976, 1977) enables real-time ground-reaction force-line visualizations. An optical beamsplitter is used to superimpose a suitably scaled vector image and the patient image during locomotion. A force plate coupled with a specially designed electronic system enables the determination of an equivalent force vector in the plane of concern. The length of the oscillographically displayed force vector is made directly proportional to the force it represents, and it is correctly oriented to reflect the orientation of the force. Real-time observations of the force vector and body positions are obtained.

Direct Measurements of Velocity and Acceleration

It would be of great value if methods were available to enable direct measurements of velocity and acceleration. For velocity measurements there are simple methods which can be used but only under very restricted conditions such as walking in a straight line. By fastening strings to appropriate body parts, small electric generators can produce voltages which are responsive to the speed of the moving string. Recording these voltages enables recordings of speed in the particular action line of concern. The disadvantage of this methodology are obvious. Accelerometers are commercially available which are compact and responsive to accelerations imposed upon them. They can be used to detect linear and angular acceleration components. Interestingly, Newton's law dealing with force, F = ma, provides the basis for the construction of the accelerometers most commonly used in kinesiology. Often, a piezoelectric crystal has attached to it a weight whose inertia produces a force which acts on the crystal to generate an electrical voltage in the crystal and proportional to the acceleration. (There are also piezoresistive accelerometers.) By using an electronic integrator whose input receives the output of an accelerometer, it is possible to arrive at a measure of the velocity of the accelerated

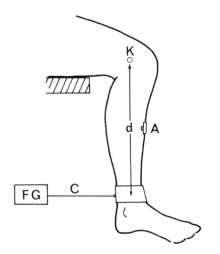

Figure 6.26. Quick release method for determining moment of inertia of lower leg.

object. A further integrating stage leads to the acquisition of displacement data.

Moments of Inertia of Limb Segments

Accelerometers, when used in the "quick-release" method, enable the accurate calculation of the moments of inertia of different body parts (Carlsöö (1972)). The method hinges on the equation $T = I\alpha$ where T is the total moment of force with respect to the fixed joint axis and is comprised of the moment of gravity and the moment of muscular force, I is the total moment of inertia incorporating that of the body segment and the gauge's attachment device to the body; and α is the angular acceleration of the body segment. Figure 6.26 illustrates the method for obtaining the value of I for the lower leg and foot. The cord, C, attached at the ankle connects to a force gauge, FG, so that the cord tension can be measured. The movement axis of the knee, K, is held fixed and an angular accelerometer, A, is attached to the lower leg. The subject is asked to extend the knee and thereby produce an isometric contraction of the quadriceps. A sudden release (quick release) of the cord attachment, without the subject's knowledge results in a forward rotation of the

lower leg as it extends. Taking moments about the axis K leads to the required result. The moment, M produced by the musculature is $M = Fd$, where d is the perpendicular distance from K to the line of pull of the cord C. It can be confirmed electromyographically that the subject is unable to relax the muscles at the instant of release. The moment of force due to gravity is conveniently made zero by ensuring that the gravitational force acts directly under the joint axis. It therefore follows that $\tau = M = Fd = I\alpha$, i.e., $I = Fd/\alpha$.

This result can be corrected for the moments of inertia of the attachments, their values simply being subtracted from I.

Timing Devices for Cinematography

Since cinematographic techniques will undoubtedly be in use in many kinesiology and biomechanics laboratories for some time to come, it is worth noting the designs of timing devices for synchronizing film views and as well obtaining the timing and duration of recorded human movements. Blievernicht (1967) presented the design of a cone-shaped rotating device for synchronizing three cameras in the three cardinal planes (Fig. 6.10). The cone, driven preferably by a synchronous motor, rotates at 60 rpm and two time scales each having 50 equal divisions were marked around the surface of the cone each at a different level. A number appeared at every fifth division progressing from 0 through 9. The two scales were marked 90° out of phase so that two floor-level cameras set perpendicular to each other recorded the same number on film. From overhead the cone appears flat. Two stationary pointers (one for each scale) set at right angles to each other ensure that pertinent readings are made from records derived from an overhead camera. Each scale division represent $\frac{1}{50}$ second and therefore by interpolation time can be estimated to $\frac{1}{100}$ second.

Figure 6.27 shows a synchronizing device based on the same principle but by

Figure 6.27. Synchronizing and timing device for two perpendicular film views with electro-myography or other related movement recordings. (Constructed by J. Moroz, Technician, Physical Education Department, McMaster University.)

design can be used for two perpendicular views only (no overhead view). This device affords a 0.04 second resolution and produces sequences of electrical pulses to enable synchronization of EMG records and the like.

CONCLUSION

In this chapter attention has been paid to methods of analysis of human move-ment. Pertinent approaches to the analyses rendered have been presented, and some indication has been given of data display and representation possibilities. A number of important technological aspects have been highlighted to show how studies might be facilitated particularly by special devices and computers. In subsequent chapters there is a further focus on muscle and the nervous system which acts to con-trol the musculoskeletal elements empha-sized thus far.

CHAPTER 7

Structure and Chemistry of Skeletal Muscle

INTRODUCTION

The animal motor system comprises three interrelated anatomical systems: the skeletal system which provides the bony levers which actually generate motion, the muscular system which supplies the power to move the levers, and the nervous system which directs and regulates the activity of the muscles. In this chapter we consider muscle.

Man has about 640 skeletal muscles of many shapes and sizes, from the tiny stapedius muscle of the middle ear to the massive hip extensor, the gluteus maximus. Muscles are situated across joints and are attached at two or more points to bony levers. Movement is produced by a shortening and broadening of the muscle which brings the lever ends into closer approximation.

Muscles differ in shape according to their functions. Some are long and slender for speed and range of movement, such as the biceps brachii; others are sheetlike to form supporting walls, such as the oblique abdominals; and some are multiple-headed to distribute and vary movement, such as the deltoid.

STRUCTURE OF SKELETAL MUSCLE

It is necessary to consider the structure of skeletal muscle before discussing its functions because structure and function are interdependent and inseparable. We consider, first, muscle structure in its gross aspects, i.e., the muscle as a discrete organ; second, the histology of the component fibers which make up the muscle organ; and finally, the ultrastructure of the contractile machinery of the fibers, the myofibrils.

Gross Structure

Muscles are discrete organs readily recognizable in any animal dissection, being attached in a manner characteristic of each species to bones or shells (with some exceptions) and crossing joints in very specific ways. Each muscle exerts a pull across the joint, and the action line of the muscle in relation to the joint determines the movement which will be produced within the limitations of joint structure.

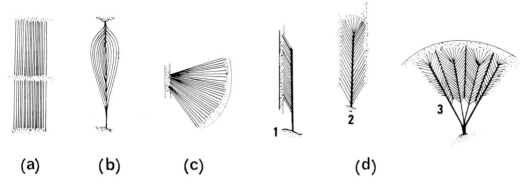

Figure 7.1. Arrangement of fibers in muscles. (a), a parallel muscle; (b), a fusiform muscle; (c), a fan-shaped muscle; (d), pennate muscles: 1, single pennate; 2, double pennate; 3, multipennate.

The Muscle as an Organ

A skeletal muscle is composed of two types of structural components: active contractile elements and inert compliant materials. The contractile elements are contained within the **muscle fibers.** Each muscle is composed of many muscle fibers, a medium sized muscle containing approximately 1 million. Fibers vary in length from a few millimeters in the stapedius to greater than 34 cm in the sartorius, and they vary in width from 10 to 150 μ. Fibers may run with their long axes parallel to the length of the muscle, as in the sartorius or rectus abdominis. These are known as **parallel** muscles (Fig. 7.1(a)). Muscles in which the fibers are arranged in the form of a spindle, as in the biceps brachii, are classed as **fusiform** muscles (Fig. 7.1(b)). There are also **fan-shaped** muscles (Fig. 7.1(c)) whose fibers fan out from a narrow area of attachment to a broad one. Examples include the pectoralis major, the anterior portion of the internal oblique of the abdominal wall, and the glutei, medius and minimus. Last, there are **pennate** muscles whose short fibers are arranged in a feather-like pattern (Fig. 7.1(d)). This type of muscle may be **single** or **double pennate** as in the forearm muscles, or the fibers may be arranged between multiple tendons as in the deltoid, the gluteus maximus, and the infraspinatus. These are described as **multipennate**. The arrangement of the fibers is related to the function of the muscle concerned. Fast-acting muscles generally have parallel fibers (biceps), while those designed for strength (gastrocnemius) are more often pennate. Coers and Woolf (1959) state that, in general, individual muscle fibers run from tendon to tendon of the muscle, that they are occasionally arranged serially, but that in man and other vertebrates they seldom anastomose with neighboring fibers.

About 85% of a muscle's mass consists of the muscle fibers themselves, the rest being composed largely of connective tissues which contain variable proportions of collagen, reticular, and elastic fibers. It is their distensibility and elasticity which assures that the muscle's tension will be transmitted smoothly to the load, and that an elongated muscle will recover its original length after being stretched. The connective tissues provide a complex arrangement of simple, essentially spring-like elements which are the **elastic components** of the muscle and which occur both in series and in parallel with the contractile elements.*

---

* Elastic components are found both in parallel and in series with the contractile elements of the muscle. Because those in parallel contribute only negligibly to passive elastic tension during stretch and become slack during contraction, they are of little interest to us. Those situated in series, however, are important factors in both passive elastic tension and active contractile tension. Therefore, throughout this section "elastic components" will be understood to refer to the series elastic components unless otherwise specified.

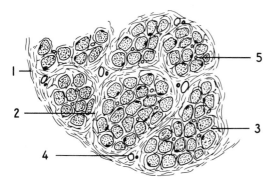

Figure 7.2. Cross-section of muscle to show its fiber bundles and the organization of connective tissues enclosing them. The epimysium (1) surrounds the entire muscle. Septa passing inward from it form the perimysia (2) which encompass bundles of muscle fibers. Note that smaller primary bundles, each with its own perimysium, are enclosed together as a secondary bundle. Delicate strands from the perimysium penetrate the bundles to form the endomysia (3) which invest the muscle fibers themselves. Arteries and veins (4) travel along the strands of the perimysia while capillaries (5) lie in the endomysia.

A connective tissue sheath, the **epimysium**, surrounds the muscle and sends septa (the **perimysia**) into the muscle to envelop bundles (fascicles or fasciculi) of muscle fibers. Larger bundles may be subdivided into several smaller bundles. From the perimysia delicate strands of fine connective tissue (the **endomysia**) pass inward to invest individual fibers (Fig. 7.2).

The total number of fibers, and hence the cross-section area of a muscle, are related to its strength requirements, but bundle size reflects the general function of the muscle. Muscles whose function it is to produce small movement increments, such as those required in manipulation, are composed of small bundles, whereas those concerned with powerful gross movements contain larger fasciculi. As a result the proportion of connective tissue is greater in the muscles which are capable of finely graded movements.

The connective tissues of the muscle blend with the collagen bundles of the tendon, forming a strong and intimate union, the **myotendinous junction.** Connective tissues of muscle and tendon are continuous. They act together as a buffer system against the possibility of too rapid development of contractile force in the muscle. Without their distensibility the muscle would be in danger of rupturing its fibers or tearing its attachments by a sudden contraction. Fascia and tendons also act to harness the pull of the muscle fibers to the bony levers. When a relaxed muscle is passively stretched or when it actively contracts, the initial tension developed is due to the elasticity of the connective tissues. In order to do work on a load, a muscle must first stretch out the elastic components until their tension is appropriate to the load before any shortening of the muscle becomes apparent. Until then, there is effectively no load on the contractile elements. Once muscle tension and load are in equilibrium, further expenditure of energy may be used to lift the load and perform external work.

Circulatory and Nerve Supply

Circulation

Muscle obtains a rich blood supply from branches of neighboring arteries. The arteries and veins travel in the epimysium, while arterioles and venules course in the perimysia and capillaries run longitudinally in the endomysia between individual muscle fibers (Fig. 7.2). An abundant circulatory network is provided by frequent transverse linkages between capillaries of adjacent fibers. Capillary anastomoses are especially well developed in the neighborhood of motor end-plates. In some instances, dilated cross-connecting vessels appear and are thought to act as reservoirs from which the muscle fibers may draw oxygen during sustained contraction, at which time capillary flow may be significantly reduced by compression of the supply vessels.

Nerve Supply

Nerves enter the muscle near the main arterial branch and divide to distribute both motor and sensory fibers to the muscle bundles. Motor fibers fall into two categories: large fibers (alpha subdivision of Group A of the Erlanger-Gasser classification) and smaller fibers (gamma subdivision of Group A). Each large **alpha motor neuron**, with its cell body lying in the ventral horn of the spinal cord, supplies a number of muscle fibers by successive bifurcation of its axis cylinder. One motor neuron and all of the muscle fibers which its axon terminals innervate constitute a **motor unit**. The number of muscle fibers per motor unit varies considerably with both the size of the muscle and the type of its function. Small muscles and muscles concerned with fine gradations of contraction have necessarily smaller motor units than do larger bulky muscles whose job is the maintenance of strong contraction. For example, motor units in the extraocular muscles of the eye consist of about five muscle fibers per motor unit, while those of the gastrocnemius may have as many as 1900 fibers activated by one motor neuron. It should be mentioned that the size of the motor unit is related to the size of the muscle fascicles, but this does not mean that a single muscle bundle is also a single motor unit. Rather, muscle fibers of a specific motor unit tend to be distributed among several fascicles in a limited area of the muscle.

The smaller **gamma motor neurons** innervate the muscle spindles, providing a means of central regulation of muscle contraction over an indirect pathway known as the gamma loop, which is discussed in detail in Chapter 11 under "Role of Spindle Innervation in Voluntary Movement."

Sensory nerve fibers are also of two general sizes: large Group I neurons whose sensory terminals lie in the muscle receptors (spindles and tendon organs) and smaller Group III fibers which probably subserve the sense of muscle pain.

Microscopic Structure

Histology of the Muscle Fiber

The muscle fiber is a syncytial mass of sarcoplasm, cylindrical in form with a bluntly tapered end and surrounded by a specialized membrane, the **sarcolemma**. The sarcolemma is a unit membrane about 100 Å thick. A number of nuclei lie peripherally just under the sarcolemma, and numerous **myofibrils**, each about 1 μ in diameter, lie longitudinally embedded in the sarcoplasm. The myofibrils are the contractile elements of the muscle. There are about a thousand in each muscle fiber.

Viewed under the ordinary light microscope the muscle fiber has a distinct cross-striated appearance of alternate dark and light areas appearing in a regular and repeating pattern. High magnification shows the striping to be a property of the myofibrils, which are oriented in register with dark regions adjacent to dark regions and light to light so that the pattern appears to extend across the whole fiber (Fig. 7.3). Dark areas are known as **A bands** because of their anisotropic (doubly refracting) effect upon polarized light. Each

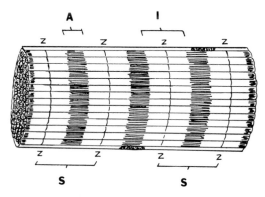

Figure 7.3. Diagram of a portion of a single muscle fiber. The drawing shows that the striations are a property of the myofibrils which lie in the fiber with light and dark bands in register. S, sarcomere; A, A band; I, I band; Z, Z line.

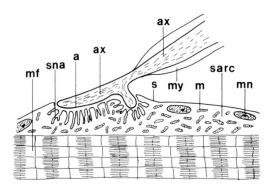

Figure 7.4. Schematic drawing of the neuro-muscular junction. a, axolemma; m, mitochondrion; s, sarcolemma, ax, axoplasm; mf, myofibril; mn, muscle nucleus; my, myelin sheath; sarc, sarcoplasm; sna, folds of the subneural apparatus. (Adapted from Couteaux, R., 1960. Motor end plate, structure. In *Structure and Function of Muscle*, Vol. 1, edited by G. H. Bourne. New York: Academic Press, p. 337.)

appears to be somewhat lighter in its mid-region, the **H zone**, which is crossed by a darker line, the **M line**. The light regions are known as **I bands** because of their more nearly isotropic effect. Each I band is clearly bisected by a dark line, the **Z disc** or **line**. The region from one Z disc to the next constitutes a **sarcomere**, and this appears to be both the structural and functional unit of the myofibril.

Neuromuscular Junction

As a terminal axon approaches a muscle fiber, its myelin sheath narrows and finally ceases a short distance before the end-plate. It has been firmly established that there is no continuity of nerve and muscle protoplasms. The nerve axon does not penetrate the sarcolemma; its arborizations lie on the surface of the muscle fiber. Axon terminals may lie under the endomysium but do not penetrate the sarcolemma. The neuromuscular junction is formed by axolemma and differentiated sarcolemma in apposition to each other. The structure of the muscle fiber in the junctional region is highly specialized. The sarcolemma is profusely folded into

troughs and grooves, presenting a spiny or lamellated appearance (Fig. 7.4). The muscle portion of the neuromuscular junction has been called the **subneural apparatus**. Many muscle fiber nuclei are seen in the vicinity of the junction, accompanied by an abundance of mitochondria. Most authorities believe that the gap between the nerve terminus and the sarcoplasm is a specialized barrier across which excitation must be transmitted by chemical means (discussed in more detail in Chapter 8 under "Synaptic Transmission").

Chemical Composition of the Muscle Fiber

The skeletal muscle fiber is composed of a semifluid sarcoplasm of water, salts, and other substances in which are suspended the nuclei, myofibrils, a reticular system of tubules, sacs and cisterns, mitochondria, glycogen granules, and lipid droplets. Water constitutes 75%, protein 20%, and other materials 5% of its mass.

Muscle proteins may be grouped into four major factions: proteins of the sarcoplasm, of the stroma, of granules, and of the myofibrils. The sarcoplasmic proteins, which occupy the space between the myofibrils, are readily extracted with water or neutral salt solutions. They include myoglobin, a pigment with a high affinity for oxygen which functions in the transfer of oxygen from the capillaries to the sites of oxidation (the mitochondria), and enzymes concerned with the glycolytic aspects of muscle metabolism.

Stroma proteins are retained in the muscle residue after extraction with strong salts and are difficult to isolate. For this reason, knowledge regarding them is limited. Some are of a collagenous nature and contribute to the structure of the sarcolemma.

Differential centrifugation is used to separate the proteinaceous granules from homogenized muscle. These granules include the nuclei, mitochondria and microsomes. In the intact fiber the mitochondria and microsomes are located among the myofibrils. Mitochondrial proteins include the enzymes of oxidative metabolism.

Myofibrillar protein consists almost entirely of actin and myosin. Small amounts of tropomyosin and troponin are also present. Actin and myosin compose about half of the total protein content of the fiber and account for most of the contractile material itself.

Glycogen granules account for about 0.5 to 1.0% of the fiber. Lipids occur in small amounts. The principal salts are those of potassium, sodium, calcium, magnesium, and chloride. Nonprotein extractives include creatine, creatine phosphate (CP), adenosine triphosphate (ATP), adenosine diphosphate (ADP), and lactic acid.

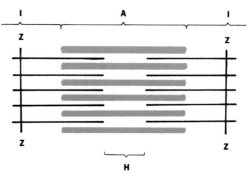

Figure 7.5. Diagrammatic representation of filament arrangement in a single sarcomere of a fibril. A, A band; I, I band; H, H zone; Z, Z line. See discussion in the text.

Ultramicroscopic Structure of Muscle

The Myofibril

Electron micrographs have revealed that each myofibril is composed of filaments of which there are two types: one thicker and shorter (in the rabbit psoas, about 100 Å in diameter by 1.5 μ in length) and another thinner and longer (rabbit psoas, about 50 Å by 2 μ). These are seen to lie longitudinally in a very definite parallel orientation. Each **thick filament** extends the length of the A band, while a **thin filament** passes from each Z line through the I band and into the adjacent A band as far as the edge of the H zone. Therefore, the denser outer portions of the A band are produced by the overlapping of thick and thin filaments for part of their lengths. The central H zone region is less dense because it contains only thick and no thin filaments. The I band is least dense because it contains only thin filaments. The arrangement of the filaments is shown diagrammatically in Figure 7.5. The bands, zones, and lines may be readily identified in the electron micrographs of Figures 7.7 and 7.10, (a) and (b).

The H zone is not homogeneous but shows variations in its density. Across its center is the dense region of the M line, on either side of which there appears a narrow region whose density is lower than that of the rest of the H zone. The lightest portion, known as the **pseudo-H zone**, maintains a constant width regardless of stretch or contraction, indicating that it is a structural feature of the thick filaments and not just another reflection of overlap. Significance of the pseudo-H zone is discussed later in this chapter.

The relation of filaments to fiber banding is clearly demonstrated by cross-sections through these areas (Fig. 7.6). If sections are taken through the denser part of the A band, the thin filaments are found surrounded by thick filaments in an orderly hexagonal array. If sectioned through the H zone of the A band, only thick filaments are present, the thin ones being absent. If sectioned through the I band, only thin filaments are present. Thus, the cross-striations of the muscle fiber seen with the light microscope are found to be due to a repeating pattern of varying filament densities along each myofibril with the patterns of adjacent fibrils in register.

Chemical Structure of Myofibrils. Chemical analysis of the fibrils shows them to be composed of about 20% protein, the rest being a watery suspension of salts and other metabolically important substances. About 80% of the protein consists of **actin** and **myosin** in a ratio of about 1:3. Actin is a low viscosity protein, molecular weight 70,000 to 76,000, while myosin is a

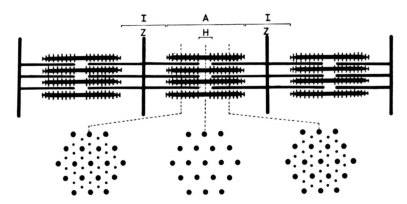

Figure 7.6. Diagrammatic representation of the structure of striated muscle. Shown are overlapping arrays of actin- and myosin-containing filaments, the latter with projecting cross-bridges on them. For convenience the figure is drawn with considerable longitudinal foreshortening. (From Huxley, H. E., 1969. The mechanism of muscular contraction. *Science 164*: 1356–1366, Fig. 1, June, 20, 1969. Copyright © 1969 by the American Association for the Advancement of Science.)

Figure 7.7. Electron micrograph of a longitudinal section through an array of fibrils in which two thin filaments lie between two thick ones. Note the projections from the thick myosin filaments connecting with the thin actin filaments. These are presumably the cross bridges. (From Huxley, H. E., 1958. The contraction of muscle. *Sci. Amer. 199*: Fig. 1, p. 71. Copyright © 1958 by Scientific American, Inc. All rights reserved.)

more viscous molecule, molecular weight 1,000,000 to 1,500,000. During contraction these two proteins combine to form a complex, **actomyosin**, which is the contractile material *per se* of the muscle fiber. Smaller amounts of two other proteins, **tropomyosin** and **troponin**, are also present. Their functions are not yet entirely clear.

Extraction methods specific for myosin remove the thick filaments and leave an electron micrograph lacking the A band. If only actin is extracted, the treated tissue shows loss of much but not all of the content of its I bands, while its A bands are unchanged. This is good evidence that the thick filaments are composed of myosin molecules and the thin filaments of actin plus other substances.

It is reasonable to assume that the association of actin and myosin, long recognized as a first step in muscular contraction, must involve some form of bonding between the two types of filaments. Electron micrographs show projections extending from the thick filaments in the dense parts of the A band (Fig. 7.7). Huxley (1958) and others have conjectured that these may represent an orderly system of cross-bridges which attach the thick filaments to each of the surrounding thin fil-

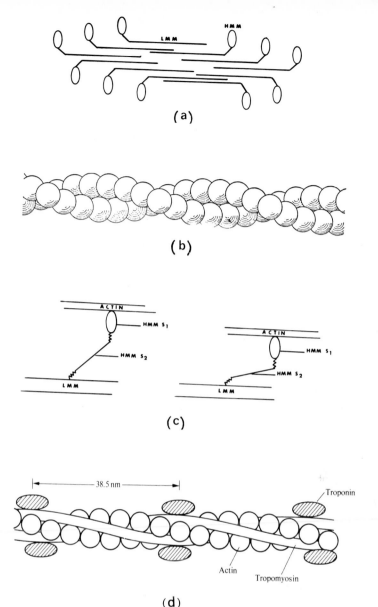

(a)

(b)

(c)

(d)

Figure 7.8. Proposed ultrastructure of myosin and actin. (a), aggregation of light meromyosin (LMM) and heavy meromyosin (HMM) into the myosin filament. LMM forms the backbone of the filament while HMM projects as the cross-bridge. (b), actin filament formed by a double helical organization of G-acting molecules. Alternate high and low points may contain the active sites for attachment of HMM cross-bridges. (c), detail of HMM subfragments: globular fragment (HMM S_1) and linear fragment (HMM S_2). HMM-S_1 is joined to S_2 by one flexible juncture, and S_2 joins the LMM backbone by another. The HMM bridges are thus capable of considerable movement to adjust, without any changes in their orientation on the myosin filament, to the variable distances between actin and myosin filaments which occur during the contraction cycle. The figure at left shows the position of the bridge when the distance is large as during contraction; the figure at the right shows the position when the separation is small as during stretch or relaxation ((a), (b), and (c) adapted from Huxley, H. E., 1969. The mechanism of muscular contraction. *Science 164*: 1356–1366, Fig. 1, June 20, 1969. Copyright © 1969 by the American Association for the Advancement of Science.) (d), the likely arrangement of actin, tropomyosin and troponin in the thin filaments of muscle. Tropomyosin appears to be arranged along the double helical organization of the actin filament. ((d) from Tribe, M. A., and M. R. Eraut, 1977. *Nerves and Muscle* (Basic Biology Course Book 10), p. 163. Cambridge, England: Cambridge University Press.)

aments. On any single thick filament, six projections form a helical pattern, each set about 60° farther around the filament and about 60 to 70 Å apart. Thus, the period of repetition is about 400 Å.

Fragmentation of myosin by trypsin digestion yields two subunits, the heavier of which has a molecular weight of about 230,000 and has been designated as **heavy meromyosin** (HMM), and the lighter, with a molecular weight of about 96,000, as **light meromyosin** (LMM). The probable organization of the myosin filaments in terms of these molecular subunits has been postulated from work carried on in Huxley's laboratories (1965). Using the technique of negative staining, in which subcellular structures are caused to stand out in contrast with their backgrounds by surrounding them with a heavy metal salt, these investigators found that the myosin molecules of the thick filaments consist of globular heads of HMM and longer tails of LMM, the latter appearing as linear strands with a strong attraction for one another and for HMM. Each thick filament is composed of about 200 to 400 myosin molecules in an overlapping array with their HMM heads projecting outward (probably as the cross-bridges) and their LMM tails overlapping in parallel to form the backbone of the filament. The molecules are oppositely oriented in each half of the sarcomere with their tails directed toward the center (Fig. 7.8(a)). As a result there is a region in the center which is devoid of HMM bridges. This is consistent with the pseudo-H zone identifiable in the striation pattern. The dense M line may result from the crossing of the LMM strands in the middle of the filament.

Actin has been shown to exist in both a globular (**G-actin**) form and a fibrous (**F-actin**) form. Hanson and Lowy (1960) noted that the actin filaments consist of a double helix of G-actin molecules with a 360 Å period, probably surrounding an F-actin core (Fig. 7.8(b)). They conjectured that the alternate high and low points of the helix are the active sites for attachment of the HMM bridges. Electron micrographs of disrupted fibrils have shown that actin filaments are firmly attached to the Z discs.

Later studies by electron microscope and X-ray diffraction have indicated that the HMM bridge is further fractionable into a globular subfragment (HMM-S_1) and a linear subfragment (HMM-S_2). It has been proposed by Huxley (1969) that the linear HMM-S_2 is connected at one end to HMM-S_1 and at the other end to LMM by flexible junctions (Fig. 7.8(c)). If this is true, then the bridges will have a rather wide range of movement and will be able to adjust to the variable distances between filaments which have been observed to exist at different muscle lengths. Interfibrillar distances are known to decrease when muscle is stretched and to increase as the muscle is returned to its rest length. In such a system the regularity of bridge attachments between actin and myosin would be determined by the constant location of the active sites on actin, rather than by exact and invariable positioning of the bridges on myosin.

Sarcoplasmic Reticulum and T System

The electron microscope has revealed another interesting and significant detail of muscle subcellular structure. The endoplasmic reticulum of the muscle fiber displays certain distinctive characteristics which have merited the use of the term **sarcoplasmic reticulum** in reference to it. First, instead of being distributed throughout the cytoplasm the membrane-limited tubules and cisterns form lace-like sleeves around the myofibrils in a distinct pattern which is repeated in a definite phase relationship to the striation bands of each sarcomere. Second, there appears to be a special arrangement of reticular elements in characteristic groups of three, known as **triads**, consisting of two cisterns separated by a transversely oriented tubule, the **T tubule** (Fig. 7.9). The triads are found at the same site in muscle tissue of any given species. In fishes and amphibians they occur at the Z lines; in reptiles, birds, and mammals, they occur at the A-I junctions (Fig. 7.10). Third, ribonucleic acid (RNA) granules, although present in the sarco-

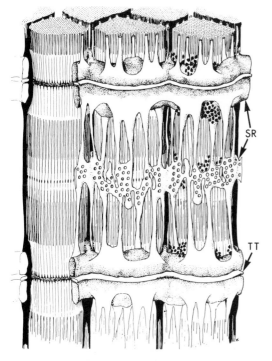

Figure 7.9. Reconstruction from electron micrographs of a portion of an adult frog skeletal muscle fiber. The sarcoplasmic reticulum (SR) is shown surrounding several myofibrils for a length of slightly more than one sarcomere. Two transverse tubules (TT) are seen extending across the figure at the centers of triads located next to the Z lines of the myofibrils. (From Peachey, L. D., 1965. The sarcoplasmic reticulum and transverse tubules of frog's sartorius. *J. Cell Biol. 25:* 209.)

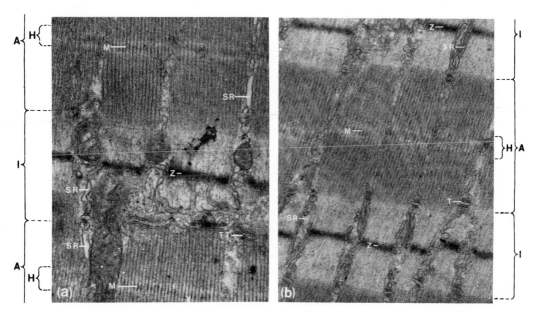

Figure 7.10. Longitudinal section of rat muscle. Two electron microscope photographs of myofibrils from a mammalian (rat) muscle. Original magnification ×46,000. (a), longitudinal elements of the sarcoplasmic reticulum (SR) lie between the myofibrils. Transverse tubules (TT) of the T system are seen at the A-I junctions. M and Z lines are conspicuous. (b), another section shows triads (T) in profile at the A-I junctions on both sides of the dark Z lines especially well in the lower half of the photograph. Longitudinal elements of the sarcoplasmic reticulum extend across the Z lines in both the upper and lower halves of the photograph. The myofibrils, lying as they do with their dark and light bands in register, show the basis for the striated appearance of the muscle fiber. (Courtesy of Drs. G. Harrison and D. Philpot, NASA Ames Research Center, Moffett Field, California.)

plasm, are absent from the reticulum itself. This suggests that the function of the reticulum may be other than protein synthesis.

The T tubules are now known to be distinct from the sarcoplasmic reticulum, although they are functionally associated with it. The **T system**, as it is called, communicates with the sarcolemma or perhaps represents invaginations of it. The content of the tubules is continuous with the extracellular fluids and contains sodium in significant amounts. The cisterns of the sarcoplasmic reticulum contain a high concentration of calcium. Walker and Schrodt (1967) have described structures between the cisterns and tubules which they interpret as interconnecting membranes. The probable roles of the sarcoplasmic reticulum and the T system in the coupling of excitation and contraction are discussed below under "Excitation-Contraction Coupling."

THE NATURE OF CONTRACTION

Excitation

Muscle may be excited and caused to contract by natural or by artificial means. Normally, excitation is accomplished only by the nervous system: nerve impulses arriving at the neuromuscular junction cause the release of a transmitter substance which diffuses across the junction and chemically excites the muscle fiber. However, whether induced naturally or artificially, excitation is evidenced by the generation and conduction of action potentials in the sarcolemma of the muscle fiber.† Action potentials travel along the fiber membrane at a speed of 1 to 3 m/sec and initiate the events which lead to shortening of the contractile elements of the myofibrils and the consequent production of tension in the muscle.

---

† The generation and conduction of action potentials in cell membranes are discussed in detail in Chapter 8.

Contraction

Muscular contraction requires the expenditure of energy obtained from chemical reactions coupled to a contractile mechanism which can use the energy to generate tension and produce external work. Some facts regarding the processes involved are well established; others still evade understanding and are as yet only theory. The following discussion is based on generally accepted concepts.

Chemistry of Contraction

The source of energy for muscular contraction is **adenosine triphosphate** (ATP), a nucleotide with three phosphate groups, two of which are attached to the molecule by **high energy bonds** (Fig. 7.11). The high energy property resides in the structure of the molecule rather than in the bond as such but, because the energy is released when the bond is cleaved, it is common practice to speak of the "bond" energy. The usual chemical bond holding phosphate to a molecule releases only 1 to 3 kcal of energy when cleaved, as compared with 7 to 12 kcal from a high energy bond (symbolized by ~P). It is this energy, made available to the muscle by cleavage of ~P bonds of ATP, that supports contraction. It is interesting to note that ATP was first identified as the energy source for muscle and later found to be the uni-

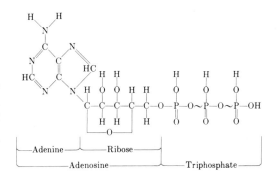

Figure 7.11. Structural formula of adenosine triphosphate. Brackets indicate the chemical components of the nucleotide.

versal form of energy currency for living organisms.

Other chemical substances in addition to ATP are essential for muscular contraction. Important among them are Ca^{++}, Mg^{++}, and actin and myosin combined to form actomyosin. The hydrolysis of ATP, to split off a ~P, also requires an enzyme known as adenosine triphosphatase (ATPase). In muscle, actomyosin functions both as an ATPase under proper conditions and also as the contractile material which utilizes the energy thus acquired. The ATPase capacity of actomyosin, which has been found to be the property of the myosin, more specifically of the $HMM\text{-}S_1$ of the myosin, requires the presence of Mg^{++} and a close association with actin to be functional.

Activation: Hydrolysis of ATP. The first overt indications of activation in the muscle are increased resistance to stretch and a rise in heat production as compared with resting muscle. This is designated the **active state**. Increase in stretch resistivity is thought to indicate the formation of actomyosin, while increased heat production is an end product of the hydrolysis of ATP. The active state is not synonymous with tension. It develops quickly, reflecting the chemical changes which lead to the mechanical changes in the muscle, but dies away before the latter are complete. Before tension can become manifested, the elastic components of the fibers must be stretched by a finite amount. Therefore, tension development lags behind the active state.

As already mentioned, the hydrolysis of ATP requires an activated and readily available ATPase. In the presence of free Ca^{++}, actin combines with myosin at the $HMM\text{-}S_1$, thus activating the ATPase capacity of myosin. ATP is probably attached to actin at the same or an adjacent site to that which is involved in its combination with $HMM\text{-}S_1$. ATPase then splits off the terminal ~P from ATP, and in some unknown manner the chemical energy thus made available is converted into mechanical energy for the shortening of the actomyosin system.

$$A + HMM\text{-}S_1 \xrightarrow{Ca^{++}} AM$$

$$ATP \xrightarrow{Mg \cdot ATPase} ADP + P_i + E \quad (7.1)$$

$$(\text{Contraction})$$

where A is actin, AM is actomyosin, ADP is adenosine diphosphate, P_i is inorganic phosphate, and E is energy. The wavy arrow indicates that the process is unknown.

Because the rate of ATP splitting closely parallels the rate of work output, it seems probable that energy is transferred directly from ATP to the contractile material without any intermediate step.

Maintenance of ATP Supply: Metabolism of Glucose. While ATP is the immediate and direct source of energy for cellular processes, the amount of ATP in the muscle cell is limited. Therefore, other reserves or energy-producing metabolic reactions must constantly replenish the ATP supply in order to support continued contraction for even a short time. The food compounds provide the ultimate energy reserves responsible for maintenance of the cell's supply of ATP. As energy is utilized and ATP is converted to ADP, energy must be supplied to reconvert the ADP to ATP. Each cell must accomplish the reconversion for itself; ATP is not transported about the body. The energy of the food compounds, of which the carbohydrate **glucose** is the molecule used most commonly and most efficiently, is released in small quantal amounts and repackaged as ~P of ATP, in which form it becomes available to the cell machinery. In most animal cells glucose is metabolized by biological oxidation.

The most economical method of extracting energy from foods requires interaction with free oxygen, the latter being supplied by the integrated functioning of the respiratory and circulatory systems. Glucose is available to all mammalian cells in the circulating blood. Under normal conditions blood glucose concentration is carefully maintained from moment to moment by withdrawals from the glycogen stores of the liver in amounts exactly equivalent to those which have been used by body cells. Muscle has an additional on-the-spot supply of **muscle glycogen** which, although essentially the same as liver glycogen, does not contribute to the blood glucose level and is depleted only by muscular activity.

It is useful to consider glucose metabolism in two parts: an anaerobic phase which takes place in the sarcoplasm and an aerobic phase which is carried on in the mitochondria.

Glycolysis, the Anaerobic Phase. Glucose undergoes a series of anaerobic chemical reactions, known as **glycolysis**, during which each 6-carbon molecule of glucose is dissimilated into two 3-carbon molecules of **pyruvic acid**. During the course of glycolysis two pairs of hydrogens with their electrons are removed and enough energy is obtained by **substrate synthesis** to convert 4 ADP to 4 ATP. However, since 2 ATP are required to drive the reactions of glycolysis, the net gain is 2 ATP per molecule of glucose for this phase. Although the process involves biological oxidation‡ it has thus far required no free oxygen.

Krebs Cycle and Related Electron Transport System, the Aerobic Phase. Each molecule of pyruvic acid then enters a long and complex process involving two sets of mutually dependent reactions known as the **Krebs cycle** (also known as the citric acid cycle and as the tricarboxylic acid cycle) and the **electron transport system** (ETS). During the reactions of the Krebs cycle 3 carbons are stripped from each molecule by decarboxylation and re-

leased as CO_2. Five pairs of hydrogens with their electrons are removed by hydrogen-acceptor **coenzymes**. A coenzyme is a molecule which acts as an essential intermediate carrier of the products of some enzyme-catalyzed reactions. Hydrogen-carrying coenzymes importantly concerned with the specific dehydrogenases of the Krebs cycle are **nicotinamide adenine dinucleotide (NAD), nicotinamide adenine dinucleotide phosphate (NADP)**, and **flavin adenine dinucleotide (FAD)**. Each coenzyme picks up 2 H, including the electrons, being thereby reduced.

$$NAD + 2H \rightarrow NAD \cdot 2H \qquad (7.2)$$

The hydrogen pairs are ultimately freed into the cell fluids as H^+ while the electrons are transmitted to other electron acceptors and passed on through the second set of reactions, the ETS.

The ETS consists of a chain of enzymes which, like a bucket brigade, receives and passes on the electrons from the coenzymes, each enzyme being alternately reduced and then reoxidized. Each of these enzymes has a built-in coenzyme called a **prosthetic group** which is an integral part of the enzyme molecule. Although not a separate molecule as is a coenzyme, the prosthetic group also carries the end product of the reaction. The enzymes of the ETS chain are known as **cytochromes**.

Electron transmission involves the flow of electrons from a higher to a lower energy level, an **electron cascade**. As they go, electron pairs lose energy in quantal amounts at three specific points in the chain. The released energy is captured and used to add inorganic phosphate (P_i) to ADP by a high energy bond.

$$ADP + P_i + E \rightarrow ATP \qquad (7.3)$$

Therefore, the energy released by one pair of electrons is sufficient to rephosphorylate 3 molecules of ADP, 1 at each energy release site.

The last enzyme in the chain donates its electrons to molecular oxygen. As each ½ O_2 accepts a pair of electrons, it becomes negatively charged ($O^=$). The negatively

---

‡ While biological oxidations may involve the addition of oxygen to the molecule, they are more usually accomplished by the removal of hydrogen and/or electrons. Often water is first added to the substrate, followed by dehydrogenation. The addition of 6 molecules of water during the reactions of glycolysis accounts for the "extra" water produced when 12 pairs of H are picked up by 6 O_2 to form 12 H_2O. As we have seen, biological oxidation may proceed anaerobically (glycolysis), or oxygen may be required as the final H-acceptor (as in ETS) to regenerate coenzyme from its reduced form.

charged oxygen attracts 2 H^+ from the surrounding cell fluids to become H_2O. There are six hydrogen pairs ($6H_2$) removed during the complete metabolism of each half of the glucose molecule and six pairs of electrons passed through the ETS. As a result, 6×3 or 18 molecules of ATP are restored by ETS for each half, or 36 for each full glucose molecule. Therefore, each molecule of glucose which is completely metabolized produces 38 ATP (36 ATP from the aerobic reactions of the Krebs-ETS phase and 2 ATP from the substrate synthesis during glycolysis).

Supplementary Means of Restoring ATP in Muscle. a. Creatine phosphate. Probably because it must be able to quickly increase its metabolic rate by 10 times or more to support vigorous exercise, muscle is supplied with a supplementary energy source in the form of another high energy phosphate compound, **creatine phosphate** (CP). CP provides a fast and direct means of rephosphorylating ADP. Although present in a concentration about 3 times that of ATP, it is still a relatively limited source as compared with the blood glucose. ADP takes up the ~P from CP to become ATP.

$$CP + ADP \overset{CPase}{\rightarrow} ATP + C \qquad (7.4)$$

where CP is creatine phosphate, CPase is creatine phosphatase, and C is creatine.

CP is later rephosphorylated by ATP derived from the oxidative chain. There is some question as to whether CP functions in an emergency capacity only or whether it is a regular intermediate step in the rephosphorylation of ADP to ATP.

b. Oxygen debt. In the absence of oxygen the Krebs-ETS reactions will cease when all of the coenzyme has been reduced, unless an alternative H-acceptor is available. In muscle, pyruvic acid can act to a limited extent in such a capacity, becoming reduced to lactic acid.

$$CH_3CCOOH + 2\ H \overset{lactic}{\underset{dehydrogenase}{\rightarrow}} CH_3CHOHCOOH$$
$$\underset{O}{\overset{\|}{}}$$

(pyruvic acid)   (lactic acid)   (7.5)

In vigorous exercise when pyruvate is formed more rapidly than it can be metabolized and reduced coenzyme is accumulating, the insufficiency of the oxygen supply (resulting from limitations of the circulatory system) can be partially offset by this mechanism. The muscle thus can continue to function at levels in excess of its oxygen supply by accelerating glycolysis and supplementing its ETS-derived ATP with the ATP provided by the anaerobic process. This is popularly known as **oxygen debt**. Eventually, however, as ADP, lactic acid, and reduced coenzyme accumulate beyond a finite limit, the process grinds to a halt and contraction ceases. Should all of the ATP be used up, rigor will result and the muscle will go into irreversible contracture.

Although in an emergency and during vigorous exercise the muscle can run for a time on glycolysis, to do so is uneconomical. While complete metabolism of glucose provides 38 ATP per molecule, glycolysis supplies only 2 ATP, or about 5%. Furthermore, when oxygen is again adequately available, as at the cessation of vigorous activity, the accumulated lactic acid and other end products must be removed. Part of the lactic acid (about one-fifth) will be metabolized in the Krebs cycle-ETS to complete the resynthesis of ADP to ATP. The remainder will be reconverted to glucose, thence to glycogen and returned to storage in muscles and liver, processes which require energy and hence use ATP.

c. Condensation of ADP. Under some conditions 2 ADP may be condensed to form a molecule of ATP.

$$2\ ADP \overset{(myokinase)}{\rightarrow} ATP + AMP \qquad (7.6)$$

where AMP is adenosine monophosphate.

d. Other food compounds. Fats and proteins can also be used to provide energy for rephosphorylation of ADP. They need not first be converted to glucose but may enter the Krebs cycle at various points (Fig. 7.12). Their preparation requires reactions which are more energy-consuming and so less economical than those of glucose. Glucose appears to be the preferred fuel for most cells, but there is some evidence that

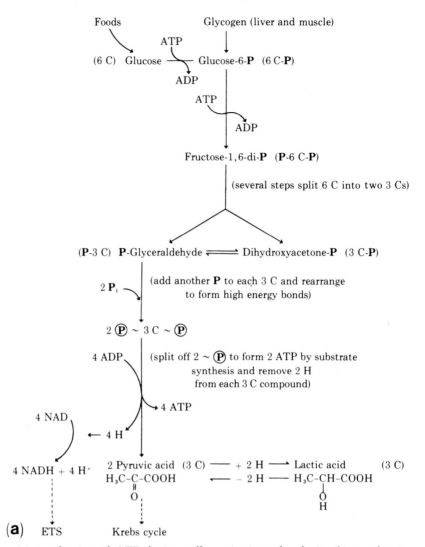

Figure 7.12. (a), production of ATP during cell respiration: glycolysis. (b), production of ATP during cell respiration: Krebs cycle. (c), production of ATP during cell respiration: ETS. In (a) and (b) the number of carbon atoms in the molecule are shown in parentheses. During cell respiration each molecule of glucose yields hydrogens with their electrons as follows:

Glycolysis	4 H	(2 pairs)
Pyruvic acid	4 H	(2 pairs)
Krebs cycle	12 H	(6 pairs)
Krebs cycle (succinate)	4 H	(2 pairs)
	24 H	(12 pairs)

All electrons except those removed from succinate are picked up by NAD and passed into the ETS. Those from succinate are taken up directly by FAD and enter the system at a lower energy level. As electrons pass through the ETS, energy is converted into ATP at three sites, each pair of electrons contributing to the formation of 3 molecules of ATP, except those from succinate which contribute 2.

Balance Sheet

	ATP used	ATP produced	ATP net
Glycolysis	2	4	2
Krebs cycle (succinate)		2	2
ETS (via NAD)		30 (10 × 3)	30
(via NAD)		4 (2 × 2)	4
		40	38

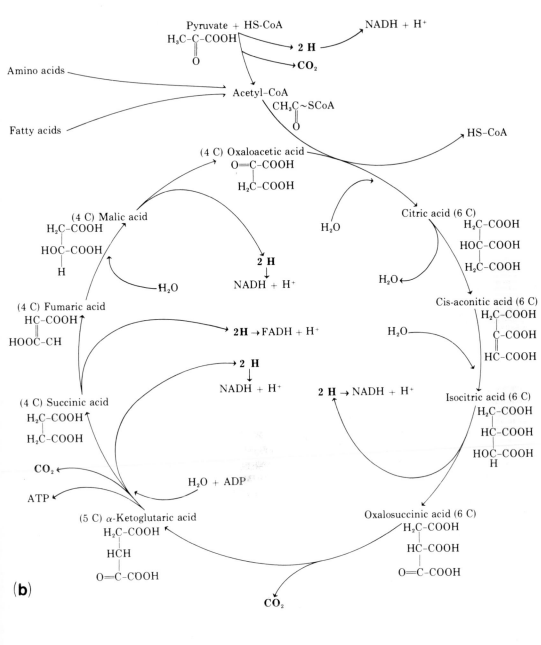

(b)

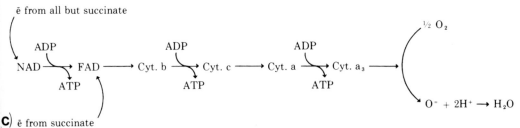

c) ē from succinate

Figure 7.12.b–c

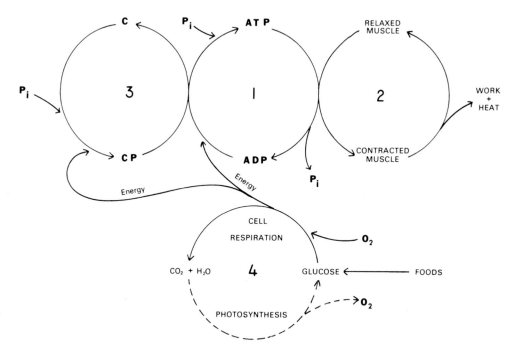

Figure 7.13. Schematic representation of metabolic interactions which maintain the muscle's supply of ATP. Circle 1: ATP is hydrolyzed to produce the energy for contraction, with a resulting production of ADP and inorganic phosphate. ATP is later resynthesized by the addition of inorganic phosphate (P_i) and of energy from creatine phosphate or from cell respiration. Circle 2: relaxed muscle becomes contracted muscle as the energy from ATP is converted into work and heat. Circle 3: creatine phosphate (CP) is hydrolyzed and contributes its high energy phosphate bond ($\sim$P) to ADP to restore ATP. Uptake of P_i and energy later reconstitutes CP. Circle 4: foods are metabolized by cell respiration, which uses O_2 and yields energy, CO_2, and H_2O. The energy is stored in the high energy bonds of ATP and CP. Photosynthesis completes the cycle by resynthesizing glucose from CO_2 and H_2O with O_2 as an end product.

a small amount of fat may be simultaneously metabolized. If the diet is deficient in carbohydrate, then fat provides the main source of fuel. Protein is not normally used, except during starvation.

Our discussion of cellular metabolism has been brief and purposely nontechnical. For further details of the chemistry of glycolysis and the Krebs-ETS reactions, the student is referred to the flow chart included here (Fig. 7.12) or to a textbook of biochemistry. Figure 7.13 presents a simple conception of the interrelations of the various chemical mechanisms mentioned.

Heat Production in Muscle

Only about 40% of the energy resident in glucose is captured as ATP-stored energy, and of the chemical energy released from the ATP during contraction only about 30% can be converted into external work. This metabolically useful energy may be called the **available energy**. The rest appears as heat and represents energy wasted as a result of the inefficiency of the chemical and physical processes.

For almost three decades the work of A. V. Hill has dominated thinking with regard to the heat production associated with muscular activity. Although some of the recent literature has raised questions concerning some points, it still is reasonable to divide the heat released by an active muscle into two major portions: the **initial heat** which appears during the contraction and the **recovery heat** which appears after relaxation. The ratio of initial to recovery heat is the same for twitch and tetanus, whether isotonic or isometric.

Initial Heat. The initial heat can be further divided into activation heat, shortening heat, and relaxation heat.

Activation Heat. The activation heat is produced upon stimulation and is associated with the appearance of the active state. It is probably related to the breakdown of ATP to initiate contraction and perhaps also to the thermal effects of the release and movement of calcium. It is the basal heat production, appearing whether the muscle shortens or not. It is independent of muscle length and tension. In tetanic contraction activation heat is sometimes called the **maintenance heat**.

Shortening Heat. Shortening heat is the extra heat produced when the muscle is permitted to shorten. It is absent in isometric contraction. Authorities seem to agree that the amount of shortening heat is proportional to the distance shortened and that the rate of its production is a linear function of the velocity of shortening. There is some disagreement as to whether or not it is also load-dependent.

Relaxation Heat. As tension subsides a portion of the initial heat can be identified as relaxation heat. It may reflect the release of the energy which was stored in the elastic components during the development of tension.

Recovery Heat. Recovery heat constitutes a larger fraction of the total heat than does the initial heat. It is produced more slowly and over a relatively long period of time following contraction. It represents heat loss during the reconstitution of ATP by interaction with creatine phosphate and by cellular respiration, and during the resynthesis of glucose and glycogen from lactic acid. It can be subdivided into an anaerobic portion related to glycolysis and a larger aerobic portion which varies with the amount of energy expended (work done) and which reflects the oxidative reactions of the Krebs cycle and electron transport system.

Regulation of Muscle Metabolism

Muscle metabolism is self-regulated. When oxygen is adequately available, glycolysis is depressed, but during vigorous muscular activity the accumulation of reduced coenzyme and other end products accelerates glycolysis. An increase in cellular levels of ADP acts as a potent stimulator for metabolism. Therefore, the rate and amount of ATP used automatically determine the rate and amount of its resynthesis.

Excitation-Contraction Coupling

For many years a gap existed in our knowledge of the means by which the electrical potentials traveling along the surface of the muscle fiber could excite the myofibrils, some of which were deep in the fiber as far as 50 μ away. The time lapse between the spike potential in the sarcolemma and contraction of the sarcomeres is not sufficient to permit a substance to diffuse such a distance (Peachey and Porter (1959)). For some time it was suspected that there must exist some form of rapid transmission to conduct excitation inward to the myofibrils. Because the triads were oriented in so definite a relationship to the cross-striation pattern, it was reasonable to think that they might be involved. Huxley and Taylor (1958) showed that differences among species regarding the location of triads correlated with the trigger points for artificially producing local shortening of sarcomeres. For example, current passed through microelectrodes produced local shortening within a frog muscle fiber only when the stimulus was applied at the region of the Z lines, and the triads were most abundant here. In crab and lizard muscle, however, contraction was induced only when stimulation was applied at the A-I junction, and this was the area of triad concentration for muscles of these species. Other species exhibited a similar agreement between local activation of the sarcomere and the distribution of triads.

Abundant evidence now indicates that the T system is separate from the sarcoplasmic reticulum and continuous with the sarcolemma and suggests that excitation is conducted inward by this channel. There is strong support for this theory. The presence of sodium within the T tubules indicates that typical action potentials may be the manner of transmission.

To link excitation and contraction, excitation must trigger the events of fiber activation which lead to tension development. Calcium, which plays an essential role in initiating the process of contraction, has been shown to accumulate in the sarcoplasmic reticulum at the site of the triads, probably having passed along the T tubules from the extracellular fluids and across the triadic junctions between tubules and sarcoplasmic reticular cisterns. Action potentials traveling along the T tubules are thought to cause the release of Ca^{++} from the cisterns into the sarcoplasm surrounding the myofibrils. The free Ca^{++} then catalyzes the formation of actomyosin; myosin ATPase is thereby activated and splits ATP. The chemical energy thus derived is converted into mechanical energy and the sarcomere contracts. As long as stimulation continues, Ca^{++} continues to be released and contraction is maintained.

Sliding Filament Theory of Muscular Contraction

When electron micrographs are taken of a **stretched** muscle, its sarcomeres are seen to be longer than in resting muscle but their A bands are unchanged. Therefore, stretching has not lengthened the thick myosin filaments. The length of the thin actin filaments also is unchanged, as shown by the constancy of the distance from the H zone of one sarcomere through the Z line to the H zone of the next sarcomere. The H zones, however, have in-

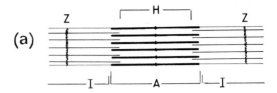

(a)

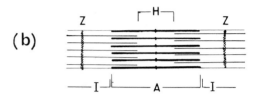

(b)

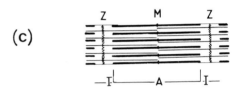

(c)

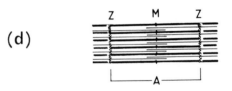

(d)

Figure 7.14. Diagrammatic representation of filament relationships in a sarcomere under various conditions. (a), stretched; (b), relaxed; (c), moderately contracted; (d), strongly contracted. A, A band; H, H zone; M, M line; I, I band; Z, Z disc.

creased and the I bands have lengthened. Apparently the extent of overlap of the two types of filaments has decreased while each filament has maintained its own integrity (Fig. 7.14, compare (a) and (b)).

In a moderately **contracted** muscle, filaments and hence A bands still retain their original lengths but changes are found in other parts of the striation pattern. Sarcomeres are shorter, and H zones and I bands have diminished. The area of overlapping of thick and thin filaments has increased (Fig. 7.14, compare (b) and (c)).

An abundance of electron microscopic evidence has led to the theory that the band pattern changes described above are due to the sliding of the filaments past one another, and that contraction is produced by the creeping of the thin actin filaments along the thick myosin filaments, the motion being mediated by chemical interactions between the filaments. This is the **sliding filament theory**, suggested practically simultaneously by two different groups of investigators, the first composed of A. F. Huxley and R. Niedergerke (1954) and the second composed of H. E. Huxley and J. Hanson (1954). The theory, which has been supported by experimental evidence of many kinds, is now accepted by most physiologists.

Definitive statements cannot yet be made regarding the interactions of actin and myosin by which the energy of ATP is translated into shortening within the sarcomere, but the process is known to be localized at the cross-bridges. It is here that appropriately directed forces are developed. It has been proposed that the myosin bridges are able to oscillate back and forth, shifting their attachments from site to site along the actin filaments. The orientation of HMM molecules in the myosin filaments suggests that bridge action exerts forces which are directionally oriented toward the center of the sarcomere. Therefore, their oscillations will draw the thin actin filaments farther into the A bands, shortening the sarcomere and exerting tension.

X-ray diffraction studies indicate that there is substantial movement of the HMM bridges during contraction, and many theories have been devised to account for bridge movement. Huxley (1969) suggests that displacement between the two halves of the HMM-S_1 may generate the mechanical force to move the actin filament a finite distance. The bridge then swings back, attaches to another site, and draws the filament farther into the A band. The cycle, repeated a specific number of times

by many bridges, results in a shortening of the contractile material and the development of tension at the ends of the muscle fiber.

As the actin filaments advance farther into the A bands, the H zones will be reduced in size. The maximal number of bridges will be able to attach when the tips of the actin filaments have reached the outer borders of the pseudo-H zones. As further contraction advances the filaments first to the M line and then beyond it, actin filaments from the two opposite sides of the sarcomere will meet and then pass by, producing a **double overlap** of thin with thick filaments (Fig. 7.14(d)). In strong contractions extreme double overlapping will cause the H zone to be replaced by an area which is denser than the outer areas of the A bands where only single overlap occurs. Electron micrographs of cross-sections taken through the sarcomere mid-region under such conditions show twice as many thin filaments as are otherwise present. Even in maximally shortened muscle, however, the pseudo-H zones are still distinguishable, remaining constant in location and dimensions and lighter than adjacent portions of the double overlap regions. This is consistent with the absence of HMM bridges in the pseudo-H zone where the tails of the oppositely oriented myosin molecules compose the A band center. In supercontraction Z lines appear to butt against the ends of the thick filaments and the thin filaments are in maximal double overlap within the A bands.

When in strong contraction the actin filaments are pulled into the sarcomeres far enough to enter the bridge areas of the opposite halves of the A bands, it is reasonable to suppose that their orientation with respect to these bridges will be abnormal. Any interaction which might occur between the filaments could not be expected to contribute to tension development and would probably oppose it. If the bridges attach, their action will be oppositely directed to that for tension production. At the least one would expect the intrusion of such "wrong way" filaments to interfere physically and/or chemically with the normal interaction of bridges with their own filaments. Double overlap with its probable consequences has been suggested as a significant factor in the observed reduction of tension capacity in a shortened muscle.

The shortening of sarcomeres adds up to produce shortening of myofibrils, which in turn sums to produce shortening of the muscle fiber. Contractile tension of the muscle, therefore, represents a summation of the short-range forces acting at multiple bridges between the myosin and actin filaments. Each cycle of a bridge contributes its small part to the production or maintenance of tension.

Action of the bridges cannot be synchronous. While some are pulling, others must be just attaching and still others detaching to shift to a new site of attachment. The peak tension at any moment reflects the average number of bridges which are active at that moment.

Relaxation

Physical Changes in Relaxation

When muscle fibers no longer receive impulses from their motor neurons, they relax. Muscular relaxation is completely passive. It is basically a cessation of tension production and may or may not be associated with lengthening of the previously shortened muscle. Muscle fibers are incapable of lengthening themselves actively. If gross lengthening occurs, it is brought about by a force outside the muscle itself, such as gravity, contraction of antagonists, or assumption of a load. However, as the bridges detach at relaxation, the internal elastic force which was built up within the fibrils during contraction is released. Recoil of the elastic components then restores the fibrils to their uncontracted lengths. The passive lengthening of the fibrils will be concomitant with the slipping of the actin filaments away from the centers of the sarcomeres.

Chemistry of Relaxation

The chemical reactions associated with relaxation are still incompletely un-

derstood. Since relaxation is a reversal of contraction, changes in the conditions which mediated contraction are required. It has been shown by various workers that relaxation requires the absence of Ca^{++}; the presence of Mg^{++}, K^+, and unsplit ATP§; the presence of a **tropomyosin-troponin system**; and the presence of a substance known as **relaxing factor**, which may be extracted from muscle microsomes. Possibly creatine phosphate and other substances are also important.

When stimulation ceases, Ca^{++} is rapidly removed from the sarcoplasm. Evidence suggests that it is returned to storage by some process which involves relaxing factor. When relaxing factor is added to muscle fibers in vitro, calcium accumulates in the sarcoplasmic reticulum.

It has been suggested that relaxing factor may act as a Ca^{++} pump, i.e., as an active carrier to transport Ca^{++} from the sarcoplasm into the cisterns of the sarcoplasmic reticulum against the chemical gradient. In the absence of Ca^{++} the bridges let go and actomyosin dissociates. Without actin, the ATPase capacity of myosin becomes inactive and in fact is then inhibited by Mg^{++}. No more ATP is split and no energy is supplied. Contraction ceases; the muscle relaxes.

ATPase in a resting muscle appears to be held in an inactive state by a complex process related to the tropomyosin-troponin system. Ebashi and colleagues (see Huxley (1969)) have suggested that tropomyosin is bound to actin and that troponin combines with the tropomyosin-actin complex (Fig. 7.15). Since troponin will not combine with myosin, its presence inhibits the attachment of cross-bridges, thus preventing formation of actomyosin and the activation of ATPase. It is suggested that myosin and tropomyosin compete with each other for some active site on the actin filament. When as a result of excitation Ca^{++} becomes available in sufficient concentration, it binds to troponin, causing a shift in the position of tropomyosin toward

§ The absence of ATP inhibits both contraction and relaxation and results in rigor.

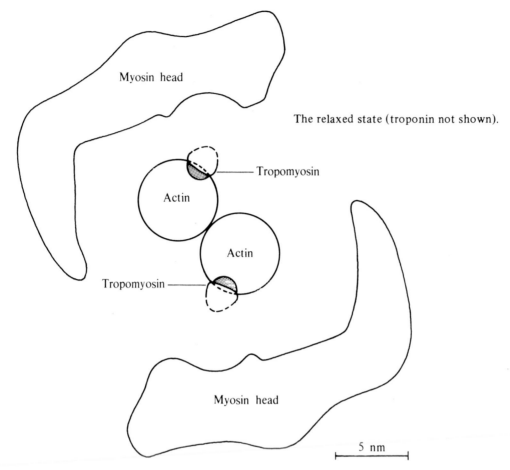

The relaxed state (troponin not shown).

Figure 7.15. The relaxed state (troponin not shown) in cross-section. (From Tribe, M. A., and Eraut, M. R., 1977. *Nerves and Muscles* (Basic Biology Course, Book 10), p. 165. Cambridge, England: Cambridge University Press.)

the groove, thus uncovering the site on the actin molecule (Fig. 7.16). Actomyosin is formed, ATPase is activated, and contraction follows.

During contraction the relaxing factor may still be actively pumping calcium back into the sarcoplasmic reticulum, but the release of Ca^{++} will exceed its storage until cessation of stimulation terminates that release. The roles of Mg^{++} and K^+ in relaxation are uncertain at present. It has been suggested (Bárány, (1967)) that, while Mg^{++} is an activator of the ATPase capacity of actomyosin, it is an inhibitor of the function in myosin alone. Potassium may function as an activator of the calcium pump mechanism.

It is interesting to note that the primary function of the tropomyosin-troponin system in the relaxed muscle fibril may be to inhibit ATPase activity. It would thus maintain relaxation by an inhibition of contraction, that inhibition being nullified by the release of Ca^{++} at the onset of stimulation. It is not unusual for a physiological process to be under such negative control. Resting heart rate and blood coagulation are two other examples. Some endocrine glands are controlled in this manner, and embryological differentiation

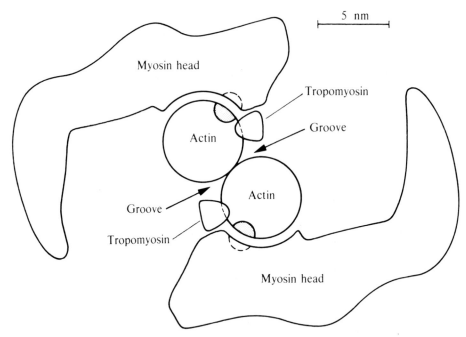

Figure 7.16. Cross-section of the muscle filaments (troponin not shown) just before the binding of actin and myosin filaments. (From Tribe, M. A., and Eraut, M. R., 1977. *Nerves and Muscle* (Basic Biology Course, Book 10), p. 164. Cambridge, England: Cambridge University Press.)

probably involves inhibition and disinhibition of specific gene systems at appropriate times.

Hypertrophy and Atrophy

Needle biopsies of human skeletal muscle are now used as common practice to identify the structural and histological differences which can be attributed to the independent variables of concern. Investigations using this technique involve a wide variety of studies such as those concerned with energy release and fatigue, recruitment patterns among the three fiber types, motor development studies, effects of immobilization, and responses to strength training. MacDougall et al. (1977) studied nine healthy volunteers "under control conditions and following five months of heavy resistance training and five weeks of immobilization. . . . " Needle biopsies were taken from triceps brachii

and analyzed for concentrations of ATP, ADP, C and CP. "It was concluded that heavy resistance training results in increases in muscle energy reserves which may be reversed by a period of immobilization-induced disuse" (MacDougall et al. (1977)).

In a subsequent paper (MacDougall et al. (1978)) studied seven healthy male subjects using the same protocol. Needle biopsies were taken from triceps brachii so that cross-sectional fiber areas could be analyzed, and overt measures of strength were recorded by a Cybex dynamometer. Training resulted in a 98% increase in elbow extension strength and significant increases in both fast and slow twitch fiber areas (Fig. 7.17). Immobilization resulted in a 41% decrease in elbow extension strength and significant decreases in fiber area for both fast and slow twitch fibers (Fig. 7.18). A range of staining intensities can be seen in Figures 7.17 and 7.18, re-

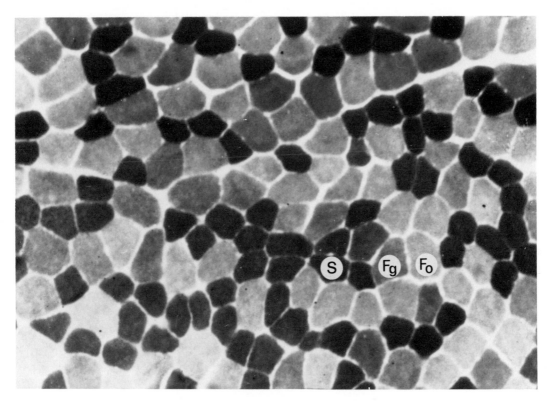

Figure 7.17. Hypertrophy of muscle: long head of triceps brachii. Magnification approximately ×200. Induced by heavy resistance weight training. Myosin ATPase, pH 4.6. Each of three fiber types is noted as follows: S, slow twitch; F_o, fast twitch oxidative; F_g, fast twitch glycolytic. (Courtesy of G. Elder, McMaster University.)

vealing the three fiber types (discussed in Chapter 4): slow twitch; fast twitch oxidative; and fast twitch glycolytic.

The histological differences displayed for hypertrophied and atrophied human skeletal muscle are of major importance to the practioner whether the application is to train the elite athlete or to rehabilitate the physically handicapped. We are only beginning to realize the implications of ultramorphological characteristics.

SUMMARY

The concepts of structure, excitation, contraction, relaxation, and hypertrophy/

atrophy which have been presented may be briefly summarized as follows.

Structure of Skeletal Muscle

1. Skeletal muscle is composed of two types of structural components: active contractile elements and inert materials. The arrangement of the fibers varies and is related to the function of the muscle concerned.

2. Muscles of every type—striated, smooth, and cardiac—contain actomyosin, and actomyosins prepared from different muscles react with ATPase in the same manner. Furthermore, no differences have been found

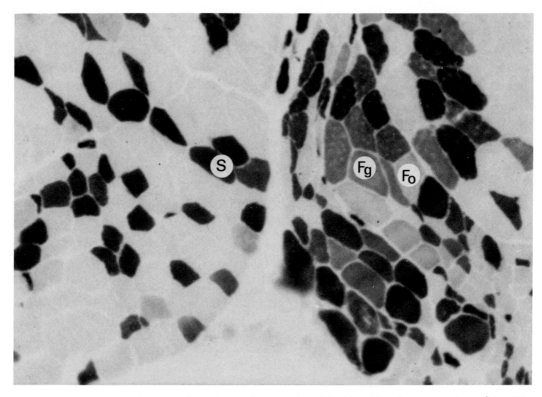

Figure 7.18. Atrophy of muscle: long head of triceps brachii. Magnification approximately ×200. Induced by 5 weeks in a cast. Myosin ATPase, pH 4.6. Normally, this reaction demonstrates only three intensities. With atrophy, the range of intensities of reaction increases. The illustration shows many fibers of all three fiber types severely atrophied. However, some fibers are relatively spared, the reason for which is unknown. Each of three fiber types is noted as follows: S, slow twitch; F_o, fast twitch oxidative; F_g, fast twitch glycolytic. (Courtesy of G. Elder, McMaster University.)

in the ATPase from different muscles or species.

3. In all types of muscle the molecules of the contractile proteins are grouped into filaments which are thick enough to be seen in electron micrographs.

4. For striated muscle there is as yet no convincing evidence that the filaments themselves shorten, but there is abundant evidence which is compatible with the sliding filament theory.

5. Tropomyosin is a constituent of the contractile systems of all muscles. Of the two kinds of tropomyosin which have been recognized, one is common to all muscles.

Excitation

6. Nerve impulses arriving at axon terminals of motor neurons cause the release of a transmitter substance which stimulates the membrane (sarcolemma) of the muscle fiber.

7. If excitation of the sarcolemma is adequate, action potentials are generated and sweep over the fiber surface.

8. The action potentials travel into the muscle fiber over the T tubules to the triad sites (the A-I junctions in mammals) of the sarcomeres.

9. At each triad, arrival of action poten-

tials triggers the release of Ca^{++} from storage in the adjacent cisterns of the sarcoplasmic reticulum, and the active state begins to develop.

Contraction

10. Ca^{++} combines with the tropomyosin-actin-troponin complex and nullifies the inhibitory action of troponin.
11. The HMM-S_1 heads of the myosin bridges attach to the active sites on the actin filaments, forming the contractile complex actomyosin.
12. Actin activates the ATPase capacity of the HMM-S_1.
13. ATPase splits ATP and energy is obtained for the movement of the bridges. The active state is now fully developed.
14. Actin filaments are pulled into the A bands for a finite distance with each bridge cycle. Shortening of the contractile material stretches out the elastic components and tension develops.
15. As long as nerve impulses continue to arrive at the neuromuscular junctions and excitation is conveyed to the triads, Ca^{++} will continue to be available, more ATP will split, the active state will persist, and contraction will be maintained by repeated cycling of the bridges.

Relaxation

16. When action potentials cease to arrive at the triads, Ca^{++} is unbound from troponin and actively returned to storage in the cisterns of the sarcoplasmic reticulum.
17. The inhibitory influence of troponin is re-established, the bridges let go, and actomyosin dissociates.
18. Since the ATPase capacity of HMM-S_1 is inactivated, no more ATP is split. No energy is supplied, and the active state declines rapidly.
19. The elastic components of the muscle recoil and draw the thin filaments back to their uncontracted state. The muscle is relaxed.

Hypertrophy and Atrophy

20. Hypertrophied muscle shows increases in muscle energy reserves; atrophied muscle demonstrates a decrease.
21. Hypertrophied muscle displays significant increases in fiber areas commensurate with increased strength; atrophied muscle shows the reverse.

CHAPTER 8

Neural Basis of Behavior

Skeletal muscles are under the control of the nervous system which determines which muscles shall contract, when, how fast, to what extent, and with what changes in force and velocity from moment to moment.

THE NEURON: MORPHOLOGY AND PHYSIOLOGY

Morphology of the Neuron

The nervous system is composed of two types of cells: **neurons** and **neuroglia**. We are concerned only with the first of these. The neuron is the functional unit of the nervous system. Each neuron possesses "in miniature the integrative capacity of the entire nervous system" (Noback and Demarest (1972)).

Neurons are usually greatly elongated cells with axon diameters ranging from 0.5 μ in small unmyelinated fibers to 22 μ in the largest myelinated fibers. Some axons exceed 1 m in length. Diameters of cell bodies range from 10 to 50 μ. Neurons are specialized to receive, conduct, and transmit excitation.

A generalized neural cell or neuron consists of four morphologically and physiologically distinct portions: a receiving pole, a terminal transmitting pole, an intervening conducting segment, and a cell body or **soma**. Each is specialized for its particular role in the cell's function. Neurons possess two types of protoplasmic processes extending outward from the nucleated soma: **dendrites** and **axons**. The processes vary in length and in the amount and extent of their branching. Dendrites are usually multiple, short, and highly branched. The space occupied by their three dimensional spread is often extensive. They constitute the receiving pole of the cell. Axons are usually single, long and, although one or more collateral branches may occur, they are relatively unbranched except at their ends. The axon is responsible for both conduction of excitation and its transmission to other cells.

An axon generates **action potentials** (nerve impulses) and conducts them from the receiving portion of the cell to the transmitting region. It is a delicate cylinder of neural cytoplasm with a limiting membrane, the **axolemma**. It varies in length and in diameter in different types of neurons. Axons are enclosed in a cellular sheath of lipid material, the **myelin sheath**, which serves to electrically insulate individual axons from one another and from adjacent neural components. The myelin sheath is formed by concentric wrappings of membranous processes from sheath cells, called **oligodendrocytes** within the central nervous system and **Schwann cells** in the peripheral nervous system. Small axons which are invested by only a single layer of sheath cell process are called "unmyelinated" fibers. Larger axons are enclosed in increasingly more numerous sheathing layers formed by more and more windings of sheath cell processes. As the folds become tightly packed together, most of the cytoplasm is squeezed out so that the sheath is ultimately composed of spiral layers of lipid-rich cellular membrane. The myelin sheath of the larger axons is segmented rather than continuous, and each segment is enclosed by a single sheath cell. The length of segments and the thickness of the myelin are quite consistent for neurons of a given caliber, larger axons having longer segments (1 to 2 mm long) and thicker sheaths. The diameter of myelineated fibers usually ranges from 1 to 20 μ. The segments are separated by short (less than 0.1 mm) unmyelinated gaps, the nodes of Ranvier. Collateral branches, when present, arise at nodal gaps and exit from the parent axon at approximately right angles (Fig. 8.1).

The area of axon outgrowth from the nucleated portion of the cell is known as the **axon hillock**. Nerve impulses are generated in the initial segment of the axon which, even in myelineated fibers, is unmyelinated. Axon ends usually divide distally into a spray of terminals, the **telodendria**, which lose the myelin sheath and end in naked tips.

The cell body or soma of the neuron is the metabolic center of the cell where, under control of its single nucleus, proteins and other metabolically important substances (enzymes, transmitter substances, neurohormones, etc.) are elaborated. The fact that materials are moved from the cell body into and along the neuronal processes has been well established. There are probably three channels for axoplasmic transport: the endoplasmic retic-

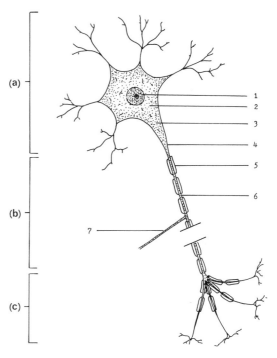

Figure 8.1. Multipolar neuron. (a), dendrites of the receiving pole. (b), axon or conducting segment. (c), telodendria and terminal arborizations of the transmitting pole. 1, nucleus; 2, cell body or soma (1 and 2 comprise the metabolic center of the cell); 3, axon hillock; 4, initial segment of axon; 5, myelin sheath segment; 6, node of Ranvier (naked membrane exposed); 7, collateral branch of the axon.

ulum; the microtubules; and the neurofilaments (McComas, 1977). When severed from the nucleated portion of the cell, a nerve process will soon degenerate because it is no longer supplied with essential materials, and a new process will grow out from the cut stump which is still attached to the cell body.

The location of the soma is the distinguishing feature among nerve cells. In vertebrates the nucleated portion of most neurons, including motor neurons and interneurons, is a part of the receiving region of the cell. The somata of sensory neurons from the skin, however, have been displaced centripetally along the course of the axon where they are better protected

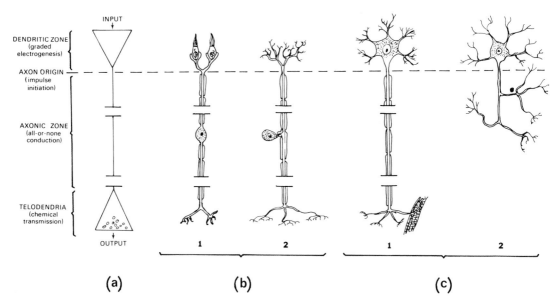

Figure 8.2. Types of neurons. (a), diagrammatic representation of the three functional portions of the neuron, showing the dendritic or receiving pole where excitation causes graded electrogenesis; the axonic or conducting segment which conducts the all-or-none impulses originating in the initial segment; and the synaptic or transmitting pole where excitation is transmitted by chemical means from the tips of the telodendria. (b) and (c): representative types of neurons. (b), **receptor neurons:** 1, special sensory neuron; 2, cutaneous neuron. Note that the nucleated portion is situated along the course of the axon. These cell bodies are located in the dorsal root ganglia or in a homologue of a dorsal root ganglion. (c), **synaptic neurons:** 1, motor neuron; 2, interneuron. These are the most common types in the nervous system. Note that the nucleated portion is located in the receptor pole of the neuron. (After Dowling, J. E., and Boycott, B. B., 1965. *Cold Spring Harbor Symp. Quant. Biol. 30:* 393.)

from injury than if situated peripherally with the receiving structures (Fig. 8.2). In these cells the dendrites communicate directly with the axon, and the cell body does not participate in the reception of excitation. In other words, the part of the fiber, often myelinated, which conducts *toward* the cell body and the portion leading *from* the cell body are both parts of the axon. Only the peripheral terminals are dendrites, while the central terminals are telodendria.

Mitochondria are present in axons, especially at the nodal areas and are numerous in the cell body and in both the receiving and transmitting portions of the neuron, being abundant in the latter. Ribosomes are mostly restricted to the cell body. Minute and unique neurofilaments are distributed throughout the cytoplasm.

Normally, excitation is conducted only from the receiving to the transmitting pole of the cell. This polarity results from nerve cells stimulated at the receiving end. An axon which is stimulated at a point along its length is capable of conduction in both directions. (Conduction of impulses in a direction opposite to the normal is referred to as **antidromic** excitation; for conduction in the normal direction, the descriptor is **orthodromic.**)

Neurons may be classified as either receptor neurons or synaptic neurons on the basis of the type of input which they receive. Receptor neurons are those which receive and transduce environmental energy such as light, sound, heat, or chemical or electrical energy. They are specialized to be excited by specific types of stimuli, and their dendritic portions are appropriately modified in structure (Fig. 8.2(b)). Synaptic neurons (Fig. 8.2(c)) receive information from other neurons by means of synaptic transmission. Their dendritic geometry may be extensive and complex, providing a wide field for reception of a great number and variety of inputs, all of which are already encoded in the manner characteristic of nervous system communication.

Physiology of the Neuron: Membrane Theory of Excitation and Conduction

Irritability is a characteristic property of all protoplasm, but types of cells differ widely in the extent to which they display the property. It is most highly developed in nerve and skeletal muscle cells. Excitation is induced in a cell by appropriate stimulation and is associated with chemical and electrical changes which spread over the cell membrane. In many types of cells the change is graded and spreads decrementally from the point of stimulation. In nerve axons and muscle cells, however, if the stimulus is adequate, the change is conducted without decrement as an all-or-none **action potential**. Adequate spread of excitation evokes the response which is characteristic of the cell.

In a muscle cell the response is contraction; in a gland cell, secretion. The essential function of a nerve cell is to transmit excitation to other cells, and it responds by releasing a chemical **transmitter substance** at its synaptic terminal. Neurons may be artificially excited by a number of different kinds of stimuli. The normal stimulus for synaptic neurons is the action upon their membranes of chemical transmitters released by other neurons. Stimulation of receptor neurons is normally provided by chemical, thermal, mechanical and electromagnetic energies. In a few instances, rare among the vertebrates, a neuron is stimulated by direct electrotonic stimulation from another neuron.

Resting Membrane Potential

Action potential generation and conduction in the neuron axolemma and the muscle sarcolemma are essentially identical. The present discussion describes these events as they occur in the neuron.

The axon membrane in the unexcited or resting state is polarized as a result of a differential distribution of ions on the two sides of the membrane. The **cations** potassium (K^+) and sodium (Na^+), the **anion** chloride (Cl^-), and certain organic anions are the ions most importantly concerned. Ions are distributed so that high concentrations of K^+ and protein ions are found intervally and high concentrations of Na^+ and Cl^- ions are located external to the membrane. The differences in ion concentrations reflect the selective permeability of the membrane. The resting membrane is at least 10 times more permeable to K^+ and Cl^- than it is to Na^+ ions. Hence, while K^+ and Cl^- pass easily through the membrane, Na^+ passes only with difficulty. Despite this fact, a small amount of Na^+ leaks in but it is promptly ejected from the cell by an active transport mechanism called the **sodium pump**.

Similarly, K^+ pumps are mainly responsible for compensating for the K^+ "leaks" so that the resting state of cell is maintained. Na^+ and K^+ pumps have been shown to be powered by the free energy from adenosine triphosphate (ATP) breakdown.

As a consequence of the semipermeability of the membrane and of the sodium pump, sodium accumulates in the intercellular fluid outside the cell in a concentration which is about 10 times greater than that inside the cell. Potassium is about 30 times more concentrated inside the cell than outside, and the chloride concentration is about 14 times greater outside than inside. This unequal distribution of an ion on two sides of a permeable membrane causes chemical and electrical forces to act on that ion. First, the chemical force results from a chemical gradient whereby ions diffuse through the membrane away from the area of their own greater concentration into the area of their own lesser concen-

inside

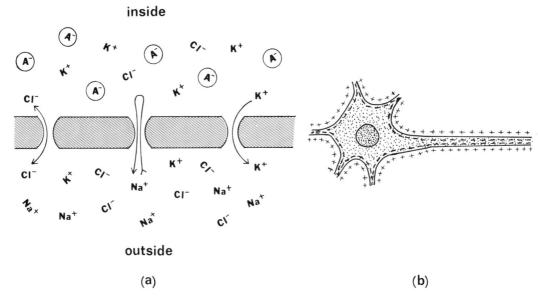

outside

(a)   (b)

Figure 8.3. Resting membrane. (a), because of permeability properties of the membrane, K^+ and Cl^- pass readily through the membrane. Na^+ passes with difficulty and is promptly ejected by the sodium pump mechanism. As a result, K^+ accummulates inside the cell and Na^+ outside. Organic anions, each indicated by A^- are too large to leave the cell. See text. (b), as a result of the differential distribution of ions, the interior of the membrane is negatively charged while the exterior is positively charged.

tration. Consequently, in the resting nerve cell the chemical gradients tend to drive sodium into and potassium out of the cell. The second force is electrostatic attraction whereby an electrically charged area at-tracts ions of opposite charge and oppo-sitely charged ions attract each other. The driving voltage for a particular ion is the difference between the value of the mem-brane potential and the equilibrium poten-tial* for that ion. In the resting condition

---

* The equilibrium potential is the electrical po-tential difference which must exist across the mem-brane to maintain the ionic concentration gradient. Its magnitude is just sufficient to equalize the influx and efflux for a particular ion and depends on the ratio of internal and external concentrations of the ion. Its value may be calculated from the Nernst equation, and is usually treated as a Cl^- or a K^+ potential, although the latter is the usual convention adopted by most researchers. The equilibrium poten-tial for Cl^- is identical with the experimentally mea-sured resting membrane potential of nerve and mus-cle cells.

the ionic currents balance each other ex-actly and the membrane potential remains constant.

In the resting cell Na^+, kept out by the action of the sodium pump, attracts and holds a considerable amount of Cl^-. Al-though Cl^-, being in higher concentration in the intercellular fluid, tends to diffuse into the cell, its inward diffusion force is balanced by the electrical attraction of the positively charged Na^+, and the chloride ions remain in equilibrium. Inside the cell, large, negatively charged protein mole-cules, which were formed within the cell and are too large to pass through the mem-brane, exert an electrical attraction on cat-ions. Therefore, since Na^+ cannot remain within the cell, K^+ is drawn into the cell and to a significant extent held there, ac-counting for the high concentration of po-tassium on the inside. The inward attrac-tion exerted on K^+ by the organic anions is opposed by the chemical gradient tending to drive it out. Under resting conditions, these two factors balance each other but there is a deficiency of positive ions inside the cell (Fig. 8.3(a)). Thus, the net charge on the inside of the membrane is negative, while the charge on the outside, where there is an excess of positive ions, is positive (Fig. 8.3(b)).

The resting potential is often called a potassium potential because it is due to the excess K^+ in the intercellular fluid. However, it is the active removal of Na^+ at a rate equal to the net rates of entry of K^+ and Cl^- that is the means by which the cell maintains its normal concentration differences. Other ions distribute themselves in a Donnan equilibrium† as indicated by their concentration ratios.

The electrical potential difference between the inside and the outside of the membrane of an unexcited cell is the **membrane or resting potential**‡ and ranges from 40 to 90 mV in nerve and muscle cells.

Excitation: Generation of Action Potential

The forces acting on sodium and potassium across the resting membrane are importantly involved in cell excitation. Any change in conditions producing an alteration in membrane permeability will result in movement of these ions in response to their driving forces, and the resting equilibrium will be upset. The time course of change in permeability of the cell membrane to a particular ion is expressed as **conductance** for that ion. Any change in conductance for one ion will appreciably affect that of other ions and hence alter the membrane potential. Conversely, any change in membrane potential will influence conductances. Stimula-

tion of a nerve axon increases membrane permeability, and hence conductance, to Na^+ at the point of stimulation. Sodium, driven by both its chemical and electrical potential gradients, passes into the cell. Because sodium is a positively charged ion, its entry decreases the negativity inside and at the same time and to a similar extent decreases the positivity outside. In other words the value of the resting potential is lowered toward zero; the membrane is being **depolarized**. A change of resting potential in the depolarizing direction constitutes a local excitatory state (l.e.s.) (Fig. 8.4, (a) and (b)).

The net influx of sodium is slow at first, but the depolarization caused by its entry produces a self-generative increase in permeability so that the rate of Na^+ influx and the rate of depolarization increase exponentially. This is an example of **positive feedback**. If the resting potential drops to a sufficient extent, it will reach a critical level which is characteristic and constant for each cell. In mammalian nerve and muscle cells critical depolarization levels range between 10 and 20 mV below the resting potential. At the critical level something happens suddenly which seems to throw the sodium gates wide open. Sodium rushes in to such an extent that the membrane potential in the stimulated area passes beyond zero and the polarization is reversed: the membrane becomes negative outside and positive inside at the point of stimulation. This almost instantaneous change in potential, which appears on the oscilloscope as a spike (Fig. 8.4(c)), is known as the **action potential**.§ The more intense the stimulus the sooner depolarization reaches the critical level and the earlier the spike appears.

The rising phase of the spike (depolarization) is accounted for by the influx of Na^+. Sodium conductance rises rapidly, in 0.1 to 0.2 msec, and then falls rapidly to a low level. Potassium conductance does not

---

† The conditions which exist when an irregular distribution of ions between two solutions, separated by a membrane, develops an electrical potential between the two sides of the membrane; e.g.,

$$\frac{[K^+]_i}{[K^+]_o} = \frac{[Cl^-]_o}{[Cl^-]_i}$$

‡ All biological "potentials" are actually potential differences.

---

§ The action potential is a sodium potential and leads to the designation of the hypothesis as the **sodium theory** of the nature of the action potential.

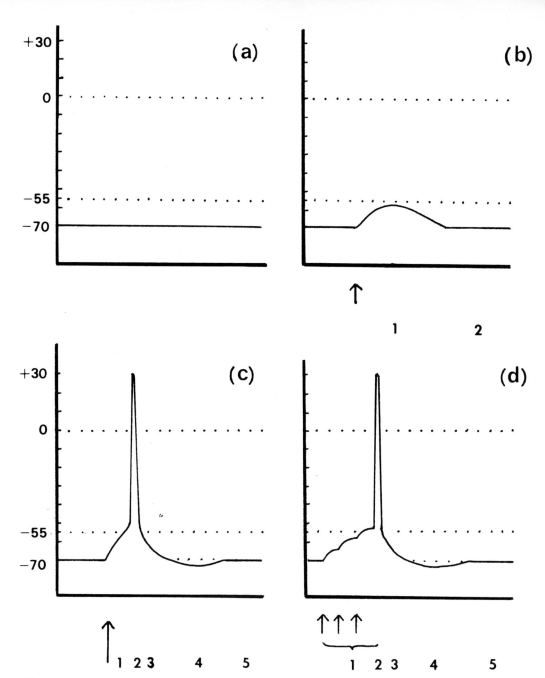

Figure 8.4. Membrane potential changes with time. Ordinate: the membrane potential in millivolts. Abscissa: time. Arrows indicate application of stimulus. (a), **resting state.** Membrane potential is −70 mV (inside). Depolarization of at least 15 mV is required to reach the critical level of −55 mV. (b), **l.e.s.** Application of an inadequate stimulus (arrow) partially depolarizes the membrane, producing a l.e.s. Since critical level is not attained (1), no spike is generated. The l.e.s. dwindles away and the resting state is restored (2). (c), **action potential.** An adequate stimulus (arrow) induces a l.e.s. (1) which quickly reaches the critical level and the membrane potential suddenly reverses to +30 mV (inside), producing an action potential of 100 mV (2). Restoration processes act promptly and start to return the membrane to its resting potential. The process slows down, producing the short negative afterpotential (3), during which the membrane is still slightly depolarized and hence abnormally irritable. This is followed by a longer lasting positive after-potential (4) which is an overshoot resulting in slight hyperpolarization and subnormal irritability. The resting potential (5) is gradually regained. (d), **summation of inadequate stimuli.** The l.e.s. of three inadequate stimuli summate (1) to reach the critical level (2), and an action potential spike is generated, followed as in (c) by negative (3) and positive (4) afterpotentials, and ultimate recovery (5).

change appreciably at first but then becomes marked as sodium conductance is inactivated. An efflux of K^+ is responsible for the falling phase of the spike (repolarization). The separation in time of these two conductance changes accounts for the change in the membrane potential which constitutes the action potential (Fig. 8.5).

The magnitude of the action potential is the algebraic difference between the resting and the active (reversed) potential values as recorded from the inside of the membrane. In Figure 8.4(c) the membrane potential which is −70 mV (inside) in the resting state is reversed to +30 mV upon adequate excitation. The action potential is therefore:

$$30 \text{ mV} - (-70 \text{ mV}) = 100 \text{ mV} \quad (8.1)$$

No action potential occurs unless depolarization reaches the critical level. If the stimulus is inadequate and does not result in the entry of enough sodium to depolarize the membrane sufficiently, the local excitatory state soon dies away (Fig.

8.4(b)). Regenerative reactions, mainly K^+ efflux, restore the resting polarization. The sodium pump gradually ejects the sodium, and eventually the resting ion distribution is regained. If, however, subsequent subthreshold stimuli are applied in such rapid succession that each succeeding stimulus evokes its l.e.s. before that of the preceding stimulus has dwindled away, the local excitatory states will summate. When the critical level is reached, an action potential will be generated. This phenomenon is designated **summation of inadequate stimuli** (Fig. 8.4(d)).

Recovery of Resting State

The action potential occupies a definite distance of the nerve fiber (5 to 6 cm in the largest mammalian neurons), lasts for a definite duration (about 0.4 msec), and then the membrane potential is rapidly returned to the resting state (Fig. 8.6). Because the action of the sodium pump is relatively slow, recovery of the resting polarization requires a faster mechanism. The outflow of K^+, as its conductance increases, brings about the almost immediate recovery of positivity outside and negativity inside. Potassium is driven outward by its chemical gradient and also by the reversal of the electrical force as the cell interior becomes positively charged. Ion distributions will be eventually restored by the slower action of the sodium pump. It has been suggested that the same carrier which actively transports Na^+ out of the cell brings K^+ back.

Conduction of Action Potential

The action potential which is generated at the site of stimulation must be conducted along the fiber to the axon terminals. Conduction is accomplished by self-propagation of the disturbance away from the site of origin. Adjacent inactive areas on the outside of the membrane are still positively charged at the resting level. Since the extracellular fluid is electrolytic, a small current, the **action current**, flows from the positively charged inactive region to the negatively charged active area where it passes in through the membrane,

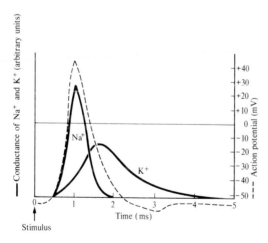

Figure 8.5. Diagram to illustrate how the sodium and potassium conductance of a nerve fiber changes as an action potential passes the recording electrodes (solid lines). The form of the action potential is given by the dotted lines (From Tribe, M. A., and Eraut, M. R., 1977. Nerves and Muscle (Basic Biology Course Book 10), p 69. Cambridge, England: Cambridge University Press.)

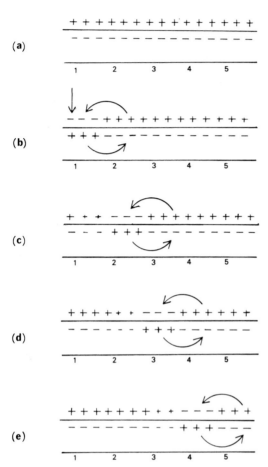

Figure 8.6. Electrical changes associated with excitation, conduction, and recovery in an unmyelinated cell membrane. (a), **resting membrane**. Resting membrane is positively charged on the outside and negatively charged on the inside. (Charges are shown on only one surface of the profile. Numbers indicate sequential areas of the membrane.) (b), **excitation**. An adequate stimulus (arrow) results in reversal of polarity (negativity) at the point of stimulation (area 1). An action current then flows from positively charged (outside) inactive area 2 to negatively charged (outside) active area 1, into and along the inside of the membrane, emerging through the inactive area. (c) to (e), **conduction**. (c), emergence of action current stimulates the membrane of inactive area 2, which then becomes depolarized. The action current now flows from presently inactive area 3 to newly active area 2. (d) and (e), the action potential is self-propagated along the fiber to areas 3 and 4. **Recovery**. In (c), (d), and (e), the area behind the action potential gradually recovers its original polarization (positively charged on the outside).

through the cell fluids, and out again through the inactive region. Although small, the current is sufficiently strong to constitute a stimulus capable of increasing membrane permeability as it emerges. The same sequence of excitatory events is repeated here: sodium moves in, polarity is reversed at this point, and the action potential has progressed along the fiber. This is repeated again and again until the action potential reaches the end of the axon terminal (Fig. 8.6). A neuron can conduct for hours without any cessation of activity to provide for restoration of the original ionic distributions.

In unmyelinated fibers such as the autonomic postganglionic neurons and the afferent fibers for dull pain, conduction is accomplished by progression of the action potential down the fiber as described above. In myelinated fibers, however, the action current skips from node to node without depolarizing the internodal portions of the axon (Fig. 8.7). This is **saltatory conduction**, and the velocity of conduction is considerably greater than in unmyelinated fibers. Because the same exchange of ions occurs less times for a given length of fiber, saltatory conduction involves fewer ions and hence requires less energy for recovery. This is an advantage which permits myelinated fibers to continue transmitting for some time even in the absence of oxygen.

Some Characteristics of Nerve Conduction

Refractory Periods. As the action potential travels along the fiber surface, it consists of a wave of negativity followed

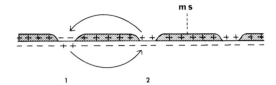

Figure 8.7. Saltatory conduction in a myelinated fiber. 1, active node of Ranvier; 2, inactive node of Ranvier; ms, myelin sheath segment. Arrows indicate flow of action current. Direction of impulse conduction is from left to right. In myelinated nerve fibers the action current skips from node to node without depolarization of internodal portions of the axon. Only a section of the surface membrane is shown in profile, with the outside of the fiber above and the inside below.

by an area of gradually recovering positivity (Fig. 8.6). While an area is in its reversed (active) state, it is *absolutely* refractory and cannot be restimulated. During recovery, the membrane is *relatively* refractory, a state which lasts many times longer than the absolute refractory period. Intense or sustained stimuli may restimulate the original site during repolarization.

The refractory periods are due to the conductance changes. Inactivation of sodium conductance decreases excitability because a greater depolarization would be required to produce further increase in Na^+ conductance to a point where its net influx would exceed the efflux of K^+. Elevation of K^+ conductance gradually restores resting polarization and hence excitability. Thus, the inactivation of Na^+ conductance and the elevation of K^+ conductance account for both the absolute refractory period and the relative refractory period.

Afterpotentials. During the relative refractory period both the amplitude and velocity of the spike are altered, reflecting changed conditions in the fiber. In some neurons the latter portion of the downward course of the spike is considerably less rapid than its rise, showing a marked concavity before reaching its initial level. This is the **negative afterpotential** because it indicates a delay in return to the resting

potential and a prolongation of some slight depolarization (hence negative). During this period of 12 to 80 msec, the membrane is hyperexcitable or supernormal and hence more easily restimulated. The recovery may continue into a hyperpolarized state, the **positive afterpotential**, which persists for a much longer time, up to 1 full second, during which the membrane is subnormal in excitability (Fig. 8.4, (c) and (d)). The positive afterpotential is probably due to a delay in the restoration of K^+ conductance to its normal level.

Frequency of Impulses. In general, natural stimuli are of sufficient duration to reactivate the membrane after the absolute refractory period. For this reason neurons normally carry trains of impulses. A single electric shock may produce a single action potential but only because its duration does not outlast the refractory period of the fiber. The stronger the stimulus the earlier it will re-excite, the shorter will be the time span between impulses, hence, the greater the frequency.

Because each action potential is followed by an absolute refractory period, action potentials cannot summate[||] but remain separate and discrete. Neurons do not conduct impulses at rates as high as the absolute refractory periods would suggest. Cognizance must also be taken of the characteristics of the relative refractory period. A fiber with a spike duration of 0.4 msec might be expected to conduct impulses at a frequency of 2500/sec but its upper limit will be closer to 1000/sec. Conduction frequencies rarely approximate their possible maxima. Motor neurons usually conduct at frequencies of 20 to 40, rarely as high as 50, impulses/sec, although, at the start of a maximum contraction, rates greater than 100 Hz have been recorded (McComas, 1977). Upper limit frequencies for sensory neurons normally lie between 100 and 200 impulses per second although auditory neurons may conduct between 800 and 1000. Information is conveyed by the presence or absence of an

---

|| The l.e.s. displays no refractoriness and hence summation is possible at subthreshold levels.

Table 8.1

Comparison of Classification Systems for Motor and Sensory Nerve Fibers

Motor*				Sensory†			
Group	Diameter	Velocity	Termination‡	Group	Diameter	Velocity	Origin (receptors)§
	μ	m/sec			μ	m/sec	
A	12–20	60–100	Muscle fibers	I	12–22	70–120	Spindle, GTO, joint
α	6–12	30–70	Axons to spindle IFs	II	6–12	30–70	Spindle, skin, joint
β	2–8	15–30	Spindle IFs				
γ	2–6	12–30	(Blood vessels?)	III	2–6	12–30	Pain
δ	1–3	3–15	ANS preganglionic				
B	0.5–1	0.5–2	ANS postganglionic	IV	0.5–1	0.5–2	Pain
C							

* Classified according to Erlanger and Gasser (1937).
† Classified according to Lloyd (1943).
‡ IF, intrafusal fibers; ANS, autonomic nervous system.
§ GTO, Golgi tendon organ.

action potential, and as well by the frequency of action potentials.

Velocity of Conduction. Velocity of conduction depends not only on myelination but, more importantly, on the diameter of the fiber. It can be fairly accurately predicted from the following equation.

Velocity in m/sec

$$= 6 \times \text{diameter in } \mu \quad (8.2)$$

Hence the largest motor and sensory nerve fibers, with diameters near 20 μ, have conduction velocities up to 120 m/sec.** In small unmyelinated fibers, velocities range from 0.7 to 2 m/sec. Large fibers not only conduct more rapidly than small fibers but characteristically have lower stimulus thresholds and larger spikes with shorter durations.

Classification of Nerve Fibers. As a result of the classic experiments of Erlanger and Gasser in 1937, nerve fibers are classified into three major groups, A, B, and C, on the basis of conduction velocities. Group C contains the unmyelinated

postganglionic fibers and group B the small myelinated preganglionic fibers of the autonomic nervous system. Group A includes the large, rapidly conducting myelinated somatic fibers. It has been further divided into four subgroups: alpha (α), beta (β), gamma (γ), and delta (δ) on the basis of velocity and diameter. The fastest fibers are those with the largest diameters. Sensory nerve fibers have been separately classified by Lloyd (1943) according to diameter into groups I, II, III, and IV, with corresponding velocities. These do not correspond exactly in size and velocity to the subgroupings of the Erlanger-Gasser class, but among afferents from the skin and muscles, group I approximates A_α and group II approximates A_β and A_γ. In order to avoid confusion, use of the alphabetical designations is restricted to efferent fibers and the numerical designations to afferent fibers. Table 8.1 provides a comparison of the two classifications as related to both motor and sensory fibers. As indicated by the columns in the table for termination of motor fibers and origin of sensory fibers, it is obvious that specific structures tend to be innervated by fibers of quite specific size and conduction characteristics.

** In mammalian skeletal muscle fibers, conduction velocity is about 5 m/sec.

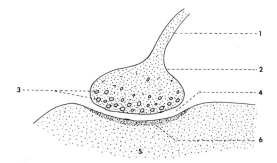

Figure 8.8. Diagrammatic representation of the components of the synapse. 1, presynaptic telodendrion; 2, bouton; 3, vesicles; 4, synaptic cleft; 5, postsynaptic neuron; 6, receptor site or subsynaptic membrane.

MORPHOLOGY AND PHYSIOLOGY OF THE SYNAPSE

While the functional unit of the nervous system is the single neuron, it is apparent that more than one neuron is traversed between receptor and effector. Usually, a chain of at least two, and more likely three or more, neurons connect receptor with effector.

Each neuron in the chain remains a separate and discrete entity. Its axon terminal ends in close proximity to the receiving structures of other neurons, but there is no protoplasmic continuity between neurons. Dendritic branches and axon telodendria interweave to form a complex network known as the **neuropil**. The region of functional contact between neurons is the **synapse**, across which excitation must be transmitted. The synapse is probably the most important aspect of neural organization; in fact, its importance cannot be overemphasized. It is responsible for the physiological continuity of conduction through the neural chains. It is also the site in the nervous system where the modification of communication occurs without which the integrated response would be impossible.

In the neuron itself, nerve impulses are transmitted in an all-or-none fashion in both magnitude and velocity, and these properties vary only with changes in the condition of the fiber. At the synapse, however, transmission is *not* all-or-none and may be amplified, reduced, or even completely blocked. As a result, the signal transmitted by a subsequent neuron in the chain may be quite different from the original input. Furthermore, blocking of some synapses and concurrent facilitation of transmission at others serve to determine the distribution of communication by directing it into specific channels.

Morphology of the Synapse

A synapse (Fig. 8.8) consists of the specialized axon terminal of the transmitting or **presynaptic neuron**, separated by a fluid-filled space (the **synaptic cleft**), from the receiving membrane of the **postsynaptic neuron**. The axon ending and the postsynaptic membrane are closely contiguous. The two apposing membranes are strongly adherent and there is evidence that they may be held together by a special **synaptic cement**.

Each telodendron terminates in a specialized unmyelinated ending which may take one of several forms. Frequently endings are bulbous swellings known as **knobs** or **boutons**, but sometimes they are diffuse arborizations which form nests, brushes, or baskets, and sometimes they are simply naked terminals which climb along a dendrite for some distance or cross it at right angles.

Presynaptic terminals contain mitochondria, neurofilaments, and numerous minute vesicles, 200 to 1000 Å in diameter, which are often clustered against the presynaptic membrane. Vesicles occur in a variety of shapes and sizes.

The synaptic cleft is continuous with the intercellular space. In width it ranges from 100 to 200 Å. It is usually occupied by vague dense material which forms a thin dark plate between the apposed mem-

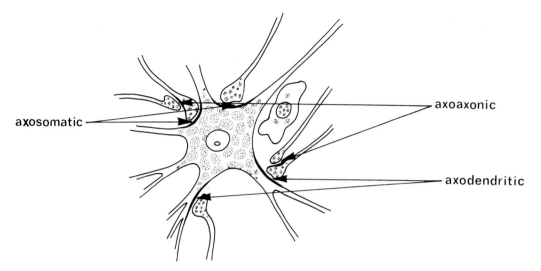

axosomatic

axoaxonic

axodendritic

Figure 8.9. Types of synapses. Synapses may be axodendritic, axosomatic, or axoaxonic (After Noback and Demarest, 1972, p. 18).

branes and is sometimes thicker on the postsynaptic side.

The contact area on the postsynaptic neuron may be called the **subsynaptic membrane** or the **receptive site**. While morphological specialization has not been clearly revealed, some electron microscope studies have shown what appear to be delicate hooklike fibrillar extensions which make contact with presynaptic fibrils in the synaptic cleft (Fernandez-Moran (1967)). Biochemical and physiological studies suggest the presence of ion-specific channels or pores and specific reactive chemical groups.

Synaptic contacts occur on dendrites and, in neurons whose nucleated portion is located within the receiving pole of the cell, on the soma. Synapses formed by contact of axon terminals with postsynaptic dendrites are **axodendritic** synapses; those formed by contact with the cell body are **axosomatic** synapses. There are also synapses in which an axon terminal makes contact with another axon terminal or even with the initial segment of another neuron; these are **axoaxonic** synapses (Fig. 8.9).

Each presynaptic neuron makes synaptic contact with many different postsynaptic neurons, often sending several telodendria to each (Fig. 8.10). Each postsynaptic neuron receives multiple axon terminals from many different presynaptic neurons: a single motor neuron in the spinal cord may have more than 1000 synapses occupying 40% of its receiving surfaces. The scope of a neuron's influence may be further extended by collateral branches of its axon which may have destinations quite different from that of the parent axon. In some cases a collateral may even turn back into the dendritic field of its own neuron as a **recurrent collateral**.

Physiology of the Synapse

Synaptic Transmission

In an active neuron, nerve impulses travel out into all of its many tiny terminal branches and into as many synapses. Abundant evidence indicates that synaptic transmission is accomplished in most instances by a chemical process. (Electrical transmission, which is known to occur in

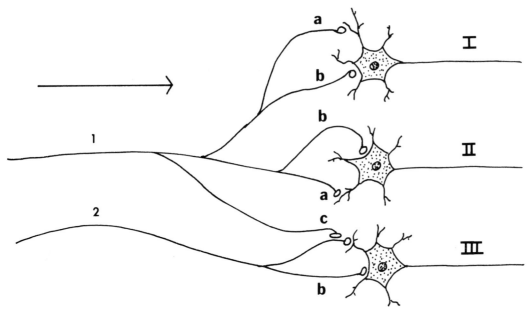

Figure 8.10. Schematic drawing of simple synaptic connections. Afferent neurons 1 and 2 make synaptic connections with interneurons I, II, and III. A collateral from the axon of neuron 1 forms synapse with a terminal of neuron 2 just before its synapse on III. Horizontal arrow indicates direction of conduction. The reader is invited to identify (a, b, or c) the synapses which are axodendritic, axosomatic, or axoaxonic.

many invertebrates such as the crayfish, squid, and annelid worm, has recently been identified in some vertebrates but is as yet unknown in mammals.) The nerve impulse itself does not cross the interneuronal gap but rather, upon its arrival at axon ends, it causes the secretion of a chemical **transmitter substance**. The minute vesicles revealed by electron microscope photographs of presynaptic terminals contain storage units of the transmitter, which may have been manufactured in the vesicles or, more likely, in the nerve cell body. Weiss (1963) and others have shown by time lapse photography and cinematography that nerve fibers are living, squirming, moving streams through which a peristaltic flow of chemical supplies is driven from the cell body at a rate of from one to a few millimeters per day. These materials may include the neurosecretions,

as well as materials to nourish and replenish the neural processes and, in motor nerves, substances which significantly influence the fast-slow response characteristics of muscle fibers.

The depolarization produced in the membrane of the presynaptic terminals by the arrival of an impulse is assumed to trigger an excitation-secretion coupling mechanism which causes the rupture of the synaptic vesicles. A quantal amount of transmitter substance is ejected into the synaptic cleft by the bursting of each vesicle. Impulses probably do not initiate transmitter release but simply accelerate a secretory process which goes on continually at a low rate. The amount released is proportional to the magnitude of the impulse, and it has been calculated that, for each 30 mV of action potential, transmitter release is increased 100-fold. An impulse probably causes all of the vesicles in immediate juxtaposition to the membrane to rupture and also mobilizes other vesicles for subsequent release by causing them to move into the strategic area.

The transmitter substance, which diffuses across the intervening space in a few microseconds, reacts with the specific chemical groups at the receptor site of the postsynaptic membrane. Bunge *et al.* (1967), and more recently Nathanson and Greengard (1977), investigating the intracellular effects of neurotransmitters, used electron micrographs to show that the transmitter substance released from synaptic vesicles is geographically associated with that part of the subsynaptic membrane possessing neurotransmitter receptors (Fig. 8.11). These sites are associated with fine channels or pores which are somehow opened by the chemical reaction to permit ions to flow through the mem-

brane at many times their normal rates. As a result, a small change occurs in the resting potential of the postsynaptic membrane at the subsynaptic site. The potential difference between this and adjacent unstimulated areas of the membrane is the **postsynaptic potential** (PSP).

Gardner (1967) derived a schematic representation of the transmitter-receptor interaction (Fig. 8.12) as it was then understood from Eccles (1964, 1965). The theme of the schematic probably still holds true today as suggested by the electron microscopy work of Nathanson and Greengard (1977), and Lester (1977) citing Heuser's work on the neuromuscular junction.

The action of the transmitter upon the subsynaptic membrane does not directly induce an action potential. The membranes of dendrites and soma (with some exceptions among dendrites of certain brain cells) are electrically inexcitable and

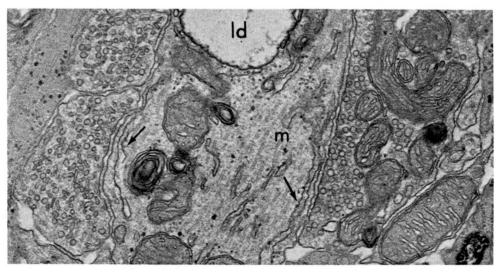

Figure 8.11. Electron micrograph illustrating synaptic junction. This electron micrograph from rat spinal cord shows a dendrite contacted by three axonal boutons. The dendrite contains microtubules (*m*), mitochondria, a large lipid droplet (*ld*) and cisterns of endoplasmic reticulum underlying its surface membrane (*arrows*). Note that the axonal terminal on the *right* contains predominantly round synaptic vesicles, whereas the two terminals on the *left* contain vesicles that are generally somewhat smaller and somewhat flattened. These differences in synaptic vesicle morphology are revealed only after primary fixation in aldehyde. × 46,000. (From Bunge, M. B., Bunge, R. P., and Peterson, E. R.: The onset of synapse formation in spinal cord cultures as studied by electron microscopy. *Brain Res. 6:* 728, 1967.)

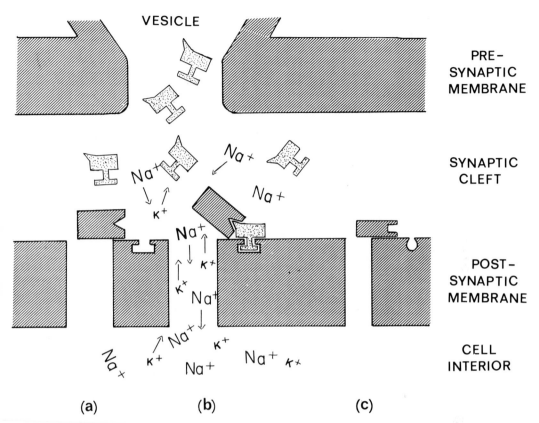

VESICLE

PRE-
SYNAPTIC
MEMBRANE

SYNAPTIC
CLEFT

POST-
SYNAPTIC
MEMBRANE

CELL
INTERIOR

(a) (b) (c)

Figure 8.12. Hypothetical explanation of synaptic transmission. A synaptic vesicle is releasing excitatory transmitter substance (stippled) which diffuses across the synaptic cleft. (a), a specific group of atoms on the postsynaptic membrane (rectangle) is so oriented that it occludes the pore which passes through the postsynaptic membrane. (b), the excitatory transmitter has already interacted with chemical groups at the receptor site, producing a change in molecular configuration which has "opened" the pore. This enables sodium to enter and, later, potassium to leave the cell. (c), a narrower channel is shown which requires a different transmitter substance, presumably an inhibitory one. (From Gardner, E. B., 1967. The neurophysiological basis of motor learning—A review. *J. Am. Phys. Ther. Assoc. 47*: 1115.)

incapable of generating action potential spikes. Therefore, the PSP is a local, graded, nonpropagated change in the resting potential which spreads *electrotonically* from the point of origin. The potential change gradually diminishes (decrements) as it spreads. The initial segment of the axon, however, is electrically excitable and has the lowest threshold of any part of the cell membrane. If the PSP is an excitatory change (depolarization) and if

its magnitude reaches the critical level of the axon membrane, an action potential will be generated in the initial segment and conducted nondecrementally over the axon. Except for the interposition of the electrically inexcitable receptor portion of the cell, the sequence of events in synaptic excitation appears to be similar to that described for direct stimulation of the axon. The chemical transmitter substances do not remain long in the synaptic cleft but are soon destroyed, each by a specific enzyme. Almost immediate destruction of transmitter is essential to neural regulation of activity because its persistence and accumulation would result in exaggerated and uncontrolled responses.

There are two types of transmitter substances, those which are excitatory and those which are inhibitory. A transmitter is excitatory if it exerts a depolarizing effect upon the postsynaptic membrane, thus bringing its resting potential toward the firing level. It is inhibitory if it decreases the possibility of firing either by hyperpolarizing the resting membrane or by stabilizing it, possibly by combining with the chemical groups of the receptor site in a way which prevents activation. A postsynaptic neuron has many synapses on its surface, some of which are excitatory and some inhibitory, and both types are often active at the same time. The constant interplay of excitatory and inhibitory activity results in a fluctuating membrane potential in the initial segment which, at any moment, is the algebraic sum of these depolarizing and hyperpolarizing influences.

Unsuccessful attempts have been made to correlate morphological differences among synapses with excitatory and inhibitory action. Some evidence indicates that excitatory synapses may have wide clefts and broad, continuous postsynaptic plates and may be located on more distal portions of dendrites, while inhibitory synapses may have narrower clefts and thinner, discontinuous plates and may be located upon dendritic trunks and soma surfaces. The situation, however, is not a simple one. Many intermediate and exceptional forms are found. In fact, some of the larger terminals show both synaptic types on the same postsynaptic dendrite. Attempts have also been made to correlate differences in the design of presynaptic endings (knobs, baskets, brushes, trails, etc.) with excitation and inhibition. At the present time no hard and fast conclusions can be drawn linking fine structure and synaptic function. However, the spatial distribution of active terminals in relation to each other and to the axon hillock may be important. Because of the decremental nature of conduction in the receptive membranes, synapses far out on dendrites should be expected to exert less influence than those closer to the cell body, and synapses on the soma near the axon hillock should be the most effective. The possibility exists, however, that large dendrites may have electrically excitable sections which could act as booster stations for their otherwise decremental conduction. Another interesting thought is that strategically placed inhibitory terminals could markedly alter the effectiveness of excitatory endings.

The chemical identity of the transmitters which act at neural junctions *outside* the central nervous system is well known. At the neuromuscular junction release of acetylcholine (ACh) by motor neuron terminals excites the end-plate membrane of the muscle fiber. Acetylcholine is released at all autonomic ganglia by the preganglionic neurons and is the transmitter at all parasympathetic and some sympathetic neuroeffector junctions. For the majority of sympathetic junctions, the transmitter is norepinephrine (nor-E).

The transmitters which operate at synapses within the central nervous system have been identified in only one instance: ACh is known to be liberated by terminals of recurrent collaterals of motor neurons at their synapses with certain cells (**Renshaw cells**) in the spinal cord. It seems certain that both ACh and nor-E may be widely involved in central nervous system transmission, and it seems equally likely that others are also concerned, especially in the brain. Candidates include γ-aminobutyric acid, histamine, 5-hydroxytryptamine (serotonin), and dihydroxyphenylalanine (dopamine), all of which are present in significant amounts; L-glutamic acid as an excitor and γ-aminobutyric acid as an inhibitor have been identified in the invertebrates.

An example of a chemical transmitter found in the brain which has a clearly defined motor function is dopamine. Victims of Parkinson's disease have long attested to the fact that their symptoms can be reversed or even eliminated when they are treated by a drug called L-dopa (*levo*-dihydroxyphenylalanine). This drug, an amino acid precursor of dopamine, when taken up by the bloodstream and converted to dopamine, supplies the amount of chemical transmitter needed by receptors in the basal ganglia. Normally, the neurotransmitter dopamine is secreted by a group of neurons which originate in the substantia nigra and project to the basal ganglia. The degeneration of this group of neurons characterizes the Parkinson patient, who displays symptoms of tremor, rigidity and delay in the initiation of movement (Nathanson and Greengard, 1977).

Variations in the shape and size of synaptic vesicles may be related to the transmitter contained. Clear, nongranular vesicles, 200 to 400 Å in diameter, probably contain ACh, and dense, granulated vesicles, 800 to 900 Å in diameter, each hold a dense spherical droplet which may be nor-E. Other vesicles differing from these may contain other transmitters.

The action of ACh is fairly well understood. Its reaction with the postsynaptic membrane produces a permeability increase which results in a rapid, localized depolarization of short duration. It is then quickly destroyed by the enzyme acetylcholinesterase (ACh-ase), which hydrolyzes it to choline and acetic acid. Destruction of the transmitter is necessary to avoid persistent and convulsive responses. Several chemical substances (e.g., eserine and neostigmine) inhibit ACh-ase, preventing destruction of ACh, and much has been learned about this neural transmitter through the use of these agents. They have also proven useful in the management of myasthenia gravis, a disease characterized by weakness and extreme muscular fatigue resulting from subnormal release of ACh by motor nerve terminals.

The classic concept of synaptic function is that each neuron releases the same kind of transmitter at all of its terminals (Dale-Feldberg law) and that the transmitter has either an excitatory or inhibitory effect on all of the postsynaptic neurons upon which it acts. The unitary nature of neuron secretion is universally accepted. There is, however, considerable evidence that the sign (+ or −) of the action of a transmitter may be determined by properties of the postsynaptic cell. In the autonomic nervous system ACh is excitatory for some effectors (for example, smooth muscle of gut and bladder) and inhibitory for others (cardiac muscle). Norepinephrine exerts both effects but oppositely in the various tissues. Furthermore, instances are known in which the effect may be reversed by hormonal influences acting on the innervated tissue. For example, the smooth muscle of the pregnant uterus is excited while that of the non-gravid organ is inhibited by ACh. Also, in some simple vertebrate nervous systems clear-cut instances have been found in which the same presynaptic neuron excites some postsynaptic neurons and inhibits others, presumably by the same transmitter (Kandel and Wechtel (1968)).

Synapses control the normal impulse traffic through the nervous system, determining the amount and pattern of information input and the consequent behavior of each neuron and group of neurons. Synaptic integrative action is based upon the interplay of antagonistic influences: facilitation and inhibition. Is it any wonder that the use of drugs is carefully monitored by our medical colleagues! We need only recognize the potent effect that results from a slight imbalance of one chemical and its predictable effect on synaptic transmission, hence, on possible behavior of the individual.

Synaptic Facilitation

Excitation in a presynaptic neuron does not necessarily result in transmission at every synapse which its terminals encounter. A certain amount of resistance is inherent in each junction and reflects the critical level of depolarization which is required to fire the postsynaptic neuron. Synaptic resistance varies from synapse to synapse and at each synapse is subject to temporary or persistant modification. If

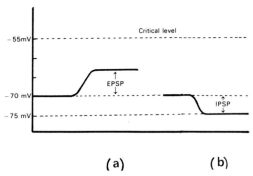

(a)   (b)

Figure 8.13. Facilitation and inhibition in synapses. The resting potential of the postsynaptic membrane is −70 mV and its critical level is −55 mV. (a), excitatory transmitter has evoked an EPSP of about 7 mV (membrane potential is now −63 mV). Further excitation equivalent to 8 mV will be required to fire this facilitated postsynaptic neuron. (b), inhibitory transmitter has induced an IPSP of about 5 mV (membrane potential is now −75 mV). Excitatory transmitter equivalent to 20 mV will be required to fire this inhibited postsynaptic neuron.

the transmitter is excitatory, the PSP which results is known to be a reduction in membrane potential, i.e., a partial depolarization. Such a decrease in the electric charge across the postsynaptic membrane is the **excitatory postsynaptic potential** or EPSP and represents a reduction of synaptic resistance toward the firing level (Fig. 8.13(a)). This is known as **facilitation**. The action of the excitatory transmitter upon the postsynaptic membrane is thought to result in a general increase in membrane permeability, an opening of all ionic pores. The most notable ion movement, however, is that of Na$^+$ because of its greater electrochemical driving force.

When a number of excitatory volleys arrive simultaneously or in close succession at several synapses of a cell, each contributes its small amount to the postsynaptic depolarization. If summation of EPSPs reaches the critical level of the neuron, a spike potential is generated in the initial segment and conducted along the fiber. The rise of the action potential wipes out the EPSP, probably by antidromic in-

vasion of the soma. However, if the total excitatory effect is in excess of the threshold level or if the presynaptic bombardment is sustained, the initial segment will be repeatedly restimulated and the postsynaptic impulse frequency will rise accordingly. The frequency of impulses in the postsynaptic axon will therefore depend upon the amount of facilitatory transmitter substance released. The greater the amount the earlier in the relative refractory period will another spike be generated.

Summation of excitatory effects occurring at a number of synapses on the same postsynaptic neuron and derived from terminals of presynaptic neurons from a variety of sources is known as **spatial summation**. The partial depolarization of the postsynaptic membrane by the concurrent subliminal inputs makes the neuron more ready to respond. As a result it may be fired by a subsequent input which alone would have been inadequate. Facilitation may also be accomplished by a high frequency of impulses arriving over a single presynaptic terminal. Such **temporal summation** is probably less effective, except perhaps at the synapses of receptor neurons. In both spatial and temporal summation, each quantum of transmitter contributes toward the possibility of ultimate firing. If a response is already ongoing, facilitatory inputs will cause amplification of the response by increasing the frequency of the postsynaptic impulse train.

Sources of synaptic facilitation include the following: multiple sensory inputs which provide summation as a result of differences in modalities and/or in topographical distribution within a single modality; proprioceptive feedback of information concerning body position or movement; and supraspinal influences from· brainstem, cerebellum, and cortex.

Synaptic Inhibition

Postsynaptic Inhibition. Although involving different transmitters, both excitatory and inhibitory synapses are presumed to have the same general manner of

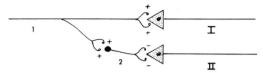

Figure 8.14. A simple inhibitory circuit. A neuron (1) synapses with an efferent neuron (I) and on a short inhibitory interneuron (2). Both of these synapses are excitatory and both neurons are activated. However, since the interneuron (2) secretes inhibitory transmitter, the efferent neuron (II) will be inhibited and will fire only if other excitatory synapses (not shown) induce sufficient EPSP to reach the critical level.

function. In both, quantal packets of transmitter are released and react at receptor sites on the subsynaptic membrane, producing a momentary increase in permeability. Eccles (1964, 1965) conjectured that the action of the inhibitory transmitter differs from that of excitatory transmitter in that it produces a selective rather than a general permeability increase by opening pores for penetration restricted to small ions. The flow of current through the membrane of an inhibitory synapse is probably due to either the outflow of K^+ or the inflow of Cl^- or both, with a concomitant increase in internal negativity (Fig. 8.13(b)). The resulting hyperpolarization is the **inhibitory postsynaptic potential** (IPSP), and it opposes the EPSP. Consequently, a greater summation of excitatory transmitter is required to lower the resting polarization to firing level. This type of inhibition is known as postsynaptic inhibition. It is the basis of reciprocal inhibition of antagonistic muscles, an essential factor in coordinated motor activity.

Excitatory input from afferent neurons is transformed into inhibition at appropriate points in the neural network by the interposition of inhibitory interneurons (Fig. 8.14). These are special short-axon neurons which release an inhibitory transmitter at their synapses, thus making it harder to fire the postsynaptic neuron. Therefore, all inhibitory pathways must contain at least three neurons and all pathways involving only an afferent and an efferent neuron (monosynaptic chains) must be excitatory. Conductive delays substantiate the inclusion of at least two synapses in even the most direct inhibitory pathways in mammals.

There are two forms of postsynaptic inhibition which merit special mention: recurrent or "surround" inhibition and disinhibition.

Recurrent or Surround Inhibition. A particular type of postsynaptic inhibition, in which active cells in sensory or motor projection systems inhibit adjacent neurons, has received considerable attention among neurophysiologists. The pathway for this inhibition involves recurrent collaterals which leave motor axons before they emerge from the gray matter of the cord. They pass back into the cord and excite special short inhibitory interneurons called **Renshaw cells**. A Renshaw cell responds to a single stimulus with a high frequency burst of impulses and the release of inhibitory transmitter, with a consequent reduction of excitability in the inciting and adjacent neurons upon which its terminals impinge (Fig. 8.15). More strongly stimulated cells exert a stronger inhibition on their neighbors than that

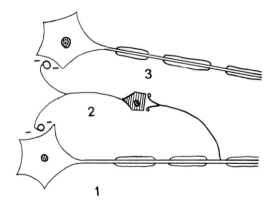

Figure 8.15. Recurrent or surround inhibition. A recurrent collateral from motor neuron (1) re-enters the ventral gray matter and synapses with a short inhibitory neuron, the Renshaw cell, (2) (cell body cross-hatched). The Renshaw cell sends terminals to the inciting neuron and to surrounding motor neurons, where its inhibitory transmitter diminishes their irritability.

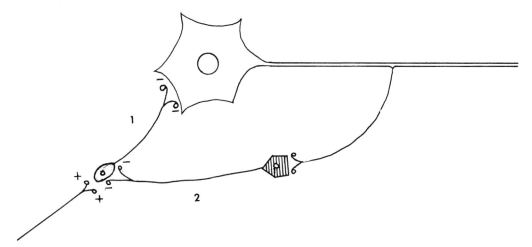

Figure 8.16. Facilitation by disinhibition. The motor neuron is under inhibition from an unknown source by way of an inhibitory interneuron (1). A collateral from the motor axon activates a Renshaw cell (2) which inhibits the inhibitory neuron. The more strongly the motor neuron is stimulated the greater is the reduction in the incident inhibitory influence. Disinhibition would thus contribute to enhancement of the muscle's response.

which they receive and hence the excitatory difference between them is exaggerated.

The exact function of this recurrent or surround inhibition is not yet clear. In motor neurons it presumably plays a role in localizing activity within a muscle and so may be of value in distributing motor unit activity for fine movements (Wilson (1966)). A similar mechanism in sensory pathways may serve to sharpen contrast (Brooks (1959)).

Disinhibition. Not only do Renshaw cells inhibit adjacent motor neurons but they may also inhibit an already existing inhibition and thereby *facilitate* neurons of the motor pool. Motor neurons are subject to a tonic inhibitory influence by some as yet unidentified interneurons, probably reticulospinal fibers. Through inhibitory synapses on these cells, the Renshaw cells depress their inhibitory action and thus release the motor neurons from the inhibition. This then is a facilitation by **disinhibition** (Fig. 8.16). The fact that this is not a usual type of facilitation is supported by both electrophysiological and pharmaco-

logical evidence: membrane potential changes are *hyperpolarizations,* and the effect is blocked by strychnine and tetanus toxin, drugs which block postsynaptic inhibitory synapses but do not affect excitatory junctions.

Normally, both facilitation and inhibition are occurring simultaneously but to different extents at the multitude of synapses of a postsynaptic neuron. The postsynaptic cell will fire whenever the algebraic sum of the two antagonistic influences is sufficient to depolarize it to its critical level; the greater the sum the greater will be the frequency of impulses generated.

Presynaptic inhibition. As the name implies, the effect of presynaptic inhibition is exerted upon the presynaptic neuron rather than upon the membrane of the postsynaptic cell. The pathway for presynaptic inhibition appears to involve neural circuits in which the inhibiting neuron synapses upon the *axon* of the presynaptic neuron close to its own termination (Fig. 8.17). The electron microscope has revealed the existence of small boutons mak-

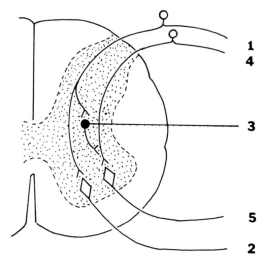

1
4

3

5

2

Figure 8.17. Presynaptic inhibition. A hypothetical circuit mediating presynaptic inhibition. An afferent neuron from a muscle spindle (1) is shown making an excitatory connection with a motor neuron (2) to its own extensor muscle. A collateral branch of the afferent neuron activates a short interneuron (3), whose terminals synapse with the axon of an afferent neuron (4) which is making an excitatory connection with an efferent neuron (5) to the antagonistic flexor muscle. As a result, excitation over the extensor afferent (1) will diminish the excitatory influence upon the antagonistic flexor motor neuron by presynaptic inhibition.

ing synaptic contact with telodendria near their large end knobs. These axoaxonic synapses are believed to be the morphological basis of presynaptic inhibition. Pharmacological evidence indicates that the transmitter substance is quite different from that of postsynaptic inhibition. First, the presynaptic inhibitory effect is not blocked by strychnine or tetanus toxin, both of which are powerful antagonists of postsynaptic inhibition, and second, it is sensitive to picrotoxin, a convulsant drug which has no action upon postsynaptic inhibition.

The distinctive characteristic of presynaptic inhibition is that EPSPs of the postsynaptic neuron are depressed without any measurable hyperpolarization of its membrane. There is good evidence that the depression is due to a partial depolarization of the presynaptic axon which reduces the magnitude of the action potential invading its terminals. For example, if a depolarization of 10 mV has been induced at the axoaxonic synapses, the spike of the presynaptic neuron will be reduced by 10 mV from its usual level. As mentioned earlier, transmitter release is proportional to the magnitude of the action potential. Consequently, when these smaller potentials reach the end knobs, less excitatory transmitter is released and the EPSP is proportionately lessened. The reduction in transmitter probably reflects a decrease in the number of ejecting vesicles, because there is no evidence that the size of individual quanta is affected. A lesser amount of transmitter results in a lower impulse frequency in the postsynaptic neuron and therefore a decreased response. When the giant axons of the squid were presynaptically depolarized by electric current, a 5% reduction in the magnitude of the presynaptic spike caused a 50% reduction in the postsynaptic response. The neurons which produce presynaptic inhibition often fire repetitively and the presynaptic spike depression may last as long as 100 msec. It is also possible that antidromic impulses traveling centrifugally in the dorsal roots may collide with the orthodromic incoming impulses and reduce their magnitude in that way.

Presynaptic inhibition provides a mechanism whereby the central nervous system can control its input by completely suppressing weak or extraneous sensory inflow and can adjust the effectiveness of signals from one part of the body in relation to conditions prevailing in another part. Most important, it can modulate or eliminate undesirable input from one specific source without altering the sensitivity of the postsynaptic neuron to input from other sources. This is in sharp contrast to postsynaptic inhibition, in which the excitability of the postsynaptic neuron is depressed.

In the central nervous system of vertebrates, presynaptic inhibition is wide-spread at all spinal cord levels, occurs commonly in the brain, and has been found in interactions between cord and brain. There is increasing evidence that all afferents entering the cord from the skin and other peripheral receptors may exert presynaptic influence upon adjacent neurons and upon themselves. Pyramidal tract cells are thought to reduce stretch reflex activity by imposing presynaptic inhibition upon spindle afferents (Fig. 8.17).

The existence of presynaptic facilitation through an increase of transmitter release by the presynaptic neuron is suspected though as yet unproven (Ganong (1965)).

Both recurrent (Renshaw) postsynaptic inhibition and presynaptic inhibition are feed-*back* inhibitions: an active neuron sends collaterals back to produce inhibition at an earlier point in the transmission pathway. In the cerebellum a feed-*forward* inhibition has been demonstrated. Basket cells and Purkinje cells are both excited by the same input, but the basket cells send terminals forward to inhibit the Purkinje cells. Presumably the mechanism limits the duration of excitation produced by any given afferent volley.

The total subsynaptic area, dendritic plus somatic, of a postsynaptic neuron is relatively enormous as compared with a single synapse and the number of presynaptic terminals impinging on a postsynaptic cell may be very large. Since both facilitatory and inhibitory synapses are represented, both effects may be exerted upon the cell simultaneously. The magnitude of the depolarizing current through the initial segment of the postsynaptic neuron will be determined by the number of active synapses and the algebraic sum of the two antagonistic influences. As long as the excitatory influence exceeds the inhibitory influence by at least the critical amount, the neuron will fire.

By selective facilitation of some synapses and inhibition of others, excitation may be directed into proper outflow channels. Muscles which should contract do so, and those which would interfere with the

movement are caused to relax by cessation of outflow to them.

Other Properties of Synapses

The properties characteristic of synaptic transmission are compatible with the accepted chemical theory. They differ in several respects, however, from the electrochemical conduction of action potentials in the nerve fiber.

Synaptic Delay. Transmission across the width (100 to 200 Å) of the synaptic cleft requires 0.1 to 0.3 msec in man and up to 1 msec in other animals. As compared to conduction velocities of over 100 m/sec in large nerve fibers, synaptic transmission is nearly 2 million times slower. Most of the delay is consumed by transmitter release, as both diffusion and chemical interaction with the postsynaptic membrane are accomplished in a few microseconds. It is obvious that, in the conduction of information through the nervous system, the more synapses in a neural chain the longer will be the time required for impulses to travel from receptor to effector.

Polarity of Conduction. Excitation proceeds only from presynaptic axon terminals to postsynaptic dendrite or soma, never in the reverse direction across the synapse. The unidirectionality of the synapse is inherent in the functional differentiation between the receiving and transmitting regions of the cell. Impulses induced artificially in a nerve axon do travel antidromically, but they die out without arousing the electrically inexcitable membranes of dendrite and soma. Furthermore, if dendrites were stimulated, they are not structured to release transmitter. Even in axoaxonic synapses excitation does not pass backward into the presynaptic terminal.

Susceptibility to Fatigue and Drugs. While the nerve fiber requires little oxygen for conduction and is practically indefatiguable, the synapse is very susceptible to hypoxia and readily fatigued. It is also more susceptible to drugs and anesthetics. During deep anesthesia synaptic transmis-

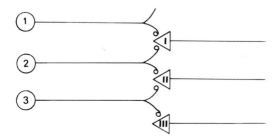

Figure 8.18. Summation requirement in synapses. Motor neuron I may be fired by impulses from (1) and (2) when firing together but will be unresponsive to either alone. Similarly, neuron II may be fired by (2) and (3).

sion is completely blocked, but nerve axons conduct as usual.

Summation Requirement. In almost every instance summation of input is required to fire a postsynaptic cell. In Figure 8.18, 1, 2, and 3 represent neurons which form excitatory synapses on motor neurons *I*, *II*, and *III*. 1 supplies only *I*, 2 supplies both *I* and *II*, and 3 supplies both *II* and *III*. If either 1 or 2 is stimulated by a single shock, an EPSP will be generated in *I* but it will not fire because the critical level has not been reached. If, however, both 1 and 2 are stimulated simultaneously or in close succession, the facilitation induced by one is reinforced by the other, and *I* may be fired by **spatial summation**. 2 will also facilitate *II* which may then be fired by 3. Repetitive stimulation of any one of the interneurons may also prove adequate to fire the postsynaptic neurons by **temporal summation**. Inhibitory synapses show similar summation effects.

Neurons with a sufficient number of active knobs are said to be in the **discharge zone**, while those which are facilitated but not fired are said to be in the **subliminal fringe**. The discharge zone increases with increased stimulus intensity, its spread being determined by the convergence of neuron activity upon the same postsynaptic cells.

Synaptic Threshold. The threshold of each postsynaptic neuron varies from moment to moment as a result of the many

influences playing upon it. However, persistent increases or decreases of threshold seem to occur as a result of use and disuse of synapses, and they may be significant in motor learning. It is possible that the phrase "lowering the resistance at the synapse," while often used and misused, may adequately describe the changes which take place at a synapse after repeated use. This phenomenon would help to explain why motor patterns, once learned, are easily "called forth" by the cortex, and also highly resistant to change and modification.

Post-tetanic Potentiation at the Synapse. Motor neurons which have been subjected to prolonged repetitive stimulation display a persistent threshold decrease which may last for hours. The effect is specific as to input, relating only to the presynaptic pathway responsible for the tetanization. Although the exact nature of the mechanism involved is unknown, there is evidence that it is presynaptic in origin and probably related to either increased transmitter release or effectiveness. Hubbard and Willis (1963) suggest that the effect is due to an increase in the ability of nerve impulses to release transmitter as a result of an increased amplitude of the presynaptic potentials associated with a hyperpolarizing effect of the repetitive stimulation. Post-tetanic potentiation represents a facilitatory effect of use and perhaps is a primitive form of learning.

Afterdischarge. Response of the postsynaptic cell may persist after stimulation has ceased. This is thought to be due to reverberating circuits which continue the presynaptic input independent of the original stimulus. Firing will ultimately be terminated by fatigue or by inhibition of the neurons responsible for the input.

Nonlinearity of Response. Postsynaptic response does not usually correspond closely to presynaptic input either in intensity or rhythm. Nonlinearity of intensity is inherent in neural organization. The convergence of presynaptic fibers on postsynaptic neurons gives rise to a phenomenon known as **occlusion**. The explanation is embodied in Figure 8.18. Assume that the

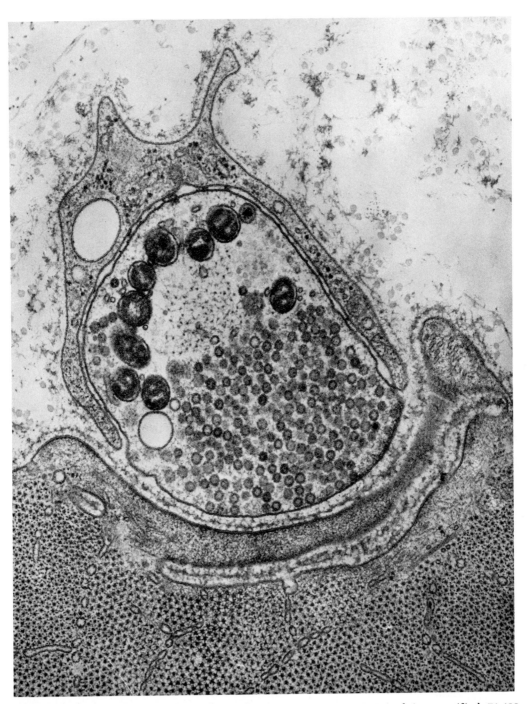

Figure 8.19. Site of acetylcholine release from a motor-nerve terminal is magnified 71,400 diameters in this electron micrograph made by John E. Heuser of the University of California Medical Center in San Francisco. The terminal is stocked with saclike synaptic vesicles containing molecules of acetylcholine; the larger dark structures are mitochondria, which generate the energy required for the activities of the nerve ending. On the arrival of an impulse the synaptic vesicles fuse with the membrane, releasing acetylcholine into the fluid-filled cleft between the terminal and the muscle cell. The molecules of acetylcholine then bind to receptors embedded in the muscle-cell membrane. Below the cleft a deep invagination in the muscle membrane, or junctional fold, is shown in partial section. (From Lester, H. A., 1977. The response to acetylcholine. *Sci. Am. 236:* 106; reproduced courtesy of Dr. John Heuser.)

rate of stimulation applied to an interneuron is sufficient to evoke maximal response of the postsynaptic motor neuron which will then elicit a maximal tension of 50 g in its motor unit. If *2* alone is stimulated, motor units *I* and *II* will both respond, yielding 100 g of tension in the muscle. *3* alone will excite motor units *II* and *III* and also produce 100 g of tension. If *2* and *3* are then stimulated at the same time, the result will be not 200 but only 150 g as the three motor units are activated. Because of the overlapping on postsynaptic neurons, the tension evoked by simultaneous stimulation of several presynaptic neurons is *less* than the sum of the tensions produced when each is stimulated alone. This is **occlusion**.

Another and more important source of nonlinearity is the modulation effected by antagonistic inputs. This is the basis of the integrative action of the nervous system.

Although the rhythm of stimulation is not usually reflected in the postsynaptic response, in certain instances a presynaptic neuron may **drive** the postsynaptic neuron, the latter responding one for one with the input over a limited range.

Neuromuscular Junction as a Specialized Synapse

The neuromuscular junction is essentially a specialized neuroeffector synapse, sharing many of the properties of neuroneural synapses but displaying other properties which are unique to it.

Generally a skeletal muscle fiber in man receives only one motor nerve terminal, rather than a convergence of numerous inputs. Multiple innervation is not uncommon in some vertebrates, however, and is in fact characteristic of slow muscle fibers in the frog.

The presynaptic nerve terminal, although discrete, is "buried" in the specialized end-plate sarcoplasm of the muscle fiber. Nerve and muscle membranes are separated by a primary synaptic cleft, and beneath the neural ending the sarcolemma is extensively folded into innumerable secondary clefts, thus tremendously increasing the membrane area exposed to the transmitter. Mitochondria are abundant in both the terminal and the end-plate region. Synaptic vesicles appear on the presynaptic side of the interface and the transmitter is known to be ACh (Fig. 8.19).

Transmitter release induces local depolarizations in the end-plate membrane known as end-plate potentials (EPPs). It was the occurrence at neuromuscular junctions in resting muscle of miniature end-plate potentials whose amplitudes were integral multiples of a minimal value that led to the concepts that transmitter was released in quantal amounts by individual synaptic vesicles and that random release occurred even in the absence of presynaptic action potentials. The amount of ACh released randomly into the junctional cleft varies directly with the concentration of Ca^{++} and inversely with the concentration of Mg^{++} at the end-plate region. Although ACh acts on the muscle fiber membrane, it is ineffective if introduced *under* the membrane into the cytoplasm.

When an action potential reaches the prejunctional terminal, the number of vesicles which rupture apparently release enough ACh to depolarize the end-plate sarcolemma to its critical level. In other words, a single impulse is able to produce a full-sized end-plate potential and to excite the muscle fiber membrane. Summation is therefore not a usual requirement for neuromuscular transmission, nor is there any synaptic inhibition at this junction. Motor inhibition must be accomplished centrally at the synapses of the motor neuron, and the muscle is then "inhibited" by the cessation of impulses traveling to it over its efferent nerve fibers.

The junction between receptor cell and afferent neuron may also be considered as a special type of synapse, at least in some instances. This is discussed briefly in Chapter 10.

SUMMARY

Briefly, the events in the functioning of neurons take place as follows.

1. Stimulation of dendrites results in excitation in the form of local graded depolarization which spreads decrementally over the dendrites and cell body.

2. If depolarization at the initial segment of the axon reaches the critical level, action potentials are generated and conducted nondecrementally along the axon.

3. Arrival of action potentials at axon terminals causes secretion of chemical transmitter substance into the synapses.

4. The transmitter substance diffuses across the synaptic gaps and chemically acts upon the membranes of the postsynaptic neurons.

5. Each postsynaptic neuron, if adequately activated, repeats the generation, conduction, and transmission of excitation. Its postsynaptic influence may be facilitatory or inhibitory, depending upon the type of its transmitter and the nature of the next postsynaptic neuron.

6. The last neuron in each chain, the efferent neuron, releases its transmitter at the neuroeffector junctions of the muscle fibers of its motor unit.

7. The muscle responds with tensions proportional to the total number of fibers activated and the frequency of impulse bombardment.

CHAPTER 9

Basic Neuroanatomy: The Central Nervous System

INTRODUCTION

The nervous system has often been described as being analogous to the electronic computer. The common features cited are the abilities to collect information, store information, compute new information, and process all information, converting it into some usable form for potential application. Like the computer, a programming unit is needed to direct the sequence of events, subroutines are called forth to perform specific operations, and automatic mechanisms are evident to respond to the input and execute the programs dictated by the subroutines and the programmer.

To appreciate the complexity of the nervous system, consider the not-too-common (fortunately) occurrence of the housewife touching the hot stove by mistake (Fig. 9.1.). Flexor muscles "automatically" respond to withdraw the finger from the noxious stimulus; hair follicles are facilitated to produce the "hair-raising" response, and sweat glands may suddenly become activated; a few milliseconds later, the housewife is *aware* of the pain and,

depending on the "severity" of the noxious stimulus, the vocal muscles may respond to produce the "ouch" response (or other perhaps unprintable word). This series of simple events occurring within a few milliseconds is mediated by the nervous system (Moore (1969)). The immediate finger withdrawal response, followed by hair and sweat gland responses, is subserved by a series of nerve pathways at the spinal cord level; the awareness of the pain and subsequent "ouch" response occur when impulses travel, via nerves, up the spinal cord, through the brain stem and finally arrive at the cerebral cortex (the "head brain" or the programmer).

Any one nerve is simply a collection of sensory (receptor) and motor (effector) neurons. Spinal nerves are composed of the neurons which "sense" the pain, the neurons which effect the withdrawal of the finger, and the neurons which effect the hair and sweat responses. Cranial nerves, such as those which activate the vocal muscles, are importantly involved, as are other nerves, ascending and descending pathways, and brain centers. It is important to recognize that the act of withdrawal from the painful stimulus occurs

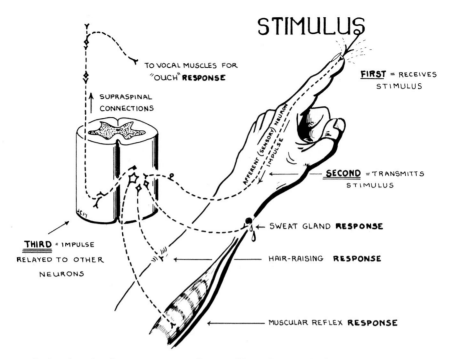

Figure 9.1. A simple stimulus-response pathway. (From Moore, J. C., 1969. *Neuroanatomy Simplified,* p. 13. Dubuque, Iowa: Kendall/Hunt Publishing Company.)

first, mediated by neuronal activity momentarily entering and exiting from the spinal cord; the recognition of the pain occurs several milliseconds later after the act of withdrawal has taken place, indicating that longer pathways and "interconnections" with brain centers are involved. It is fortunate for the housewife that the withdrawal occurs immediately and is not dependent on her perception of the pain; a few milliseconds of additional contact with the noxious stimulus could result in a more serious burn due to the longer duration of the stimulus.

It is apparent that, even in a very simple stimulus-response series of events, all levels of the nervous system have the potential of becoming involved. This and subsequent chapters are concerned with the basic components of the somatic nervous system, highlighting those which are importantly involved in motor behavior. (The autonomic nervous system is not discussed.) The nervous system is a highly organized, extensive network designed to integrate internal reactions of the individual and correlate these with the external environment so that appropriate responses are made. After all, it is the internal mechanisms which ultimately make it possible to take advantage of the "external" mechanisms which have been discussed in the first six chapters of this book. The nervous system is arbitrarily divided into two large divisions, the central nervous system and the peripheral nervous system. The central nervous system, the subject of this chapter, is composed of the brain and spinal cord. The peripheral nervous system, composed of the nerves, ganglia, and end organs which connect the central nervous system with all other parts of the body, is discussed in subsequent chapters.

To aid the reader in understanding the complexity of the central nervous system, a general overview of the major structures will be presented. The descriptions are by no means complete but are intended to

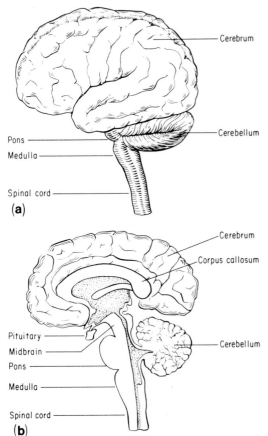

Cerebrum

Cerebellum

Pons

Medulla

Spinal cord

(a)

Cerebrum

Corpus callosum

Pituitary

Cerebellum

Midbrain

Pons

Medulla

Spinal cord

(b)

Figure 9.2. Drawings of lateral (a) and medial (b) views of the human brain. (From Thompson, R. F., 1967. *Foundations of Physiological Psychology*, p. 85. New York: Harper and Row.)

help the reader understand the functions attributed to each major brain and spinal cord structure and become familiar with common terminology used in most textbooks and literature.

CEPHALON MORPHOLOGY AND PHYSIOLOGY

The brain or **cephalon** is divided into five major sections, termed **telencephalon** (or "head brain"), **diencephalon** (or "prim-itive-brain"), **mesencephalon** (or "middle brain"), **metencephalon**, and **myelenceph-alon** (or "hind brain"). Man has one of the most sophisticated telencephalons whereas lower forms of life have extensive diencephalons and limited telencephalons (Fig. 9.2). An outline of the brain, its major sections, and the principal anatomic units associated with each follows.

Telencephalon:
  Cerebral cortex
Diencephalon:
  Basal ganglia
  Thalamus
  Reticular formation (part of)
  Internal capsule
Mesencephalon:
  Midbrain (tectum and tegmentum)
  Reticular formation (part of)
  Red nucleus and substantia nigra
Metencephalon:
  Pons
  Reticular formation (part of)
  Cerebellum
  Vestibular apparatus
Myelencephalon:
  Medulla oblongata
  Reticular formation (part of)

Telencephalon

The telencephalon is composed of the **cerebral cortex** which is primarily a vast information storage area. "Approximately three quarters of all the neuronal cell bodies of the entire nervous system are located in the cerebral cortex" (Guyton (1972)). All information is collected, stored, computed, and processed in the cerebral cortex so that such information can be called forth at will to control motor behavior.

Certain areas of the cortex perform specific functions although it has been shown that the cortex has not been as discretely divided as some might believe (Guyton (1972)). Figure 9.3 shows some of the functional areas of the human cerebral cortex. The map, originally constructed by Penfield and Rasmussen in the 1950's suggests that some areas (those labeled) have specific functions while the other areas perform in a more general way. The **pri-**

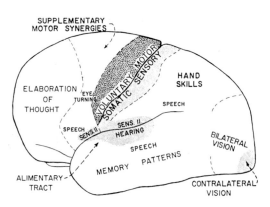

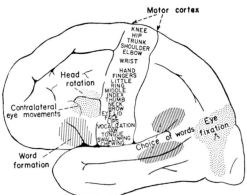

Figure 9.3. Functional areas of the human cerebral cortex. (From Penfield, W. and Rasmussen, T., 1968. *The Cerebral Cortex of Man.* New York: The Macmillan Company.)

Figure 9.4. Representation of the different muscles of the body in the motor cortex and location of other cortical areas responsible for certain types of motor movements. (From Guyton, A. C., 1972. *Structure and Function of the Nervous System,* p. 190. Philadelphia: W. B. Saunders Company.)

mary sensory areas receive signals from the lower centers of the brain and spinal cord and transmit the results of the "analyses" back to the lower centers and to other regions of the cortex. The **motor areas** of the cortex send impulses to specific muscle groups either directly or indirectly via other brain structures. Representation of the different muscle groups for all body parts within the motor cortex is shown in Figure 9.4. Generally, the degree of representation of body parts correlates with the complexity of movements required (Fig. 9.5). The **pre-motor** area is believed to be concerned primarily with the acquisition of specialized motor skills; in addition, control of complex eye and mouth movements is located here. Some of the **somatic sensory** and motor areas overlap so that sensations and muscular contractions in the same area of the body are monitored by the same area of the cerebral cortex.

In 1909, Brodmann classified the functions of the cortex according to structurally distinctive areas (Fig. 9.6); his classification is used by most neuroanatomists and is found in most modern-day texts. Brodmann's areas 1, 2, and 3 (of the **postcentral gyrus**) represent the somatic sensory region while areas 4 and 6 (of the **precentral gyrus**) represent the motor and premotor areas, respectively.

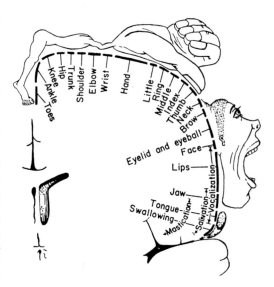

Figure 9.5. Degree of representation of the different muscles of the body in the motor cortex. (From Penfield, W. and Rasmussen, T., 1968. *The Cerebral Cortex of Man.* New York: The Macmillan Company.)

The reader should consult a textbook concerned with the structure and function of the nervous system for details concerning all areas of the cerebral cortex. For the

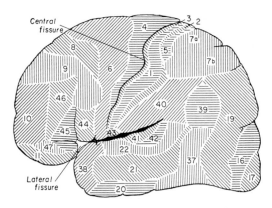

Figure 9.6. Structurally distinctive areas of the human cerebral cortex. (Modified from Brodmann.) (From Everett, N. B., 1971. *Functional Neuroanatomy*. Philadelphia: Lea and Febiger.)

purposes of this text, it is sufficient to say that movement can be initiated by the motor cortex, that patterns of human movement can be learned and then stored in the motor cortex (sometimes called an **engram**) to be called forth at will, and that motor cortex signals can be modified by impulses reaching the sensory cortex, or other brain structures, as the signals descend to the **spinal cord** (effector) level.

Diencephalon

That part of the brain called the diencephalon is composed of structures specific to modern man's needs, but old in terms of the evolution of the species. Some of the structures of the diencephalon dominate the human body during early stages of ontogeny and will persist in directing adult body functions. They will also direct motor behavior under special conditions of stress and reduced cerebral cortex activity. The principal structures mentioned here are the **basal ganglia, thalamus**, part of the **reticular formation**, and the **internal capsule**.

Anatomically, the basal ganglia are composed of the **caudate nucleus, putamen, globus pallidus, amygdaloid nucleus** and **claustrum**. The latter two are not directly concerned with motor function and

will not be discussed. Numerous pathways exist between the motor cortex and the caudate nucleus and putamen (collectively labeled the **striate body**); the latter sends numerous fibers to the globus pallidus, which communicates back to the motor area of the cortex via the thalamus. These circular pathways operate as a kind of feedback loop or servo-control mechanism (Guyton (1972)). The functions of the basal ganglia are generally associated with the **extrapyramidal system**, one of the systems by which the cortex communicates with the final common pathways to muscle. If the cortex were destroyed, discrete movements of the body, especially the hands, would be impossible, but gross movements of a subconscious nature would still be possible; i.e., walking and controlling equilibrium (Guyton (1972)).

It is important to recognize that the basal ganglia function as a total system. Therefore, assigning specific functions to each portion will not completely describe the physiology of the system. As a whole, the basal ganglia are generally recognized as centers capable of inhibiting muscle tone throughout the body (Noback and Demarest (1972)). The striate body seems to initiate and regulate gross movements performed unconsciously. The globus pallidus is usually ascribed the ability to provide background muscle tone for movements that are initiated either by the striate body or by the cortex. Thus, the basal ganglia, together with other centers, modulate motor activities through circuits that feed back to the cortex.

The thalamus serves many important functions other than that related to motor control. Along with the **subthalamus, substantia nigra**, and **red nucleus**, the thalamus operates in close cooperation with the basal ganglia to exert influences on motor activity. Also, impulses from the **cerebellum** are transmitted back to the motor cortex via a specialized nucleus of the thalamus, the **ventral lateral nucleus**. The **subthalamic nucleii** are importantly involved in the total circuitry which provides the background discharge necessary for the success of fine coordinated movements.

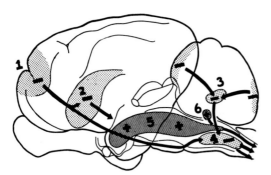

Figure 9.7. Outline view of the brain of the cat showing facilitatory and inhibitory areas. Inhibitory: 1, portion of cerebral cortex; 2, basal ganglia; 3, cerebellum; 4, reticular formation (medulla). Facilitatory: 5, reticular formation (diencephalon and midbrain); 6, vestibular nucleii. (From Magoun, H. W., 1958. *The Waking Brain*, p. 16. Springfield, Ill.: Charles C Thomas.)

The reticular formation is a vast network of neurons and nucleii which anatomically is located throughout most of the brain stem. A portion of it is located in the diencephalon and it extends caudad as far as the myelencephalon. Those functions concerned with motor control are the primary focus of this textbook although the reticular formation plays major roles in arousal and consciousness and various states of sleep and relaxation. It exerts powerful influences on phasic and tonic motor activities. That portion of the reticular formation located in the diencephalon is generally ascribed the ability to exert facilitatory influences on spinal motor discharge. Thus, flexor or extensor reflexes, decerebrate rigidity and responses evoked from the motor cortex are facilitated (Magoun, 1958).

An appreciation of some of the facilitatory and inhibitory influences initiated in the diencephalon structures can be gained by studying Figure 9.7, an outline view of the brain. Areas yet to be discussed such as other portions of the reticular formation and some of the cerebellar and **vestibular nucleii** are also shown.

The internal capsule is actually a massive bundle of nerve fibers which links the cerebral cortex with other portions of the central nervous system. If the reader can imagine the diencephalon as a structure shaped something like a fist and inserted upward "inside" the cerebral cortex, it is easier to understand that the major routes to and from the cortex travel through the diencephalon. Ascending fibers projecting from subcortical nucleii to the cerebral cortex and descending fibers projecting from the cerebral cortex to lower centers of the brain and spinal cord are massed together and include both sensory and motor pathways. Generally, the internal capsule is divided into **anterior** and **posterior limbs**, each of which has distinct pathways associated with them. For example, the **corticospinal tract** (cerebral cortex to spinal cord) passes downward through the rostral portion of the posterior limb of the internal capsule and brain stem, crossing to the opposite side at the **medulla.** The caudal half of the posterior limb of the capsule contains various projections from the many pathways of the thalamus to the cortex (Noback and Demarest (1972)). The anterior portion of the capsule includes mostly the many fibers which connect the several portions of the brain stem with each other and with portions of the cortex. Overall, the internal capsule, elaborated at the diencephalon level, can be thought of as the great "elevator" or "escalator" system of the human body because it represents the *only* means by which nervous system impulses may descend from or ascend to the cerebral cortex.

Mesencephalon

The term mesencephalon is often used synonymously with the term **midbrain.** The mesencephalon is the most anterior extension of the brain stem which still appears to have the basic structural characteristics of the spinal cord. The dorsal portion of the midbrain (**the tectum**) is easily identified by the two pairs of relay nucleii which subserve the visual and auditory systems. The ventral portion of the midbrain (**tegmentum**) contains nucleii for the third and fourth **cranial nerves**, all of the ascending and descending tracts mentioned earlier in the section concerned with internal capsule, and a portion of the reticular formation. Two important centers are also located here: the substantia nigra and the red nucleus (Thompson (1967)). The function of the substantia nigra is not well known but apparently it is the major center for excitation of the gamma loop associated with the **neuromuscular spindle** (see Chapter 11 on the spindle). Its importance is recognized because it apparently activates the **gamma efferent system** even before the **alpha motor neurons** to muscles are activated, and provides the background muscular tone so that discrete and highly coordinated movements can be performed. The red nucleus is primarily concerned with gross body movement especially as the body deviates from the standing upright posture.

Metencephalon

The subdivision of the brain called the metencephalon includes the **pons** and part of the reticular formation. The cerebellum and vestibular apparatus are usually associated with the metencephalon because of location although they are not considered to be a part of this subdivision.

The pons, like the midbrain, contains ascending and descending tracts and a large mass of transverse fibers on its ventral aspect. The literal meaning of the word pons is "bridge" which implies its function: it interconnects the two sides of the cerebellum and brain stem as well as the fibers connecting the cortex with the spinal cord. Also, several cranial nucleii are located in the pons, notably the main motor nucleus of the fifth nerve and the nucleus of the seventh nerve.

That portion of the reticular formation located at metencephalon level is concerned with facilitation and inhibition of lower spinal cord neurons.

The cerebellum is often dubbed "the great motor coordination center." It over-lies the pons and presents a convoluted appearance with numerous fissures. Its specialty as a center for sensory-motor coordination is noted by the many afferent and efferent fibers associated with it. Sensory input is received from the vestibular system, spinal fibers, auditory and visual systems, reticular formation, and various regions of the cerebral cortex. In turn, it sends efferent fibers to the thalamus, reticular formation, and other parts of the brain stem. The cerebellum is often subdivided anatomically, and this type of organization appears to have some functional significance. The oldest (phylogenetically) part is the flocculus which projects to the vestibular nucleii. The next oldest is the **medial region (vermis)** which projects to the **medial (fastigial) nucleus** and the vestibular nucleii. The intermediate portion projects to one of the three cerebellar nucleii known as **interpositus**, and the lateral portion projects to the **lateral (dentate) nucleus**. The lateral portion, principally the dentate nucleus, is the part markedly developed in man and other mammals (Thompson (1967)).

An extremely important aspect of the cerebellum is its extensive connections with many parts of the brain stem and cerebral cortex. The anterior lobe of the cerebellum projects to the primary motor area of the cerebral cortex via the dentate and interpositus nucleii and the ventral lateral nucleus of the thalamus. The vermis and intermediate regions are functionally associated with the visual and auditory areas of the cerebral cortex. In addition to projections mentioned above, the cerebellar cortex also projects, via the appropriate **cerebellar nucleii**, to the reticular formation, the pons, the midbrain, the red nucleus, the basal ganglia, and the motor cortex via the ventral lateral nucleus of the thalamus. As mentioned in the section concerned with the thalamus, the cerebellum is involved in a large number of circles or loops so that it receives sensory information, projects to many parts of the brain including the cerebral cortex, and is importantly involved in the descending tracts to the spinal cord (Thompson (1967)).

Myelencephalon

The **medulla oblongata** is the main structure composing the myelencephalon. It contains ascending and descending fiber tracts and a number of cranial nerve nucleii. It is easily distinguished from other parts of the brain stem at the point where descending tracts cross and continue to the spinal cord on the contralateral side of the body (called **decussation of the pyramids**). The medulla houses the most caudad portion of the reticular formation which, at this level, exerts inhibitory influences on spinal cord neurons.

CORONAL AND CROSS-SECTIONAL ANATOMY OF THE CENTRAL NERVOUS SYSTEM

It has already been mentioned that the means of communication between the various parts of the brain and the spinal cord involves a complex series of nerve networks which travel in specific pathways or tracts. To facilitate the understanding of the "routes" fibers take in ascending and descending within the central nervous system, selected coronal and cross-sectional views of the central nervous system

are shown. A coronal section through the cerebrum is shown in Figure 9.8; note the location of the internal capsule and the caudate nucleus. A cross-section of the midbrain is shown in Figure 9.9 and reveals the location of the **cerebral peduncles**, the pons, and the origination of some of the cranial nerves. A cross-section at the level of the pons (Fig. 9.10) demonstrates the location of the reticular formation, one of the cranial nerves, the **cerebellar peduncles** and some of the **pontine nucleii.** Figure 9.11 displays the medulla in cross-section so that some of the nucleii, cranial nerves, medial pathways, and the pyramid decussation can be seen.

At the spinal cord level, a cross-section at the cervical level (Fig. 9.12) reveals the characteristic gray formation usually described in the shape of an "H" or butterfly and the outer white portion. The gray portion represents the several neuron cell bodies in this part of the spinal cord while the outer part is white in color because it is totally made up of axons of nerve fibers, both ascending and descending. The "white" ascending and descending fiber tracts are called **fasciculi** and are organized into three columns: dorsal, ventral and lateral. The "gray" matter is divided into dorsal and ventral horns: the cell bodies in the dorsal horns receive sensory information from the periphery via afferent neurons and transmit to the brain and

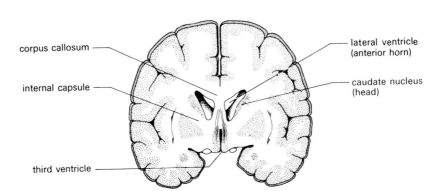

Figure 9.8. Coronal section through the cerebrum. (From Sage, G. H., 1977. *Introduction to Motor Behavior: A Neuro-psychological Approach*, p. 24. Reading, Mass.: Addison-Wesley Publishing Company.)

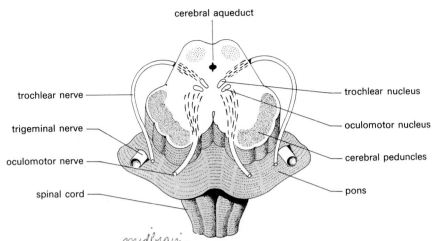

Figure 9.9. Cross-section of the mesencephalon. (From Sage, G. H., 1977. *Introduction to Motor Behavior: A Neuro-psychological Approach*, p. 26. Reading, Mass.: Addison-Wesley Publishing Company.)

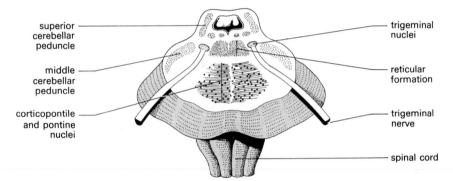

Figure 9.10. The pons in cross-section. (From Sage, G. H., 1977. *Introduction to Motor Behavior: A Neuro-psychological Approach*, p. 27. Reading, Mass.: Addison-Wesley Publishing Company.)

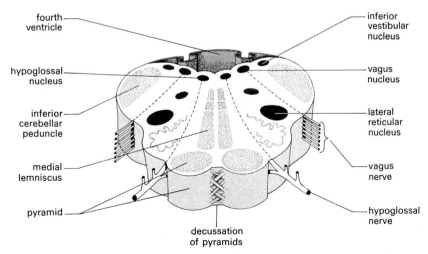

Figure 9.11. Medulla oblongata in cross-section. (From Sage, G. H., 1977. *Introduction to Motor Behavior: A Neuro-psychological Approach*, p. 28. Reading, Mass.: Addison-Wesley Publishing Company.)

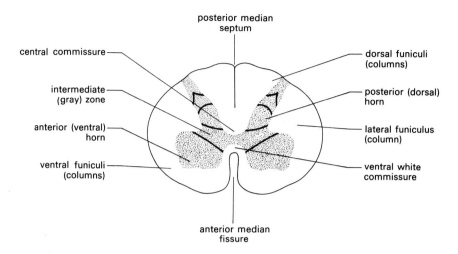

Figure 9.12. Section through the spinal cord at the cervical level. (From Sage, G. H., 1977. *Introduction to Motor Behavior: A Neuro-psychological Approach*, p. 31. Reading, Mass.: Addison-Wesley Publishing Company.)

other levels of the spinal cord; the ventral horns are composed of cell bodies which are the "**final common pathway**" to muscle (or to other effectors), and thus are referred to as efferent neurons. The terms, "final common pathway" (an old Sherrington term), alpha motor neuron, and **lower motor neuron** are often used synonymously to refer to the final nerve cell which innervates striated muscle.

PATHWAYS

All parts of the brain and nervous system are linked together in an elaborate network of pathways. Those pathways entering and exiting from the cerebral cortex are of interest here.

Descending Pathways

The means of communication from the cerebral cortex to spinal cord level, terminating on lower motor neurons, is functionally divided into two systems: the **corticospinal system** (also called the **pyramidal system**) and the extrapyramidal sys-

tem. Other descending efferent projections are used to communicate with other parts of the brain and are named for the parts with which the fibers interconnect. For example, the **corticorubro**, **corticopontine** and **corticostriate** projections are those pathways which carry fibers from the cerebral cortex to the red nucleus, pons, and striate body, respectively.

Corticospinal System (Pyramidal System)

Roughly 60% of the fibers arise from the motor cortex (sensory fibers comprising the remainder) and descend through the internal capsule, crossing over at the medulla level and continuing down into the **lateral white columns** of the spinal cord. These pathways supply the motor nerves of the contralateral side of the body. The corticospinal system is often regarded as a very fast system because one synapse exists in the circuit between the motor cortex and the effector, at the juncture with the alpha motor neuron (Fig. 9.13). The sensory fibers project to the dorsal horn of the spinal cord for the purpose of modifying information which is entering from the periphery.

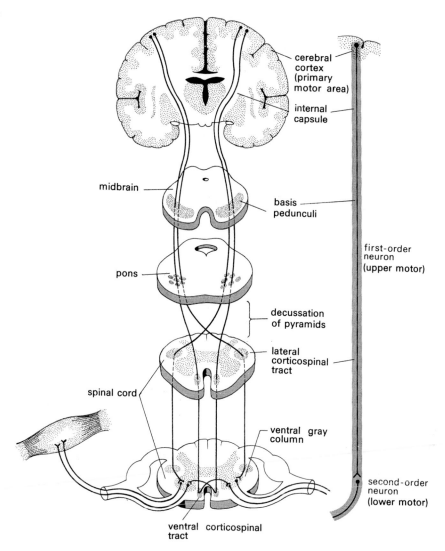

Figure 9.13. The pyramidal motor pathways. (From Sage, G. H., 1977. *Introduction to Motor Behavior: A Neuro-psychological Approach*, p. 126. Reading, Mass.: Addison-Wesley Publishing Company.)

Extrapyramidal System

The second means of communication from the motor cortex (as well as other cortical areas) to the final common pathway to muscle is via a series of "side trips" which involve other parts of the brain so that they can modify the signal which finally arrives at the spinal cord level. None of these pass through the pyramids of the medulla; they can travel ipsilaterally or contralaterally. These pathways are also named according to the parts of the central nervous system with which they communicate. Thus, the names **rubrospinal**, **vestibulospinal**, **reticulospinal** and **tectospinal** immediately convey to the reader the fact that the fibers originate in

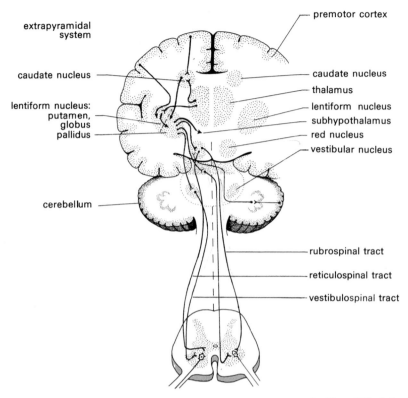

Figure 9.14. Some extrapyramidal motor pathways. (From Sage, G. H., 1977. *Introduction to Motor Behavior: A Neuro-psychological Approach*, p. 128. Reading, Mass.: Addison-Wesley Publishing Company.)

the red nucleus, vestibular nucleii, reticular formation, and the roof (tectum) of the midbrain, respectively, and all terminate at the spinal cord level (Fig. 9.14).

The rubrospinal pathway, which originates in the red nucleus of the midbrain, descends in the lateral white column and innervates distal muscles. The vestibulospinal pathway, which originates in the lateral vestibular nucleus, descends in the **ventral white column** of the spinal cord. The reticulospinal pathway which originates in the reticular formation at the pons level, descends in both the ventral and lateral white columns of the spinal cord. Both the vestibulospinal and reticulospinal pathways terminate on proximally located musculature. The tectospinal pathway, which originates in the tectum of the mid-

brain, descends in the ventral white column and innervates "neck" muscles.

Vestibulospinal and Reticulospinal Tracts

The functions associated with two of these extrapyramidal systems are further amplified by associating them with the specific tracts they follow as they descend to the spinal cord. Vestibulospinal pathways are divided into **medial** and **lateral tracts**. Vestibular nucleii located in the upper medulla and lower portion of the pons travel in the **medial vestibulospinal tract**, project to the spinal cord, and end in the cervical and thoracic level of the cord. The **labyrinthine response** (explained later in Chapter 11), which corrects head position with respect to gravity, is mediated by the vestibular nerve which feeds into this tract and modifies neck and upper extremity muscle activity. **Lateral vestibular nucleii** are somatotopically organized and influence postural mechanisms. They project down to the spinal cord via the **lateral vestibulospinal tract** and are generally known to facilitate extensors and inhibit flexors in order to maintain the upright position.

The **reticulospinal tracts** are not somatotopically organized. They are divided into **medial** and **lateral reticulospinal tracts**: the medial is often called a **reticulofacilitatory tract** because it facilitates extensor reflexes and inhibits flexor responses; the lateral, the **reticulo-inhibitory tract** is involved with inhibiting stretch reflexes in extensor muscles and facilitating flexor responses. The medial reticulospinal tract assists the lateral vestibulospinal tract with problems of balance, although the vestibular pathways are more substantially involved. Both of the reticulospinal tracts act on the alpha and **gamma motor neurons** for the purpose of modifying and coordinating reflexes at the spinal cord level.

Cerebellum Feedback Loop

It was mentioned earlier that the cerebellum is regarded as the "great motor coordination center." To explain its many interconnections in a chapter of this scope is impossible. There are approximately three times as many afferent fibers as efferent emerging from the cerebellum which is why it is regarded as a **somatic afferent organ** or, as Sherrington called it, the "head ganglion." The feedback loop to and from the cortex involves the following structures: cerebral cortex, pons, cerebellum, dentate nucleus, red nucleus, thalamus, and cerebral cortex; or, if you prefer, the loop would be called the **cortico-ponto-cerebello - dentato - rubro - thalamo - cortico** servo-mechanism. The largest cerebellar projection emerges from the dentate nucleus, proceeds to the red nucleus, continues to the thalamus and ultimately to the cortex. Other efferents travel from **deep cerebellar nucleii**, via the **superior peduncles** to the red nucleus and thalamus

or to the vestibular nucleii and reticular formation. The afferent pathways are usually divided into three parts: the inferior, middle and superior. The **dorsal spinocerebellar tract**, a very complex path, is an example of an afferent projection belonging to the inferior division. An afferent from the pons (originating in the cortex first) is an example of a projection belonging to the middle division. The superior division is primarily an "output" center, consisting mostly of efferent pathways. In summary, the many cerebellar projections emphasize the fact that the cerebellum is important in the coordination of muscle activity but does not initiate activity; it acts as a monitor for other brain centers and modulates muscle activity.

Ascending Pathways

The means of communication from the spinal cord to the many brain centers, i.e., sensory information, can be divided into two functional units: those which contribute to conscious awareness and those which may not reach the cerebral cortex but play an important role in coordination. Most of the ascending pathways travel via the thalamus, especially those concerned with conscious perception; the "unconscious" pathways terminate in the cerebellum.

Conscious Awareness

Those pathways concerned with **conscious perception** usually cross to the contralateral side at some point in their "journey" to the sensory cortex. In most cases, three neurons are involved between the spinal cord and the cortex. Those ascending in the dorsal columns (called the **medial lemniscus**) are associated with specific information such as the highly specific sensations which arise from touch, two-point discrimination, and **kinesthesia**. The first order neuron terminates in the **dorsal column nucleii** in the medulla. Two fasciculi of the medulla, the **cuneatus** and **gracilis** are the termination points for the upper and lower extremities, respectively. The second order neuron crosses to the

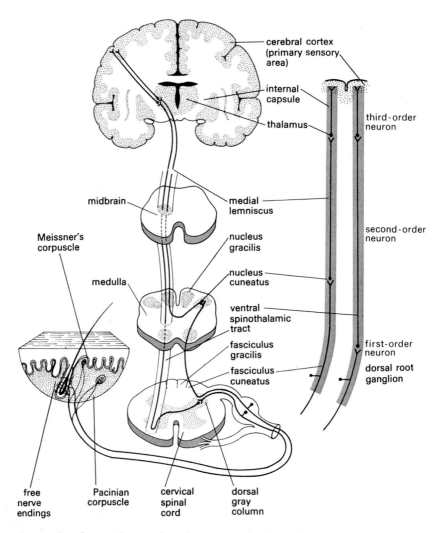

Figure 9.15. Sensory pathways for touch and pressure: the fasciculi cuneatus and gracilis and the ventral spinothalamic tracts. (From Sage, G. H., 1977. *Introduction to Motor Behavior: A Neuropsychological Approach*, p. 85. Reading, Mass.: Addison-Wesley Publishing Company.)

contralateral side, terminating at the **ventrobasal complex** of the thalamus. The third neuron projects to the primary sensory area of the cerebral cortex (Fig. 9.15). A second ascending system subserving conscious perception travels in the **anterolateral system** and also involves three neurons. It carries nonspecific information from small fiber systems such as those for pain, temperature and crude touch. The first order neuron synapses in the dorsal horn of the cord; at the same level of the

spinal cord, the second neuron crosses to the contralateral side and ascends in the **anterolateral tract** (also called the **ventral spinothalamic tract**) to the ventrobasal complex of the thalamus. The third neuron usually goes on to the cortex (Fig. 9.15).

Proprioceptive Pathways

The morphology and physiology of proprioceptors is discussed in detail in

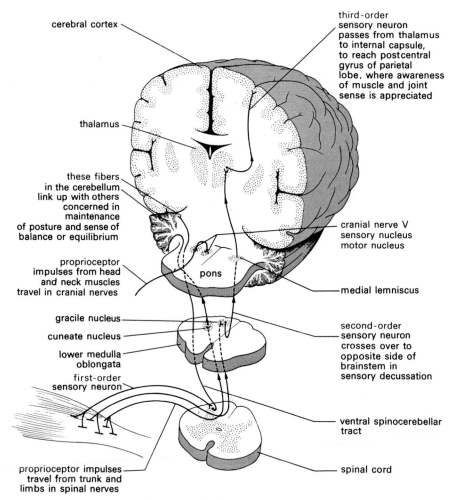

cerebral cortex

third-order
sensory neuron
passes from thalamus
to internal capsule,
to reach postcentral
gyrus of parietal
lobe, where awareness
of muscle and joint
sense is appreciated

thalamus

these fibers
in the cerebellum
link up with others
concerned in
maintenance
of posture and sense of
balance or equilibrium

proprioceptor
impulses from head
and neck muscles
travel in cranial nerves

pons

cranial nerve V
sensory nucleus
motor nucleus

medial lemniscus

gracile nucleus

cuneate nucleus

lower medulla
oblongata

first-order
sensory neuron

second-order
sensory neuron
crosses over to
opposite side of
brainstem in
sensory decussation

ventral spinocerebellar
tract

proprioceptor impulses
travel from trunk and
limbs in spinal nerves

spinal cord

Figure 9.16. Proprioceptive pathways to the brain. (From Sage, G. H., 1977. *Introduction to Motor Behavior: A Neuro-psychological Approach*, p. 96. Reading, Mass.: Addison-Wesley Publishing Company.)

Chapter 11. The intention of this section is to describe the pathways used between the receptors and the cerebellum. The **dorsal spinocerebellar tract** carries information from the neuromuscular spindle and the **Golgi tendon organ** via two neurons which travel ipsilaterally. The first order neuron terminates in the cord for lower extremities and in the medulla for the upper extremities. The second order neuron enters the cerebellum via the **inferior cerebellar peduncles** and terminates there.

The **ventral spinocerebellar tract** carries information from joint and cutaneous receptors. The first order neuron ascends ipsilaterally to the medulla. The second order neuron crosses to the contralateral side and enters the cerebellum via the **superior cerebellar peduncles** (Fig. 9.16). Projections to the cerebral cortex for awareness of position and movement follow a slightly different path, the third order neuron passing from the thalamus to the **postcentral gyrus** of the cortex.

Other Pathways

Ascending pathways subserving other functions include the **spinotectal** which is associated with head and eye movement, the **spino-olivary** which is presumed to be associated with movement and the **spino-vestibular** which is concerned with postural reflex mechanisms.

REFLEXES

By definition, a reflex is an automatic act such that a specific stimulus will produce a predictable response. Psychologists and physiologists have interpreted the term reflex in various ways, the former tending to confine the reflex to spinal cord level. Some physiologists have taken a broader connotation of the term and have classified reflexes according to levels of the central nervous system involved: third level reflexes are confined to the spinal cord, such as the simple stretch reflex; second level reflexes include those automatic responses involving levels of the brain stem; first level reflexes involve the cerebellum and cortex. The motor learning implications of reflex acts are germane to this discussion and have been reserved for that portion of Chapter 11 where reflexes are discussed in greater detail, as they are manifested by proprioceptors.

SUMMARY COMMENT

The salient features of the central nervous system, its organization and function, have been highlighted. The chapters to follow attend to some elements of the peripheral nervous system, focusing on integration of the sensory-motor systems and proprioceptive neuromuscular constructs.

CHAPTER 10

Basic Organization of the Neuromuscular System

The mammalian system is organized so that changes both in the external environment and its own internal environment can be monitored and acted upon. Receptors are necessary to detect the changes; effectors, if action is to be taken, need to be stimulated into activity. Thus, a centralized control system is necessary to coordinate all of these monitoring and decision-making activities (Tribe and Eraut (1977)).

The individual components of the neuromuscular system, the receptors, neurons, and muscle fibers, are organized into functional units which make possible the integration of activity of body parts into purposeful response patterns. Three major divisions may be recognized: the sensory or input system, the motor or output system, and the integrative system. The first includes receptors with their afferent neurons, and the second includes effectors (muscles) with their efferent neurons. The integrative system is composed of a complex arrangement of interneurons whose synapses determine the strength and direction of the signal which is transmitted from the input to the output system.

THE SENSORY SYSTEM: MORPHOLOGY AND PHYSIOLOGY OF RECEPTORS

The survival of the organism, its normal functioning, and its behavior depend upon its ability to detect and respond appropriately to changes in the external and internal environments. Receptors are the first elements in the communication pathway leading ultimately to the effectors. They function as transducers to convert stimulus energy into neural excitation.

Receptors are strategically situated over the body surface and within and among the muscles, bones, and viscera, with sensitivities attuned to specific and pertinent types of stimulus energy. Axons of afferent neurons enter the nervous system at all levels: spinal cord, brain stem, cerebellum, and cerebral cortex. Many branch divergently and, as they synapse upon interneurons with a variety of destinations, they establish the basis for extensive distribution of the input signal. Even the simplest motor behavior, such as the

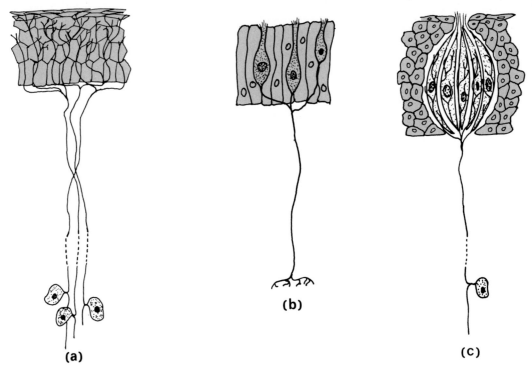

Figure 10.1. Types of receptors. (a), relatively unspecialized afferent neurons acting as receptors: free nerve endings. (b), neural cell with highly specialized receiving pole and axon serving as afferent fiber: olfactory cells. (c), specialized receptor cells having no axonal segment. These cells act to trigger dendrites of the afferent neuron closely associated with them: a taste bud.

automatic response to a noxious stimulus derives from conduction of action potentials over chains of neurons.

General Morphology of Receptors

Types of Receptor Structures

Receptors are specialized cells or organs highly sensitive to specific stimuli. There are three general types of receptors:

1. Relatively unspecialized **afferent neurons** which branch freely at their receiving poles. Examples are the free nerve endings in the skin (Fig. 10.1(a)).

2. Special neural cells known as **receptor cells**. Their receiving poles are highly specialized so as to present a low threshold for some particular type of environmental energy. Some resemble ordinary neurons except for the specialization of their dendritic terminals: their axons serve as afferent fibers (Fig. 10.1(b)). Others do not resemble typical neural cells, having little or no axonal segment. Their function is to set up the physicochemical conditions which will trigger an afferent neuron in close association with them (Fig. 10.1(c)). Examples of receptor cells are the taste buds, olfactory cells, the rods and cones of the retina, and the hair cells of the cochlea and labyrinth.

In numerous sensory systems one afferent neuron serves many receptor cells which may be spatially distributed over several square centimeters.

3. Complex **sense organs** consisting of specialized non-neural auxiliary or accessory structures surrounding or otherwise

closely associated with receptor cells and afferent neurons. Examples are the eye, ear, and neuromuscular spindle. The auxiliary structures assist the receptor cells in dealing selectively with their particular kind of stimulus energy. They analyze the incident stimulation and distribute it properly on the sensory surface or among the groups of receptor cells. The accessory structures of the eye are the cornea, iris, and lens which focus light on the retina. In the ear the tympanum, ossicles, lymph-filled canals, and the basilar membrane conduct and distribute sound vibrations to the auditory receptor cells. The structures associated with the neuromuscular spindle are discussed in Chapter 11.

Receptors may be classified in a number of ways: (a) according to the specific type of stimulus to which they respond: chemoreceptors, photoreceptors, pressoreceptors (baroreceptors), nociceptors, etc.; (b) according to the type of sensation aroused: visual, auditory, cold, etc.; (c) according to anatomical criteria such as morphology, location, cortical termination, etc.; (d) according to the type of response evoked: posture, movement, visceral function. Because the response of the organism is often determined by a combination of such criteria, a generally useful scheme classifies receptors into four groups as follows.

1. **Exteroceptors:** receptors located peripherally and responding to stimuli from the external environment. The organism's responses are sensation and/or movement.

2. **Interoceptors:** receptors located within the body in association with the viscera and responding almost exclusively to stimulation from the internal environment. The organism's response is a change in visceral function.

3. **Proprioceptors:** receptors located in muscles, tendons, joints, and the labyrinth of the inner ear and responding to stimuli arising from some aspect of posture and/or movement. The organism's response is change in activity of appropriate muscles. Some exteroceptors, especially those of the skin also evoke proprioceptive responses.

4. **Nociceptors:** receptors distributed throughout the body and responding to injury. The organism's response is usually withdrawal movement (flexion).

Physiology of Receptors

The Sensory Unit

A single sensory axon and its receiving pole with all of the receptor cells associated with its peripheral terminals constitute a **sensory unit**. For some units the receptive field may be very large. Generally the area supplied by one unit overlaps or interdigitates with areas supplied by other units. Consequently, stimulation of one small site may involve several sensory units (Fig. 10.4).

Excitation of Receptors

In a sensory system, special receptor cells may or may not be present. The afferent neuron is the essential element, and the generation of nerve impulses by depolarization of its initial segment is the significant feature common to all. In the absence of receptor cells, the stimulus may act directly upon the receptive pole of the neuron by mechanical deformation of the nerve terminal itself. In systems where receptor cells are present, the stimulus acts on the receptor which then fires the afferent neuron. Most receptor cells function like dendrites: they are electrically inexcitable and when stimulated produce a local graded potential, the **receptor potential**, which spreads electrotonically. The receptor potential may directly depolarize the initial segment of the afferent neuron or it may induce the secretion of a chemical mediator at a receptor-neural junction. The chemical mediator then acts upon the dendrites of the afferent neuron in a manner similar to that of synaptic transmitter. It induces in them a graded electrotonic potential, known as the **generator potential**, which is similar in every way to the postsynaptic potential at the synapse. If the generator potential is of sufficient magnitude when it reaches the initial segment, the neuron will fire and the frequency of

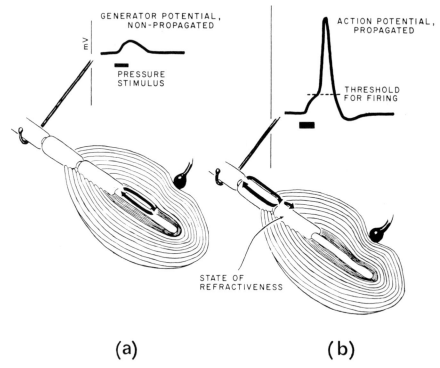

Figure 10.2. Generator potential vs action potential. Distinction between the generator potential and the action potential is illustrated by the Pacinian corpuscle. (a), the generator potential is a local, graded, nonpropagated shift in potential across the membrane of the sensory axon terminal within the corpuscle, induced by light pressure. The currents set up by the generator potential spread electronically and reach the first node of Ranvier, shifting the membrane potential toward depolarization. (b), as a result of stronger pressure, the membrane potential at the node reaches the critical level and an all-or-none spike develops. The spike will travel in a saltatory manner along the myelinated axon. (From Eldred, E., 1967, Peripheral receptors: Their excitation and relation to reflex patterns. *Am. J. Phys. Med.* 46: 71.)

impulses will be related to the intensity of the stimulus (Fig. 10.2). The fact that the generator potential and the impulse arise separately or are subserved by different mechanisms is proven by a variety of experiments. For example, various drugs abolish nerve impulses but leave the generator potential unaffected. In instances in which the receptor potential itself stimulates the initial segment or when the stimulus acts directly upon the afferent neuron, the receptor potential and the generator potential are one and the same.

In more complex sensory systems which include both specialized receptor cells and auxiliary structures, each sense organ comprises two intimately related mechanisms: one responsible for its specific sensitivity and the other responsible for transduction of the stimulus into a form of energy capable of discharging the neuronal terminals.

In sensory systems which include receptor cells distinct from the afferent neuron, the following sequence of events has been postulated in the transmission of information to the central nervous system.

1. The external stimulus acts upon the receptor cell, sometimes directly, sometimes through interposed auxiliary structures, and sets up a graded receptor potential in the receptor cell.

2. The receptor potential spreads electrotonically over the cell and at a receptor-neural junction causes secretion of a chemical mediator.*

3. The chemical mediator acts upon appropriate regions of the dendritic receiving pole of the afferent neuron, evoking a graded generator potential.

4. The generator potential spreads electrotonically over the dendrites and cell body to the initial segment. If it is sufficient to depolarize the membrane to its critical level, all-or-none nerve impulses are generated and conducted along the axon.

5. At the axon's terminals, chemical transmitter is secreted at synapses of the first central interneuron.

6. If sufficient summation of input occurs, excitation continues on through the central nervous system.

Regulation of Sensory Input

Threshold. It is well established that a single sensory neuron has a threshold requirement which must be satisfied by the stimulus in order to induce the all-or-none neuron response, that each neuron has a maximal frequency limit, and that each sensory system has a limited number of neurons which may be recruited by increased stimulus intensity. Therefore, there are limits inherent in the system below which no response is evoked and above which no increased response is achieved by increased stimulus intensity.

Adaptation. Nerve fibers readily adapt to the flow of directly applied constant current. After delivering one or a few impulses, the fiber becomes silent. Adaptation is also characteristic of receptor cells. The generator potential is at first proportional to the stimulus intensity but declines gradually during steady stimulation. The

rate of decline appears to depend upon mechanical coupling between the stimulus and the sensor element of the receptor and to be related more to mechanical properties of adventitious tissues than to the electrical properties of the sensor. Impulse frequency in the afferent neuron declines with the same time course as the generator potential and is probably related to depression or inactivation of the mechanisms responsible for the generator potential.

Adaptation and negative afterpotential have been shown to be related as to time course in some receptor neurons. Repetitive activity in the neuron may result in an accumulation of Na^+ or K^+ or both because of decreased activity of the sodium pump. As a result prolonged negative afterpotential is produced with concomitant reduction of irritability and decrease in the impulse frequency evoked by the continuing stimulus. In other words, adaptation may be related to electrogenic pumping (Goldberg and Lavin (1968)).

Some receptors adapt slowly and continue to maintain impulse frequencies with little decrease for as long as they are subjected to appropriate stimulation. For example, the muscle spindle afferents in posture (antigravity) muscles continue to fire as long as the muscles are under stretch. Such slowly adapting receptors mediate **tonic responses**. Rapidly adapting receptors are characterized by a sharp decline in frequency with sustained stimulation. They are concerned with **phasic responses**. For example, when the skin is in constant contact with clothes, receptors of hair follicles, which are movement detectors, quickly adapt to the slight pressure on the skin.

Some types of sensory systems include receptors which exhibit a range of adaptability as a function of time or some other parameter of the stimulus. For example, some sensory receptors in and around joints are excited by acceleration or deceleration of joint movement but adapt quickly when the movement ceases. Others adapt very little, maintaining a discharge rate related to the stationary angle of the joint.

---

* It should be remembered that not all receptors produce a receptor potential and not all receptor potentials cause the release of a chemical mediator. In fact, some authors believe that chemical transmission is the exception rather than the usual occurrence at receptor-neural junctions.

Central Regulation of Sensory Input. Responses of complex sense organs are determined not only by the fundamental properties of individual receptor cells but also by the influences which they exert on each other and the control exerted over them by the central nervous system.

Regulation by Presynaptic and Recurrent Inhibition. Strongly stimulated afferent neurons inhibit themselves and adjacent neurons by presynaptic inhibition. The most strongly stimulated cell exerts the greatest inhibitory effect. Therefore, since its neighbors exert less inhibition upon it, its impulse frequency is exaggerated in comparison with that of its neighbors. For example, the perception of contrasting temperatures in the skin of a limb partially immersed in hot (or very cold) water is greatest at the water-air boundary. The water feels much hotter here than where it contacts the skin below the surface. This may be explained by the supposition that presynaptic inhibition stemming from sensory neurons in the immersed skin is being exerted upon afferent neurons originating in both the strongly stimulated receptors beneath the water and the less stimulated receptors above the water. Because the more strongly stimulated cells exert the greater inhibitory effect, inhibition of the cells above the water line will be relatively greater than their excitation and their impulse frequencies will be much more reduced than will those of the cells in the skin areas contacted by the water. The difference in inputs from the two adjacent groups of receptors will thus be enhanced by the inhibition. Furthermore the neurons serving completely submerged receptors will be subject to inhibition from all sides, while those of water line receptors will receive inhibitory influences only from the submerged side. Combination of these two effects will lead to enhancement of perception localized at the water line. In other words, contrast has been sharpened.

It has been suggested that presynaptic inhibition functions by decreasing the magnitude of the generator potential. This appears to be true in both light and sound receptors.

Pain fibers are subject to inhibition by large afferents from skin, muscle, and the major sense receptors. Melzack and Wall (1965) proposed a "gate control theory" to explain this. Figure 10.3 illustrates their theory and shows that presynaptic inhibitory mechanisms are involved. The substantia gelatinosa exerts presynaptic inhibition on the axons of the large afferents synapsing with the interneuron transmitter cells. The stimulation "image" may also be sharpened by contrast induced by postsynaptic recurrent inhibition involving Renshaw cells.

Regulation by Efferent Neurons. The sensitivity of some receptors is controlled by the central nervous system through efferent neurons. The best studied example is the muscle spindle, discussed in detail in Chapter 11, whose sensitivity can be "set" by the central nervous system so as to maintain a particular rate of discharge under changing conditions. Visual, auditory, and olfactory receptors are also subject to central regulation by efferent fibers.

Receptor Specificity

Anatomical Basis of Specificity

Receptor cells are distinguished by their greater irritability to some particular type of stimulus, a property described as specificity. Specificity is based on one or more of the following factors which either restrict responsiveness to certain forms of environmental energy or contribute to a high level of sensitivity to one particular kind.

Structural Design. Most receptor cells and accessory organs are so adapted that their sensitivity is restricted to one type of stimulus. The eye is designed as an organ to handle light effectively, and the rods and cones are specialized to respond to the intensity and wavelength of the incident light energy. The cochlea is designed to receive and conduct sound vibrations, and different areas of its basilar membrane respond to different wave lengths.

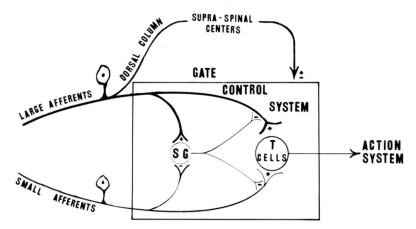

Figure 10.3. Diagram of the gate control theory of pain perception and response. SG, neurons in the substantia gelatinosa; T, first interneuron transmitter cell; +, excitatory synaptic action; −, inhibitory synaptic action. Note that SG exerts presynaptic inhibition on input to T cells. (After Melzack, R., and Wall, P.D., 1965. Pain mechanisms: A new theory. *Science 150*: 971.)

Location of Receptors. The site of a receptor often determines the stimuli to which it will respond. Auditory hair cells lie deep within the skull where they are inaccessible to stimuli other than sound waves. Skin receptors, located at the body surface, are exposed to many forms of stimulation. Stretch receptors in the chest wall are strategically placed to respond specifically to inflation of the thorax at inspiration. Pressure receptors in the carotid arteries are stimulated only by pressure in those blood vessels supplying the brain.

Topographical Arrangement of Pathways. Afferent pathways in the nervous system isolate transmission lines for one kind of receptor or one body region from those of others. Discrete cranial nerves and pathways serve the major senses. For example, information regarding position and movement of the legs travels in nerve bundles which are distinct from those serving the arms.

Central Terminations. The neural path for each sensory system has a definite destination in the central nervous system. The impulses aroused by pain reach neurons to flexor muscles. Impulses aroused by muscle stretch terminate on efferent neurons to motor units of the stretched muscle. Impulses from the retina travel to the visual areas in the occipital lobes of the cortex, while impulses from the cochlea reach the auditory areas of the temporal lobes.

In a number of cases specificity cannot be accounted for by any of the above. For example, there are several histological types of cutaneous receptors but specificity is poorly defined. Some temperature-sensitive afferents arise from receptors which appear to be specific for either coldness or warmth, but for others the response seems to be related not to receptor type but to axon diameter. The skin of the ear has few morphologically distinct receptor endings yet it has all of the cutaneous sensations. Itching seems to be a distinct sensation but no special receptors have been found. Sexual sensations do not arise from stimulation of histologically specific endings. Different responses arise from receptors which appear similar. Touch, pressure, and pain sensations can vary so much in quality that little is known regarding the types of endings and axons concerned. In some instances sensations seem to be dependent on *patterns* of stimulation and may involve two or more types of receptors.

It is possible that further study may reveal physiological specificity in receptors which at present appear identical. Vibration-specific receptors, once thought to be nonexistent, have recently been identified.

Sensitivity Range

Each sensory system transmits to the central nervous system information about stimuli which fall between minimal and maximal stimulus amplitudes of its particular modality. Some receptors respond over the full range of their sensitivity, with corresponding frequency variations. Others respond only to a limited portion of the range. For example, some cold receptors display optimal response at a particular temperature, and a whole population of receptors is required to cover the full excursion of a joint.

Some receptors appear to change their specificity at different regions within the stimulus range. There are some temperature receptors which respond to cold at low temperatures, become unresponsive as temperature rises and then, beyond a certain level, paradoxically become heat receptors.

There are receptors which fire only when the stimulus is applied ("on" receptors); some only fire as the stimulus is removed ("off" receptors). Others fire in both instances ("on-off" receptors). Examples are the Pacinian corpuscles specific for pressure, the touch receptors of the hair follicles, and certain visual receptors.

Specificity of Sensory Perception

Some but not all sensory stimulation results in perception. Although perception depends upon psychological factors related to past experience, there are certain parameters of sensation which have morphological or physicological bases.

Modality of Sensation. Different kinds of stimuli are distinguished as distinct modalities of sensation: light, sound, taste, smell, etc. Because all impulses in all neurons are qualitatively the same, in order to be correctly recognized as a particular sensation the impulses aroused by each particular type of stimulus must reach its special area of the cerebral cortex. Conversely, impulses arriving at a specific area of the cortex will be interpreted as having been aroused by that particular type of stimulus (or referred to a particular receptor location, as in the "phantom limb" phenomenon). For example, impulses arriving at the visual cortex, regardless of source, will be interpreted as light; those arriving at the auditory cortex, as sound. If the optic and auditory nerves could be cut and successfully cross-sutured, we would "hear" the lightning and "see" the thunder.

Quality of Sensation. Within a single modality we can identify different qualities: light consists of colors; smell of odors; taste of flavors; sound of tones; cutaneous sensations of touch, pressure, warmth, cold, pain. We cannot describe qualities of sensation except by comparison with some other quality; they are basically psychological. In some cases, however, the differentiation of qualities has an anatomical or physiological basis, such as the further specialization of receptors. Examples are the three types of cones in the retina, the different types of taste buds, and the maculae and ampullae of the labyrinth. Other bases may involve the overlapping of receptor fields, or certain combinations of afferent neurons from spatially separated receptors. Our knowledge, although extensive in some modalities, is still inadequate to explain fully all of the qualities of sensation.

Intensity of Sensation. Intensity is a parameter of sensation whose physiological basis is well known. Generally the areas supplied by sensory units overlap and interdigitate (Fig. 10.4). A weak stimulus activates only the receptors with the lowest thresholds but, as the intensity is increased, less irritable receptors are also activated, involving more sensory units. Some of the newly recruited receptors belong to the already active units and as a result their impulse frequencies rise. With further increase the influence of the stim-

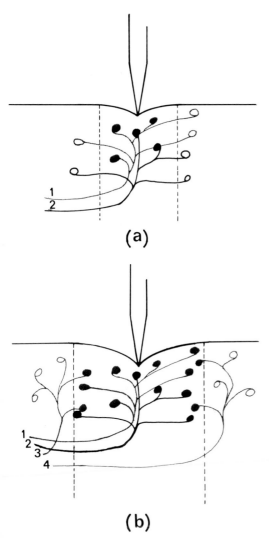

(a)

(b)

Figure 10.4. Recruitment of sensory units. Pressure exerted upon the skin surface by a probe activates receptors in the affected area. ●, active receptors; ○, inactive receptors. (a), light pressure activates five receptors served by the sensory units 1 and 2. Three of unit 1's six receptors and two of unit 2's five receptors are active. The sphere of stimulus influence is bounded by broken vertical lines. (b), stronger pressure widens the area of influence and activates more receptors in units 1 and 2 (now six in 1 and five in 2) and recruits two more units: 3 (two receptors) and 4 (two receptors). The greater number of active receptors in 1 and 2 will result in greater generator potentials and hence higher impulse frequencies in their sensory nerve fibers.

ulus tends to spread over a larger area, activating sense organs beyond the actual sphere of contact and adding to the afferent inflow. In short, the stronger the stimulus the greater will be the number of active sensory neurons and the greater the impulse frequency in each. The more intense the bombardment of cortical centers the stronger the sensation.

THE MOTOR SYSTEM: MOTOR UNITS

The impulses aroused in sensory units traverse the neural pathways of the central nervous system. Upon reaching cortical centers, some impulses produce conscious sensation or evoke discriminative thought processes which may ultimately result in movement. Following shorter routes, others directly induce muscular response without cortical intervention. Movement is the most common response to stimulation.

Like the sensory units, which are composed of receptors and afferent neurons, the units of the motor system include efferent neurons and muscle fibers. The cell bodies of the motor (efferent) neurons lie in the ventral horn of the gray matter in the spinal cord or in the motor nuclei of the cranial nerves. All of the cell bodies of the efferent neurons serving a particular muscle are gathered together topographically and constitute the **motor neuron pool** for that muscle. Their axons leave the cord through ventral roots and travel peripherally with the appropriate spinal nerve to the specific muscle nerve. Because a muscle nerve contains fewer motor neurons than there are muscle fibers in the muscle, it is obvious that each motor neuron must supply a number of muscle fibers. **A motor unit**, therefore, consists of one motor neuron, its axon branches, and all of the muscle fibers which they innervate (Fig. 10.5).

In each muscle, motor units vary in size within a characteristic range. The mean size may comprise as few as three to six muscle fibers, as in the extrinsic eye muscles, or more than 1700 as in the gas-

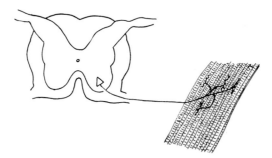

Figure 10.5. The motor unit. A single motor neuron is shown sending its axon via the ventral root to a muscle where its telodendria innervate four muscle fibers distributed within a restricted area of the muscle.

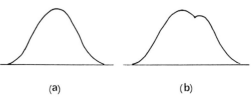

(a)   (b)

Figure 10.6. The motor unit is not all-or-none. (a), a myogram of a twitch response to a single maximal pulse. (b), two stimuli were delivered in such rapid succession that the second fell during the absolute refractory periods of the muscle fibers responding to the first pulse. The second peak is due to recruitment of fibers for which the first stimulus was inadequate. They are now responding to summation of the two stimuli. (After Ralston, H.J., 1957. Recent advances in neuromuscular physiology. *Am. J. Phys. Med.* 26: 94.)

trocnemius muscle.† Mean motor unit size is related to the function of the muscle. Muscles concerned with strength or endurance, such as the trunk and leg muscles, are composed of large motor units, while those involved in fine or manipulative movements, such as the muscles of hand, face, and eyes, have small units.

The various fibers of a single motor unit are not aggregated into one bundle but are scattered among a number of muscle fascicles within a localized area of the muscle belly, interdigitating with fibers of other units. Impulses traversing the motor neuron reach all of the neuromuscular junctions where its axon forms synaptic end-plates with the muscle fibers. All muscle fibers of the unit whose thresholds are reached will respond together. However, since muscle fiber thresholds are not identical, the neural discharge may fail to activate some fibers. The muscle's tension will be determined by the total number of active muscle fibers, which in turn will vary with impulse frequencies in the motor neurons. In other words, the motor unit

is *not* all-or-none in its response. This can be demonstrated by applying two stimuli in such rapid succession that the second falls during the absolute refractory period of any fibers responding to the first stimulus. The twitch myogram will show a second peak, indicating that some fibers which were unresponsive to the first stimulus have been recruited by summation of the two stimuli which alone are inadequate to trigger additional fiber responses (Fig. 10.6). A further indication is found in the fact that the motor unit potentials‡ recorded from the muscle surface by the electromyograph increase in amplitude with increase of excitation or stretch tension. Recruitment of lazy fibers may be implicated in the greater tension of tetanic contraction as compared with a single twitch. It is also possible that the stretch response may be due to some favorable effect of stretch upon the irritability of

---

† Motor unit size may be calculated by dividing the total number of muscle fibers, estimated from the muscle's volume and architecture, by the number of afferent neurons in the muscle nerve. Examples: platysma, 25; first lumbrical, 95; tibialis anterior, 609; medial gastrocnemius, 1775. An average figure for all muscles of the body is about 180.

---

‡ A motor unit potential is the deflection on the electromyogram produced by the summation of the muscle fiber action potentials of one motor unit. Each appears as a spike on the oscilloscope or graphic record, the size of the spike being proportional to the size of the motor unit. A motor unit potential is recognized by its size and regularity of occurrence, both of which are relatively constant for any given set of conditions.

muscle fibers, perhaps related to an increased release of Ca^{++}.

Although the motor unit does not respond all-or-none, it does represent the indivisible functional unit of neuromuscular function. The greater the stimulation the more motor neurons respond and the greater their firing rates. As more motor units become active and their frequencies rise, muscle tension increases as each motor neuron arouses its own group of muscle fibers to tetanic contraction. The contractile tension exerted by a muscle is determined by the total number of muscle fibers which are contracting, each to its all-or-none capacity. Tension increases progressively in a graded manner as more and more motor units are recruited. The increments of gradation are directly related to the mean size of the motor units composing the muscle. Muscles with relatively small motor units are capable of small variations in tension and hence of delicate movements, while the greater increments of muscles with large motor units are suited to strong, gross movements.

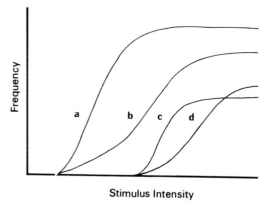

Stimulus Intensity

Figure 10.7. Motor unit curves, showing thresholds and frequencies. (a), a motor unit with a low threshold, rapid rise in firing frequency, and a high maximum. (b), a unit with a low threshold but a slower rise and lower maximal frequency. (c), a unit with a high threshold, rapid rise, and low maximal frequency. (d), a unit with a high threshold, slower rise, and moderate maximal frequency. Note (1) that, while (a) and (b) have the same threshold, their rise rates and maxima differ; (c) and (d) likewise; and (2) that, while (a) and (c) (and (b) and (d)) are shown with similar rise rates, their thresholds and peak frequencies differ.

Response Characteristics of Motor Units

Threshold and Frequency

Some motor units begin to respond at low stimulus intensities and show a rapid increase in frequency which then levels off (Fig. 10.7, curve (a)). Others may have equally low thresholds but a slower change in frequency with increased stimulus intensity (curve (b)). High threshold units show similar differences in frequency rise patterns (curves (c) and (d)). Frequency peaks also vary among neurons with the same thresholds and the same frequency increase rates (compare curves (a) and (b) with (c) and (d)). In general, low threshold units show greater discharge rates than high threshold units, but the latter produce motor unit potentials of higher amplitude. The motor units with high thresholds, large spikes, and relatively low frequencies tend to be recruited only when high tensions are demanded of the muscle. Frequencies in motor neurons generally range between 25 and 40/sec, rarely exceeding 50/sec and reaching the range of 40 to 50/sec only when tension is over 75% of maximal voluntary contraction.

Discharge Patterns

Motor unit discharge frequencies, as indicated by electromyograph potentials, are not always rhythmic. Strong stimuli may evoke bursts at peak frequency, alternating with intervals of lower frequency or even silence. Some respond to any stimulation with intermittent bursts in a regularly repeated pattern.§

A motor unit may fire only over a

---

§ Similar neural response patterns exist among interneurons in the central nervous system and are considered to constitute one aspect of the information coding which is essential for the discriminative and integrative functions of the nervous system.

specific tension range, becoming silent as tension rises above its peak and resuming as declining tension returns to its response range. Such silent periods during strong stimulation may be due to a Renshaw-like inhibition imposed autogenically or by higher threshold neurons.

Fast and Slow Motor Units

Among motor units within the same muscle, differences of response are observed which are related to the sensitivities and response characteristics of motor neurons. The ventral (efferent) roots of motor nerves show a heterogeneous population of nerve fibers which fall into two subdivisions, alpha and gamma, of the class A category of the Erlanger-Gasser classification. (See Table 8.1 for Erlanger-Gasser classification.) More than half (about 70%) of the motor fibers have diameters and conduction velocities which place them in the A_α division, and these are known as **alpha motor neurons**. Their characteristic sizes and speeds, however, separate into a bimodal distribution with peaks toward the upper and lower ends of the alpha range. Therefore, two groups of alpha motor neurons are recognized: one containing the largest diameter fibers with high velocities, and another group of smaller fibers with velocities which are

distinctly lower but still within the alpha range. The former have been designated as **fast** motor neurons and the latter as **slow** motor neurons.

The descriptive terms **phasic** and **tonic** are also applied to these alpha motor neurons. Table 10.1 compares phasic (fast) and tonic (slow) motor neurons. These two groups of alpha motor neurons constitute the neural components of motor units, and their differences are reflected in differences in the response characteristics of the motor units.

The remainder (about 30%) of motor neurons in the muscle nerve fall into the gamma division of the Erlanger-Gasser group A and are hence known as **gamma motor neurons**. They do not innervate the muscle fibers but supply the muscle spindles, performing an important function in the regulation of muscle activity which is discussed in considerable detail in Chapter 11.

In terms of twitch time course, motor unit responses fall into two classes: **fast** and **slow**. Muscles composed of a mixture of muscle fiber types contain both fast and slow motor units. Each single motor unit is homogeneous, containing either fast or slow muscle fibers but not both. No instances of heterogeneous motor units have been found. The fast units are innervated

Table 10.1
Comparison of Phasic and Tonic Motor Neurons

Characteristic	Phasic Motor Neurons	Tonic Motor Neurons
Fiber size	Large	Smaller
Conduction velocity	Rapid	Slow
Threshold	Low to electricity	High
	High to physiological stimuli (i.e., stretch)	Low
Impulses		
Size of spikes	Large	Small
Spike duration	Short	Longer
Afterpotential (hyperpolarization)	Short	Prolonged
Response to muscle stretch		
Frequency	Low	High
Pattern	One or two discharges	Prolonged discharge
Rest state	Electrical silence	Tonic activity

by phasic (fast) alpha motor neurons and the slow units by tonic (slow) motor neurons.

Role of Motor Neuron in Determining Properties of Its Motor Units. In many mammals, all of the limb muscles are slow at birth, although nerve conduction velocities are already either fast or slow. In the adult animal, however, the muscles also have differentiated into fast and slow. Study of the course of postnatal differentiation has shown that the distinction between the two types of muscles is brought about by a relative increase in the shortening speed of the muscles which are genetically predetermined to be fast muscles, with little change occurring in the genetically slow muscles.

If a normally slow muscle is denervated soon after birth and then allowed to be reinnervated by a nerve which normally supplies a fast muscle, the slow muscle is changed into a fast muscle. The opposite effect can also be accomplished and what should have been a fast muscle is converted into a slow one.

In the adult the normally slow soleus muscle can be transformed into a histochemically and physiologically fast muscle by anastomosing its nerve to the nerve that originally served a fast muscle, such as the peroneus. Conversely, the normally pale and fast gracilis can be made red and slow by cross-innervation with the normally slow crureus muscle. In both instances reinnervation by its own nerve produces no significant change, indicating that surgery is not responsible.

A large number of such cross-union and hetero-innervation experiments have demonstrated that metabolic and contractile changes are induced by a "foreign" nerve, leaving little doubt that motor neuron influence is an important, and perhaps the prime, factor in determining the nature and properties of the muscle fibers. This conclusion is in agreement with the concept of homogeneity of the muscle fiber composition of motor units.

It has been suggested that a motor neuron influences differentiation of its muscle fibers trophically. Chemical sub-

stances that are moved by axoplasmic flow down the motor neuron and across the neuromuscular junction are postulated as mediators. Axoplasmic flow has been well established, and its presence is associated with the time of differentiation both in normal postnatal development and during reinnervation.

Several theories have been proposed to explain conversion of one type of muscle to the other. Some workers think that the change from slow to fast may be produced by an alteration of the time course of the fiber's active state. A second hypothesis implicates the mean frequency discharge of the motor neurons, for there seems to be an inverse relationship between impulse frequency and contraction speed of the muscle fibers. Relatively high frequencies are associated with slow motor units and *vice versa*. Experimental results have provided support for both hypotheses. Low frequency stimulation by artificial means or absence of stimulation, as assured by appropriate surgical techniques, have been found to transform the characteristics of slow muscles in the direction of fast muscle, while continuous or high frequency stimulations have converted fast muscles into slow ones. These relationships are consistent with the theory that the tonic activity of slow muscles is due to their constant postural stimulation by gravity. These experiments also show that skeletal muscle is capable of responding adaptively to the type of activity demanded of it, a factor of considerable medical importance and one which may be significant in training and in the development of motor skill.

"Catch" Mechanism in Slow Motor Units. A study by Burke et al., (1970) of motor units in the cat triceps surae (the medial and lateral gastrocnemius muscles and the soleus) has revealed a phenomenon in the slow motor units resembling the "catch" mechanism of crustaceans and mollusks by which certain muscles are able to exert prolonged tensions without a large energy expenditure. It was found that in slow motor units interruption of a background stimulation frequency by introducing an additional stimulus produced a marked prolongation of normal tetanic enhancement (Fig. 10.8). Apparently the shortening of one interstimulus interval in the rhythmic frequency extended the range of output tension that could be produced by a given unit without a large change in the mean firing frequency of its motor neuron. The background frequencies which were optimal for the phenomenon were found to be in the same range observed to be characteristic of slow type motor units when activated by maintained stretch, but it was not clear whether such a pattern of firing is actually utilized by motor units participating in normal behavior.

Three Types of Motor Units. Since the identification of three types of muscle fibers (A, B, and C), evidence has been accumulating for the existence of three types of motor units. These have been designated as motor units A, B, and C or, by some authors, as types I (fast), intermediate, and II (slow). The behavior of each unit suggests that it is homogeneous for one type of muscle fiber. The type A motor units represent the classic fast or phasic units discussed above. Types B and C represent two distinguishable types of slow or tonic motor units, with type B intermediate between A and C in some of its properties and similar to C in others.

The flexor digitorum muscle of the rat is homogeneous for type A units. The rat soleus muscle, although containing only slow units, is heterogeneous for both slow types, B and C. Some muscles contain all three types of motor units in various proportions. The cat and rat gastrocnemius muscles include types A, B, and C, with the fast type A units predominating (Close (1967); Wuerker et al. (1965)). These structural differences between the gastrocnemius and soleus muscles appear to be correlated with their functional roles. Although both plantar flex the ankle, sharing a common tendon, the gastrocnemius is designed for uneconomical rapid, powerful contraction, and the soleus for endurance. The power and speed of ankle plan-

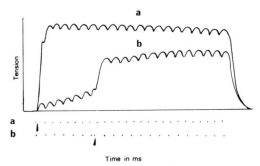

Figure 10.8. Effect of alteration in interstimulus interval. Curve (a), initial enhancement by a double stimulus (indicated by arrow on frequency trace (a) below curves). Curve (b), no initial enhancement; later enhanced by one short interstimulus interval (indicated by the arrow below the stimulus trace for (b)). Curves represent tension responses of a slow motor unit in the cat gastrocnemius to interruption of a basic rhythmic stimulation. Tension traces (solid lines) are labeled (a) and (b) and their pulse sequences (dotted lines) are similarly designated. Intervals of the train were about 100 msec. The arrow at the first pulse in (a) indicates a double stimulus with a very short (about 10 msec) interstimulus interval used to produce strong initial mechanical enhancement (twitch fusion). The arrow in (b) denotes the introduction of an added pulse, resulting in one interstimulus interval which is shorter than that of the basic train (but longer than the 10 msec interval of the initial summation). In (b) the initial double stimulation was omitted; hence there was no large initial enhancement. (Adapted from Burke, R.E., Rudomin, P., and Kajac, F.E., 1970. Catch property in single mammalian motor units. *Science 168*: 122.)

tar flexion which are required for running and jumping can be supplied by the gastrocnemius but this muscle cannot appreciably shorten owing to the shortness of its fibers and their oblique arrangement in the muscle. Also it fatigues quickly. The soleus, with longer parallel fibers and greater endurance, "covers" for the inadequacies of the gastrocnemius. It can shorten enough to accomplish full ankle plantar flexion and has sufficient endurance to oppose gravity for prolonged periods. Henneman and Olson (1965) have suggested that separate heads were developed in response to need, each designed for its own special purpose and each complementing the other; nature had discovered that two heads were better than one!

The three types of motor units, when all are present, seem to participate in contraction in a definite sequence, each being recruited at its own tension level and contributing to a particular tension range. The cat gastrocnemius muscle provides a good example. It contains all three types of motor units, with type A predominating. Weak stretch was found to arouse the low threshold tonic motor neurons and their type C motor units became active. With increased stimulation, other units were activated as their thresholds were reached. The order of recruitment was inversely related to velocity, type C units being followed by types B and A at respectively greater intensities. As more and faster units were added, the previously active units dropped out. Because new motor units are recruited before previous ones have reached their maxima, smoothness of contraction is assured.

Henneman and Olson (1965) have pointed out the probable significance of the recruitment order. First, the C-B-A order makes possible relatively small tension increases at all tension levels. At low tensions the small-fibered type C motor units provide small progressive increments. At higher tensions, although the motor units are composed of larger muscle fibers, their tension increments still contribute only a small percentage of increase in the total tension already present at the time of their recruitment. As a result, fine control is possible at all tension levels. Second, because the small-fibered type C units act first, they are subjected to the most frequent use. Their slower contractions and oxidative metabolism are thus a distinct advantage, consistent with the early and frequent demand placed upon them by their recruitment primacy.

THE INTEGRATIVE SYSTEM: NEURAL CIRCUITS

Essentially the integrative system is composed of neural circuits interconnected in all possible combinations. Circuits range from simple two-neuron pathways to devious networks of infinite complexity. The brain receives an extensive input of information compounded from all of the stimuli playing upon the body's receptors. By some means not yet understood, incoming signals are analyzed, appraised, and interpreted against a background of past experience. In some remarkable manner a decision is made as to the appropriate response, and the outgoing signal is correspondingly fabricated so as to regulate all the muscles concerned in that response. Limitless variation in impulse transmission is possible by means of the simple expedient of varying synaptic facilitation and inhibition within the circuitry.

Integrative Function of the Neuron

The neuron itself is undoubtedly capable of some degree of integrative action. Experimental evidence is accruing which suggests that the neuron is much more than a passive conductor of information. Rather it probably has far greater potentialities for the analysis and abstraction of the afferent patterns impinging upon it than was thought. Apparent integrative capacity has been noted in individual neurons of sense organs, interneurons, and motor neurons. Multiple trigger zones, rather than a single trigger area restricted to the initial axonal segment, have been identified in some types of neurons. Bullock (1959) suggested that as many as four or five different kinds of circumscribed loci in various parts of the neuron may function to exercise some evaluative action, so that the cell does not pass on the information it receives merely in a one-to-one ratio.

Dendrites, which for some time were

thought capable only of nonpropagated electrotonic response to stimulation, have been shown to have definite trigger zones in some instances. Considering the extensiveness and complexity of dendritic trees, especially in the brain and cerebellum, multiple trigger zones and propagated impulses in this region of the nerve cell would greatly increase the potentiality for integrative processing of afferent input than is possible in the case of a single-trigger cell. Furthermore, the pattern of synaptic firing then becomes a parameter for integration. Still other parameters are input frequency, number of active endings, pattern of firing bursts, and coincidence or succession of firing of different pathways, all of which must be considered in a complete analysis of the neuron's integrative role.

Coordinative Centers

During embryological development, nerve cell bodies migrate to form nerve centers strategically located within the central nervous system so that specific types of sensory input may readily be translated into motor outflow which will produce appropriate responses. Centers are interconnected to permit the coordination of responses and to allow for variation as may be required by changing environmental conditions. In the reflex centers of the spinal cord, extensive and important integrative activity occurs, and the fundamental interregulation of individual muscles is accomplished automatically there. Centers of the brainstem, cerebellum, and subcortical nuclei integrate activity in the various parts of the body and adjust muscle tone for balance and equilibrium. Conscious centers of the cerebral cortex are capable of imposing voluntary modification upon any part of and at any point in a movement pattern by suitably altering the descending signal. The reader is referred to Chapter 9 for information on the organization and function of the central nervous system.

So far as is presently known, the complex circuitry of the integrative system provides the mechanism for all of the functions of the nervous system, both conscious and unconscious, even including the more abstract faculties such as thought, judgment, sensation, and emotion.

Types of Neural Circuits: Number of Neurons

Two-Neuron Circuit

While most circuits include at least three neurons (an afferent neuron, an interneuron, and an efferent neuron), a simple spinal circuit consisting only of an afferent and an efferent neuron plays a significant role in regulating muscular contraction. This is the pathway of the passive stretch reflex and includes only the afferent neuron from the muscle spindle stretch receptor and the efferent (motor) neuron to the contractile elements of the same muscle fibers, the two neurons synapsing in the spinal cord without an intervening interneuron.

Three or More Neuron Circuits

Except for the stretch reflex circuit, pathways in the spinal cord and brain contain all three types of neurons. Circuits of only three neurons, one of each type, are the exception rather than the rule, however. Generally the central portions of the circuit, the interneurons, are multiple, forming chains and networks. As a result, basically simple circuits are converted into complex pathways.

There are essentially three basic circuit types: (a) divergent circuits, by which a single receptor may influence many effectors; (b) convergent circuits, by which a number of different receptors may influence the same effector; (c) repeating circuits, by which a single input is multiplied a number of times.

Divergent Circuits

Figure 10.9(a) shows diagrammatically a simple **divergent circuit**. The axon of the afferent neuron (1) branches to synapse

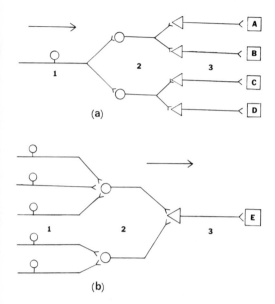

(a)

(b)

different lengths and with a variety of branching patterns so that a stimulus impinging upon a single receptor organ, e.g., the eye, can evoke responses involving many parts of the body. Further illustrations include a noxious skin stimulus which evokes a mass withdrawal response, or a sudden loud sound which may produce a total body response.

Convergent Circuits

Figure 10.9(b) presents a simple **convergent** pattern. Axons of several afferent neurons (1) synapse with two interneurons (2) whose axons in turn synapse with a single efferent neuron (3) serving the effector (E). As a result, impulses arising in a number of receptor organs converge upon a single effector, thus amplifying the signal to the muscle. Again the diagram is over-simplified but serves to present the basic principle.

Repeating Circuits

Two basic types of repeating circuits are known, reverberating and parallel circuits, in both of which a single input results in repetitive firing of efferent neurons.

Figure 10.10(a) illustrates a **reverberating circuit**. The afferent neuron (1) synapses with a chain of interneurons (2, 3, and 4), with the fourth neuron transmitting the signal to the efferent neuron (5). Impulses traveling this circuit, however, reverberate through an axon collateral from one of the chain (3) to restimulate a neuron (2) situated earlier in the chain. Repetitive firing will continue to activate the effector (F) until terminated by fatigue or by inhibition imposed through another circuit.

The **parallel circuit** is shown in Figure 10.10(b). A linear chain of interneurons (2, 3, and 4) connects the afferent (1) with the efferent neuron (5). Collaterals from these interneurons (2 and 3) also project to synapses of the efferent neuron, however, so that in the case illustrated a single input will stimulate the efferent not once but

Figure 10.9. Basic circuits: divergent and convergent (diagrammatic). (a), a simple divergent circuit. The axon of an afferent neuron (1) branches to synapse with two interneurons (2) which send collaterals to synapse, each with a different efferent neuron (3). As a result, impulses originating in the single afferent neuron reach four different effectors, A through D. (b), a simple convergent circuit. Axons of several afferent neurons (1) synapse with two interneurons (2) whose axons in turn synapse with a single efferent neuron (3) serving effector E. As a result, impulses arising in a number of receptors converge upon a single effector amplifying the signal. These schematic diagrams are intended merely to convey the basic principles of circuitry within the nervous system. Arrows indicate the direction of transmission through each circuit.

with two interneurons (2) which also send collaterals, each synapsing with a different efferent neuron (3). As a result, impulses originating in the single afferent reach four different effectors, A, B, C, and D. The circuit pictured is overly simplified and geometric in form. The central components, namely the interneurons, characteristically occur not singly but in chains of

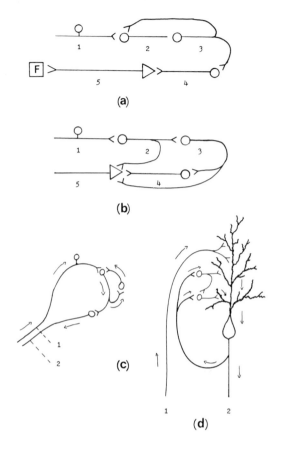

Figure 10.10. Basic repeating circuits (diagrammatic). (a), a simple reverberating or oscillating circuit. An afferent neuron (1) synapses with a chain of interneurons (2, 3, and 4) with the last (4) transmitting the signal to an efferent neuron (5). Impulses traveling through the circuit, however, reverberate through an axon collateral of one of the interneurons (3) to restimulate a neuron (2) situated earlier in the chain. Repetitive firing will continue to activate the effector F until terminated by fatigue or by inhibition imposed through another interconnected chain. (b), a simple parallel circuit. A linear chain of interneurons (2, 3, and 4) connects afferent (1) with efferent neuron (5). Collaterals from the interneurons also project forward to synapse on the efferent, however, so that in the case illustrated a single input will stimulate the efferent neuron not once but three times. Because of synaptic delay the arrival of impulses from collaterals 2 and 3 will precede the arrival of 4. (c), a reverberating reflex circuit. A simple reverberating circuit which may be responsible for continuation of a response after the stimulus ceases (the dog's scratch reflex, for example). Such circuits require damping by cerebellar function to obviate oscillation. 1, afferent neuron; 2, efferent neuron. (d) a reverberating cerebral circuit. An ascending fiber from the thalamus (1) excites a large pyramidal cell whose descending corticospinal fiber (2) gives off a collateral branch. The collateral stimulates two or more interneurons which have excitatory terminals synapsing upon this same pyramidal cell. Interconnection between the interneurons further increases the repetitive firing of the pyramidal cell. As a result, a single ascending volley may be multiplied many times in a second. The number and complexity of such circuits may be important in the rate of learning of which an individual is capable. ((c) and (d) from Gardner, E.B., 1967. The neurophysiological basis of motor learning—A review. *J. Am. Phys. Ther. Assoc. 47:* 1115.)

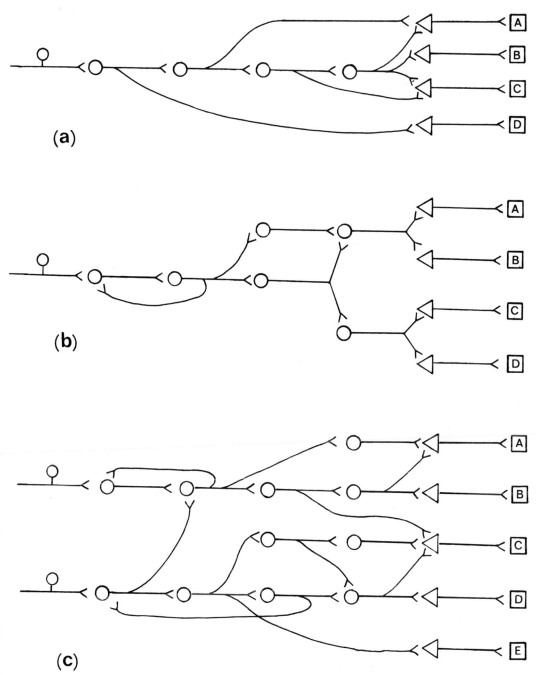

Figure 10.11. Some hypothetical combinations of basic circuits. These relatively simple diagrammatic combinations suggest the manner in which complexity may be introduced in central nervous system circuitry. The student should determine the response which will be induced in each effector and the order of sequential responses. The possibilities of combinations are obviously limitless. Further modification by interposition of appropriate facilitatory and inhibitory circuits provides an adequate basis for the observed complexity of reflex response patterns.

successively (three times), with synaptic delay determining the order of their arrival. Complexity is readily introduced into the repeating circuits by collaterals supplying additional efferent pathways or interconnecting with other circuits of the same or different type. Interposition of inhibitory neurons assures suitable direction of excitatory flow.

Combination of Circuits

The possibilities of combination are obviously limitless. Consider the relatively simple combinations suggested by the diagrams in Figure 10.11 and determine the response which will be induced in each effector. With further modification by appropriate facilitation and inhibition, these combinations of simple circuits provide a basis for the observed complexity of response patterns.

SUMMARY

The neurons of the nervous system are arranged in a complex but orderly manner. Afferent and efferent neurons are combined to form the peripheral nerves which connect receptors and effectors with the central nervous system. The interneural chains, which complete the circuits from receptor to effector, compose the spinal cord and brain. Cell bodies serving specific functions are aggregated into clusters (nuclei or centers) from which axons travel in groups (tracts) to distant destinations. Collateral branches leave the longer tracts to provide interconnections with centers at various levels of brain and cord. Interconnections between circuits, properly modified by facilitatory and inhibitory influences, provide the framework for coordinated movement patterns.

CHAPTER 11

Proprioceptors and Allied Reflexes*

INTRODUCTION

All animals, including man, are born with genetically built-in neural circuits which are preprogrammed by modification of synaptic transmission to produce stereotyped response patterns useful to the species. These are not learned. They are present at birth or appear as the developing nervous system progresses to completion. They represent the heritage of the species. Genetic material prescribes and ontogeny builds the components for the types of movement characteristic of human behavior. These components include not only the bones, joints, muscle attachments, and nerve supply, both afferent and efferent, but also appropriate interconnections and patterns of facilitation and inhibition. A child is born with a repertory of a few hundred movements which compose

the raw material of motor learning. The modification and recombination of these in all possible ways results in the acquisition of additional motor patterns, some of which are very different from inherited patterns. These are learned motor skills.

Stereotyped responses in the form of simple human reflexes are well known, such as the stretch reflex; withdrawal (flexion) reflex; extensor thrust and the positive supporting (extension) reflexes; crossed extensor reflex; righting reflexes; placing reactions and others. Some of these appear to be very simple; others must be amazingly complex. Some are fully formed at birth, while others develop as natural expansions of these during the maturation of the neuromuscular system. It should be noted that, to the term **reflex**, a variety of meanings have been ascribed. Neurophysiologists tend to include automatic patterns mediated at all levels of the central nervous system; psychologists argue that the "true" reflex is spinal cord mediated and all other motor patterns, mediated at higher levels, cannot truly be called reflex. The former description most

---

* Some of the material and several figures in this chapter have appeared previously in Quest, Monograph XII, 1969, pp. 1–25 (authored by Elizabeth B. Gardner), and are reproduced here with permission of the publishers.

adequately conveys the authors' use of the word.

Each of these reflex patterns consists of a coordinated combination of several to many joint movements, and each joint movement further consists of a coordinated combination of muscle actions: contraction of prime movers, relaxation of antagonists, and supportive contractions of synergists and stabilizers. All of these must be precisely regulated in regard to their intensity, speed, duration, and sequential changes in activity from the beginning to the end of the movement. This requires a considerable amount of integrative function which is largely automatic and unconscious.

Muscle-to-muscle integration is accomplished by basic reflex reactions which are initiated by receptors strategically located to feed back information to the central nervous system. Information must be received continuously regarding body position, muscle length and tension, speed, range and angle of movement, acceleration of the body or its parts, and balance and equilibrium. This information must then be integrated by cord and lower brain centers and converted into a suitable modification of the impulse outflow to produce immediate adjustment of each muscle concerned. As the state of a muscle changes, the information input will also change, evoking re-modifications in never-ending succession. Much of the information also becomes available to centers in the conscious areas of the brain where it may be sorted, analyzed, interpreted, and converted into an outflow of signals to modify voluntary body movements appropriately as occasion demands.

Voluntary movement requires a foundation of automatic responses which assure a proper combination of mobility and stability of body parts. Since activity occurs in many muscles simultaneously or sequentially, precise regulation is essential. Fortunately, neural control of muscles, whether activity is unconscious or deliberate, is mostly involuntary: muscles are smoothly regulated by reflex mechanisms. The voluntary contribution to movement is almost entirely limited to initiation, regulation of speed, force, range, and direction, and termination of the movement. Volition does not normally include control of individual muscles, although the human capability of doing so and even of controlling single motor units has been amply demonstrated. For example, reaching for an object is voluntarily prescribed as to direction, speed, and the object sought; but the functional features of shoulder girdle fixation, elbow extension, wrist stabilization, and finger movement are regulated by subcortical mechanisms.

Two contrasting neurophysiological hypotheses exist regarding the subconscious regulation of the many muscles concerned in the behavior subserved by neural circuits; the central control hypothesis and the peripheral control hypothesis. The **central control hypothesis** postulates that an unspecific input, or even the central nervous system itself, triggers activity in a nerve net which has been genetically structured (in terms of mutual excitatory and inhibitory influences) so that, once activated, the total pattern proceeds automatically. In invertebrates numerous instances are known in which specifically identifiable interneurons drive "follower" cells to produce relatively complex coordinated movements. Because in these cases the interneurons must be continually active for the response to proceed, it is obvious that they must be under the control of a preceding triggering stimulus. The stimulus may be a single brief event. Once initiated, the movement sequence proceeds without requiring further regulatory input. In vertebrates some instances are known in which stimulation of selected hypothalamic or cortical regions elicits complex motor acts or sensory experience, and the response often greatly outlasts the stimulus. However, little is known regarding the neurons or pathways involved (Willows and Hoyle (1969)).

The **peripheral control hypothesis**, in contrast, attributes sequential activation of various component circuits of the nerve net to specific inputs from **proprioceptors**, each of which triggers either excitatory or inhibitory output to appropriate muscle pathways. Most neurophysiologists have favored the latter hypothesis because proprioceptive loops have been so convincingly demonstrated. It seems highly probable that in man there is no movement pattern controlled purely by either one of these methods but rather that features of both are involved. While man's basic motor patterns are probably genetically coded in the species and laid down during ontogeny as specifically structured interconnected nerve circuits, activity within the nerve nets appears to be adjusted to the moment-by-moment changing conditions of environment and/or body orientation (changes which could not be anticipated genetically) by proprioceptive feedback.

Although little is known regarding the exact structure of vertebrate nerve nets, a large amount of information has accumulated regarding proprioceptors and their structure, mode of function, and reflex effects upon the activity of muscles. Some of this knowledge has been derived from clinical and laboratory studies on humans, but the greater portion and the most detailed information has come from the study of other mammals, especially monkeys and cats. It is not wise to transpose indiscriminately from one species to another but, where interspecies similarities and instances of parallel evidence exist, speculation regarding the operation of the same mechanisms in man is justified, especially as the basis for the formulation of hypotheses and for the design of experiments.

According to Sherrington (1906), proprioceptors are those end organs which are stimulated by actions of the body itself. They are somatic sensory organs located so as to secure inside information and to bring about cooperation and coordination among muscles effectively. The nervous system uses these sensory receptors to modify and adjust muscle function so that peripheral automatic (subconscious) regulation will dominate in most of our so-called voluntary or volitional movements. When proprioceptors are stimulated by movement or position, impulses traverse neural chains to act upon muscles in diverse and interrelated ways. By exciting various proprioceptors, contraction of any muscle tends to organize other muscles to cooperate with it. In the parlance of the electronic engineer, these reflexes operate as negative feedback loops by means of which motor activity becomes in large measure self-regulating. In other words, aspects of the movement process such as muscle tension, absolute muscle length, velocity of change in muscle length, joint angle, joint movement, head position, and contact with surfaces act as stimuli to initiate signals in nerve fibers which are fed back into the central nervous system. In some way as yet unknown, this information is compared with the desired pattern which nature, past experience, or conditioning has established. If the afferent signal indicates any divergence from this pattern, centers in the nervous system modify efferent signals so that the activity of the proper muscles is appropriately increased or decreased to correct the difference.

Proprioceptors may be conveniently classified into three groups: the muscle proprioceptors, the proprioceptors of the joints and skin, and the labyrinthine and neck proprioceptors.

MUSCLE PROPRIOCEPTORS

The muscle proprioceptors are the **neuromuscular spindle** and the **Golgi tendon organ**, both of which are incorporated into the gross structure of the muscle itself. Powers (1976) also includes the **Vater-Pacinian corpuscle**, located in muscle.

Neuromuscular Spindles

The term, neuromuscular spindle, is the one usually employed as being most descriptive of the receptor, although the terms, muscle spindle or spindle, are often seen and are equally acceptable.

Neuromuscular spindles are highly specialized sense organs which are distributed among the bundles of contractile fibers in the muscles. They are found

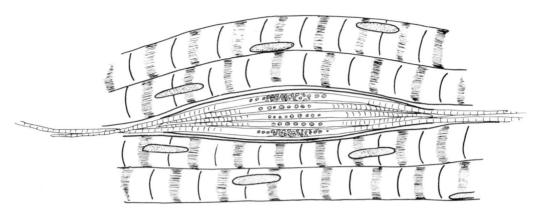

Figure 11.1. The neuromuscular spindle. A diagrammatic representation of a muscle spindle *in situ*, longitudinal section, lying in parallel with the extrafusal or contractile muscle fibers. A spindle consists of a fluid-filled capsule containing small intrafusal muscle fibers of which there are two major types. The two outer intrafusals represent nuclear bag fibers which are percapsular. The three lying centrally represent nuclear chain intrafusal fibers and are intracapsular. Innervation is omitted. (From Gardner, E. B., 1969, Proprioceptive reflexes and their participation in motor skills. *Quest XII*: Fig. 1-B, p. 5.)

throughout the mass of the muscle but tend to be more concentrated in the central portion. There are more spindles in man's phasic muscles than in his tonic (postural) muscles, as would be expected since the former require more precise control. The neuromuscular spindle is probably the most important and surely the most complex of the proprioceptive receptors. One usually expects, and finds, structural complexity associated with functional complexity. In the case of the muscle spindle, its structure presents an outstanding duality which is also reflected in its function.

The muscle spindle is sensitive to length and, when stretched, responds to both constant length, as in maintained position or posture, and changing length, as during movement. The firing of the sensory neurons of the spindle reflect both the rate of change in length (**phasic response**) and the ultimate length finally achieved and maintained (**tonic response**). Both aspects of muscle length are signaled by variations in the firing frequency of the afferent neurons serving the receptor.

Morphology of the Spindle

Structural details vary slightly from spindle to spindle, depending upon the particular function of the muscle in which the spindle lies. In general, each consists of a fluid-filled capsule 2 to 20 mm long and enclosing 5 to 12 small specialized muscle fibers (Fig. 11.1). These are known as **intrafusal fibers** (i.e., within the spindle capsule), to distinguish them from the contractile or **extrafusal fibers** of the muscle. The latter, when stimulated by their large alpha motor neurons, contract to produce the muscle's tension.

Intrafusal fibers differ from contractile muscle fibers in several ways. Their diameters range from one-tenth to one-fourth the diameter of the contractile fiber, their nuclei are concentrated in the central or **equatorial region** rather than being distributed throughout the fiber, and the contractile material is restricted to the **polar ends**. There are many sizes of intrafusal fibers (Fig. 11.2). Each spindle contains one to three large intrafusal fibers ranging

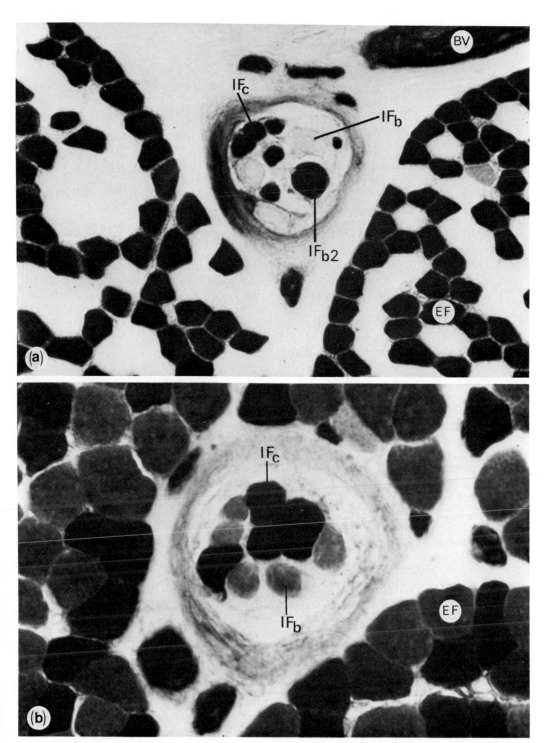

Figure 11.2. The neuromuscular spindle in cross-section. Demonstrates three intrafusal fiber types as follows: nuclear bag fiber (IF$_b$), low ATPase pH 10.3 activity; nuclear chain fiber (IF$_c$), high ATPase pH 10.3 activity; second type of nuclear bag fiber (IF$_{b2}$), high ATPase pH 10.3 activity. (Courtesy of G. Elder, McMaster University.) (a), magnification approximately ×200. The equatorial region of the spindle is seen surrounded by its multilayered capsule. Extrafusal muscle fibers (EF) and a blood vessel (BV) are identified. (b), magnification approximately ×400. A second neuromuscular spindle, demonstrating that some chain fibers have larger diameters than bag fibers.

from 12 to 26 μ in diameter and one to eight smaller intrafusal fibers with diameters ranging from 4 to 12 μ. These two types of receptor cells differ not only in diameter but in length. The smaller fibers are usually contained entirely within the spindle capsule (**intracapsular fibers**), while the larger fibers pass well beyond the capsule (**percapsular fibers**). In the large fibers the centrally located nuclei are aggregated into a swollen baglike region in the equatorial portion and hence these are called **nuclear bag fibers**. There may also be a single-file projection of several nuclei known as the **myotube**, extending outward from the bag region on either side. The smaller intrafusal fibers contain only a single column of nuclei through their equatorial region. These are known as **nuclear chain fibers**. The large bag fibers, extending well beyond the capsule, attach to the connective tissues and endomysia of the contractile muscle fibers. In some instances, a single nuclear bag fiber may pass through several capsules each having a nuclear bag. In such cases the structure is known as a **tandem spindle**. Although the significance of tandem spindles is not yet clear, they are probably concerned with the intricate role of the spindle in muscle regulation. The chain intrafusals, fully contained within the capsule, attach to the inner surface of the capsular connective tissue at either end. The two types of intrafusal fibers also vary in viscosity, the bag fibers being more viscous than the chain fibers. Viscosity may determine the type of contraction of the intrafusal fibers, a characteristic which has an important influence upon its function. The innervation of the two types of intrafusal fibers also differs.

Only recently, a few investigators have chosen to distinguish three. Thus, intrafusal fibers, like the actual contractile elements of muscle may be classified into three types: two types of bag fibers and one type of chain fibers (Boyd et al. (1975)). Very little data are available on this but close examination of Figure 11.2(a) reveals a second type of nuclear bag fiber. It can also be seen (Fig. 11.2(b)) that the popular

view that bag fibers are large and chain fibers are smaller in diameter is not necessarily true (Elder (1978)).

Spindle Innervation

Afferent Innervation. Afferent (sensory) neuron endings are intimately associated with the intrafusal fibers and are stimulated mechanically when the fibers are stretched. Impulses then pass over the axons and enter the spinal cord by the dorsal roots. Within the spinal gray matter they distribute to a number of pathways, rostrally, caudally, and contralaterally. Most prominent, however, is their influence upon their own muscle group. In general, spindle afferents exert an excitatory effect upon the muscle in which they lie, a facilitatory effect upon synergistic muscles, and an inhibitory effect upon antagonistic muscles. An important exception is discussed later.

Most muscle spindles receive two types of afferent innervation, designated **primary** and **secondary**. The two types of afferent neuron are distinguished by differences in sensory ending and by differences in axon size. The primary afferent neurons terminate in **annulospiral** endings which coil around the nuclear regions of the intrafusal fibers while the secondary afferents terminate juxta-equatorially, i.e., farther toward the striated polar regions, either in smaller coils or in **flower-spray** endings. Axons of primary afferents are large group I fibers (known as Ia), while the secondaries fall into group II of the Lloyd classification.

The two types of afferent neurons distribute differently to the two types of intrafusals (Fig. 11.3). Each spindle has only one primary afferent. It enters the spindle and branches to supply an annulospiral ending to each of the intrafusal fibers of the spindle, both bag and chain. Each spindle receives one to five of the smaller secondary afferents. Their endings are restricted almost entirely to the chain fibers, and their axons rarely branch, the axon to ending ratio being essentially 1:1.

The two types of afferents differ in sensitivity, the primary afferents having

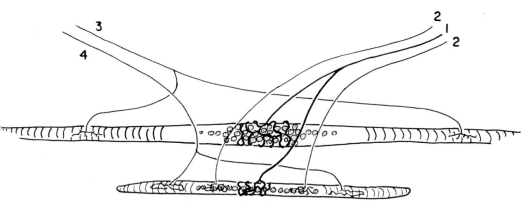

Figure 11.3. Intrafusal fibers and their innervation. A large nuclear bag intrafusal fiber is shown above with equatorial region filled with nuclei and with contractile polar ends extending beyond the limits of the picture. A nuclear chain fiber is shown below. The nuclear chain fiber is smaller in diameter and shorter, with the characteristic single row of nuclei in its central region. Afferent innervation is pictured on the two types of intrafusal fibers. The single large primary afferent neuron (1) ends in coiled terminals (annulospiral endings) on the nuclear region of each intrafusal, while the smaller secondary afferents (2) have branched endings (flower sprays) located on the outer parts of the nuclear region and appear only on the chain intrafusals. Efferent innervation is also shown. The gamma (fusimotor) neuron (3) ends in gamma plates located distally on the polar regions of the nuclear bag fiber. The nuclear chain fiber receives another type of gamma neuron (4) which terminates in gamma trail endings situated closer to the equatorial region. (From Gardner, E. B., 1969, Proprioceptive reflexes and their participation in motor skills. *Quest XII*: Fig. 1-B, p. 5.)

much lower thresholds to stretch than do the secondaries. Only a few millimeters of stretch per second are sufficient to activate the primaries. Furthermore, the primary afferents signal both phasic and tonic stretch, while the secondaries signal tonic length only. The primary afferent neuron signals the phasic length state of the muscle by changes in its impulse frequency *during* stretch. This is the phasic response. The frequency reflects, not length as such, but *rate of change* in length, i.e., **velocity** of the stretch. When stretching is completed, the frequency of discharge drops to a constant level appropriate to the new tonic length. This is the tonic response (Fig. 11.4). When a small stretch is rapidly imposed, the phasic response frequency rises sharply and then drops markedly when stretching ceases. The difference between the maximal frequency attained during the phasic portion of the stretch and the level to which frequency settles in the tonic response is called the **dynamic index**. The tonic value is taken at 0.5 second after the final position has been reached. The contrast between the responses of primary and secondary spindle afferents during a slow stretch is illustrated in Figure 11.5.

Efferent Innervation. Intrafusal fibers are supplied with motor innervation in the form of small gamma-sized neurons known as **gamma motor neurons** or **fusiform neurons**, whose cell bodies lie in the anterior horn of the gray matter of the spinal cord. Axons of these neurons leave the ventral root, travel to the muscle in the appropriate spinal nerve, and terminate in motor endings on the contractile polar end regions of the intrafusal fibers. Each spindle receives 7 to 25 (average 10 to 15) efferent neurons. Impulses traveling over these fusimotor neurons evoke contraction

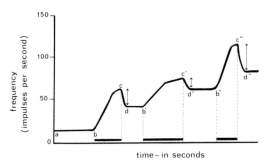

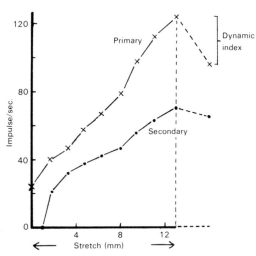

Figure 11.4. Phasic and tonic responses of the spindle primary afferent neuron to interrupted stepwise increases in muscle length. Each stretch, denoted by the solid bars on the time scale, involved the same amplitude of stretch but was imposed at a different velocity as indicated by the time. After each stretch the new length was maintained. a, the initial frequency before stretch was applied; b, the beginning of the stretch; c, frequency at the completion of the stretch; d, frequency after the new length was attained; b', c', d' and b", c", d", responses to subsequent stretches. Note that frequency of impulse discharge increased during each stretch (b to c, b' to c', b" to c", the phasic response) but dropped back to a new constant level consistent with each new length attained (d, d', and d", the tonic response). The small double-ended vertical arrows represent the dynamic index for each stretch. Hypothetical. (Based on Matthews, P. B. C., 1968. Central regulation of the activity of skeletal muscle. In *The Role of the Gamma System in Movement and Posture*, Revised edition. New York: Association for Aid of Crippled Children.)

Figure 11.5. Responses of primary and secondary spindle afferents. Upper curve: the phasic frequency response of a primary spindle afferent neuron during and directly following a slow stretch of the muscle. Lower curve: response of a secondary spindle afferent under same conditions. Broken vertical line indicates point at which stretch was completed, and the muscle was held at this new length. The last point on each curve is the frequency recorded after the new length had been maintained for 0.5 second. Note the marked drop (dynamic index) in the response of the primary afferent, indicating that it is responding to velocity of stretch, while the secondary afferent has responded only to absolute length. (Adapted from Harvey, R. J. and Matthews, P. B. C., 1961. The response of de-efferented muscle spindle endings in the cat's soleus to slow extension of the muscle. *J. Physiol. (Lond.)* 157: 370. Cf. Matthews, P. B. C., 1968. Central regulation of the activity of skeletal muscle. In *The Role of the Gamma System in Movement and Posture*, Revised edition, Fig. 17, p. 32. New York: Association for Aid of Crippled Children.)

of the polar ends of the intrafusal fibers just as impulses in the large alpha motor neurons evoke contraction in the large contractile fibers. Contraction of the intrafusals exerts no detectable influence on muscle tension. Since an intrafusal fiber is connected at both ends either to the interior wall of the capsule or to the connective tissues of the extrafusal muscle fibers, the contraction of its polar ends imposes a stretch upon the nuclear region (Fig. 11.6). Afferent endings are activated just as they are during passive stretch of the whole muscle. Stretch produced by gamma innervation may be referred to as **internal stretch**, while stretch imposed by gravity, an outside force, or shortening of an antagonistic muscle is designated as **external stretch**. Impulses generated in afferent neurons, whether by internal or external

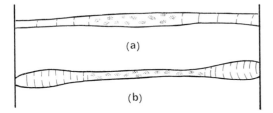

Figure 11.6. Internal stretch of an intrafusal fiber. The vertical bars on either side represent tendons to which the intrafusal fibers are attached. (a), a nuclear bag fiber in the neutral state with polar ends uncontracted; (b), the same fiber under gamma (fusimotor) stimulation. Polar ends have contracted, putting the nuclear region under stretch.

stretch, traverse the usual neural circuits to the muscles.

The structural duality of the muscle spindle is also reflected in its motor innervation: there are two types of gamma neurons (Fig. 11.3). Existence of the two types is supported by anatomical, physiological, and pharmacological evidence. *Anatomically* there are two types of endings differentially located on the intrafusal fibers. There are endings found on the polar regions of nuclear bag fibers which resemble smaller versions of motor end-plates, and these endings are known as **gamma plates**. On the nuclear chain fibers, more diffuse endings occur known as **gamma trails**. Gamma trails are situated more centrally on the polar regions, i.e., closer to the equatorial region, sometimes overlapping the secondary sensory endings, while the plate endings occur more distally. Furthermore, in the muscle nerve there occur two distinct types of gamma-sized axons: one type is thickly myelinated, the other thinly myelinated. The two types show no separable differences in their axon diameters, overlapping over the whole range.

Physiologically, some gamma neurons conduct rapidly and may be called **fast** fusimotor neurons, while others conduct more slowly (**slow** fusimotor neurons). There is little overlapping in conduction

velocities between the two groups. Indirect evidence suggests that the thickly myelinated axons are the fast-conducting ones, although this needs further substantiation.

Furthermore, the gamma neurons display two different types of influence on the primary afferents. Some gamma neurons, when stimulated concurrently with external stretch of the muscle, appear to produce an increase in the dynamic index. In other words their action exaggerates the phasic response of the primary endings. Such neurons are called **dynamic fusimotor neurons**. Other gamma neurons markedly *decrease* the dynamic index. These are known as **static fusimotor neurons**. They may also increase the tonic response somewhat, but this effect is less noticeable than is the reduction of the phasic response (Fig. 11.7).

Pharmacologically, evidence for two separate gamma systems is found in Rushworth's (1968) investigation of drug effects on phasic responses as compared with tonic responses. He found that barbiturates depressed the tonic response but left the phasic responses unaffected, and that procaine completely suppressed the tonic response and reduced the phasic response. Moreover, Rushworth found that the phasic response was exaggerated in cerebellar disease, while the tonic response was absent. Other studies have shown that the anterior lobe of the cerebellum normally inhibits dynamic gamma activity, suggesting that in disease of the cerebellum this inhibition is reduced or absent. Such differential effects on the two types of response support their correlation with separate structural properties.

It is not known for sure which gamma axon is related to which type of motor ending and, coincidentally, which type of intrafusal fiber. Indirect evidence seems to relate the dynamic gammas to the plate endings, and therefore the nuclear bag fibers, and the static gammas to the trail endings and the chain fibers. For example, the primary afferent neurons have their endings on both types of intrafusal fibers and are influenced by both gammas; however, the secondary afferents, whose end-

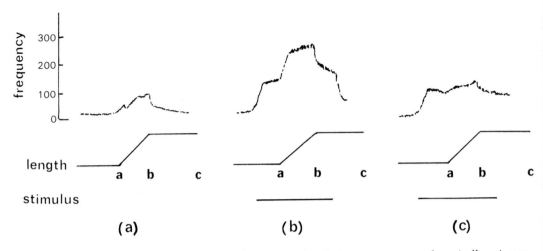

Figure 11.7. Effects of dynamic and static fusimotor stimulation on response of a spindle primary afferent to stretch. (a), no gamma stimulation; (b), stimulation of a dynamic gamma neuron; (c), stimulation of a static gamma neuron. The application and duration of stimulation is indicated by the bars (under (b) and (c)) in the lower trace. The middle trace shows the time course of the stretch: a, beginning of stretch; a to b, phasic stretch; b to c, static stretch. The top trace shows the frequency response in the spindle primary afferent neuron under the three conditions. (a), without gamma stimulation, the frequency rise of the phasic response closely paralleled the phasic stretch (a to b) and was followed by the tonic response (frequency decrease) as the new length was maintained. (b), under dynamic gamma stimulation the resting frequency in the primary afferent was markedly increased even before the stretch was begun. The phasic response during the stretch was almost twice that in (a). When stimulation ended, the tonic response frequency returned to about the same level as in (a). (c), the initial or resting frequency was increased, as in (b), by stimulation of the static fusimotor neuron. However, activity of this gamma neuron markedly reduced the phasic response as compared with (a). In fact, it was hardly greater, if at all, than the initial gamma background discharge. (Adapted from Matthews, P. B. C., 1962. The differentiation of two types of fusimotor fibres by their effects on the dynamic response of muscle spindle primary endings. *Quart. J. Exp. Physiol. 47:* 324. Cf. Matthews, P. B. C., 1968. Central regulation of the activity of skeletal muscle. In *The Role of the Gamma System in Movement and Posture,* Revised edition, Fig. 24, p. 46. New York: Association for Aid of Crippled Children.)

ings are restricted almost entirely to chain fibers, are activated only by static gammas. Examined from another viewpoint, if dynamic gammas end in the plate endings on the nuclear bags as suggested, they should influence the primary afferents but not the secondaries, which is the actual experimental finding. And if static gammas end on chain intrafusals, they should affect both types of afferent ending, and, in fact,

the static fusimotor neurons can **drive**† both primary and secondary afferents. So it seems likely that the dynamic fusimotor neurons distribute to the nuclear bag intra-

---

† A neuron is considered to drive another neuron when it causes the other neuron to respond one for one to its frequency over a limited number of cycles.

fusal fibers and the static fusimotors to the chain intrafusals.

Evidence is accumulating that the type of contraction of the two kinds of bag and the chain intrafusal fibers differ. This is significant because the nature of intrafusal contraction may determine the afferent response, while the particular type of gamma neuron simply activates a particular type of intrafusal fiber. Smith (1966) has shown that bag fibers contract slowly in a local, graded manner. This would be consistent with the phasic response in which frequency increases with the rate of stretch. The smaller chain fibers, however, contract in a faster, twitchlike manner, and complete tetanus can be evoked in them by a stimulating frequency of about 15 impulses/sec. This would be compatible with the tonic response to a maintained stretch, the extent of tetanus in the intrafusal fiber determining the frequency in the afferent neuron. Since the primary afferents serve both types of intrafusal fiber, they would be expected to signal both phasic and tonic stretch, while the secondaries, associated mainly with the chain fibers, would signal only tonic length; this is in fact the case. Further, if it is the dynamic type of gamma which serves the bag fibers and evokes a contraction appropriate to phasic response as suggested, it is not surprising that stimulation of dynamic gammas during external stretch enhances the phasic response, increasing the dynamic index as it does (Fig. 11.8). Finally, if bag fibers contract at both "slow" and "fast" rates, *perhaps* some bag fibers act like chain fibers, indicating only tonic length.

After gamma denervation, chain fibers atrophy quickly, whereas bag fibers are slow to degenerate. This suggests that the chain fibers have a greater dependency upon gamma innervation, which is in agreement with tonic sensitivity, while the phasically sensitive bag fibers should be more dependent upon external stretch. Greater sensitivity to stretch is consistent with the fact that the nuclear bag fibers are percapsular.

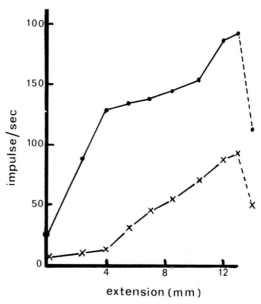

Figure 11.8. The influence of fusimotor stimulation on the response of a spindle primary afferent neuron. Both curves represent the dynamic response of the same primary afferent neuron. Upper curve (●), ventral roots were intact; therefore the spindle was under fusimotor influence. Lower curve (×), ventral roots were cut, and as a result the spindles were deefferented. The last point on each curve (joined by a broken line) shows the frequency recorded at 0.5 second after completion of the phasic part of the stretch. Note the early increase in frequency which occurred with gamma firing. In this case the dynamic index was also increased. (After Jansen, J. K. S., and Matthews, P. B. C., 1962. The central control of the dynamic responses of muscle spindle receptors. *J. Physiol. (Lond.) 161:* 357. Cf. Matthews, P. B. C., 1968. Central regulation of the activity of skeletal muscle. In *The Role of the Gamma System in Movement and Posture,* Revised edition, Fig. 21, p. 42. New York: Association for Aid of Crippled Children.)

Evidence has indicated that spindles also receive branches from the alpha motor neurons to the contractile fibers. These have been designated as **beta axons.** The presence of skeletomotor as well as fusimotor axons has been demonstrated histologically by Adal and Barker (1965) and electrophysiologically by Bessou et al. (1963). Beta innervation is relatively rare in cats and humans but more common in rats, in whom it seems to derive from slow alpha motor neurons. When present, beta axons terminate in typical extrafusal-type end-plates on nuclear bag fibers close to the ends of the spindle poles, and they appear to have a weak dynamic effect. While the function of these beta axons is still uncertain, such a contribution to intrafusal motor innervation would obviously be useful in the integration of muscle activity.

Physiology of Spindle Innervation

Functions of Primary Afferent Neurons. More is known of the effects of the firing of the primary endings than of the secondary endings. Firing of the spindle primary afferent neurons reflexly excites motor units of the muscle in which the spindle lies, simultaneously facilitating synergists and inhibiting antagonists. Figure 11.9 shows diagrammatically the simple divergent circuit presumed to mediate the effects produced. It has been well established that the primary afferent enters the dorsal root of the spinal nerve. Its cell body lies in the dorsal root ganglion and its axon passes into the gray matter of the dorsal horn, makes it way to the ventral horn, and synapses directly upon an anterior horn cell of the motor pool of its own muscle. In other words, the reflex arc from spindle to muscle consists of only two neurons: a **monosynaptic** pathway. When stimulated by stretch of its endings, the primary afferent neuron fires signals into the central nervous system which evoke a contraction just sufficient to relieve the stretch. In any muscle contraction it is essential not only that the prime mover be activated but that the activity of the muscles of its functional group (those which operate upon the same joint) be appropriately modified. Synergists must support the activity of the prime mover, and antagonists must not oppose it. Collateral branches from the axon of the primary

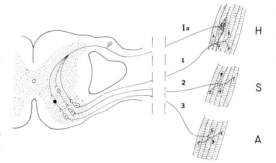

Figure 11.9. Circuits of the spindle primary afferent neuron. Ia, spindle primary afferent neuron: 1, alpha motor neuron to homonymous muscle (H); 2, alpha motor neuron to synergist (S); 3, alpha motor neuron to antagonist (A). The primary afferent neuron (Ia) of the spindle is shown entering the spinal cord (seen in cross-section) by the dorsal root. In the ventral horn its axon makes monosynaptic connection with the alpha motor neuron (1) to its own muscle (H). A collateral connects with the alpha motor neuron (2) to the synergist (S) through an excitatory interneuron (shown with a cell body as an open circle). Another collateral connects through an inhibitory interneuron (shown with a cell body as a filled circle) with the alpha motor neuron (3) to the antagonist (A). These circuits assure proper cooperation among the muscles of the functional group.

afferent make connections through interneurons with the motor neurons of the synergists. These are excitatory synapses, and the synergists are facilitated. Other collaterals synapse through inhibitory interneurons with antagonists, producing reciprocal inhibition. There is some indication that the inhibition of antagonists may be accomplished by presynaptic inhibition exerted upon their own primary afferents (see Fig. 8.17). This seems reasonable, since the contraction of any muscle will automatically put its antagonists on stretch and, if the antagonist should respond to its own stretch reflexes, it would oppose the contraction of the prime mover muscle and produce co-contraction. Connections with synergists and antagonists are probably disynaptic, since the time consumed in transmission from spindle afferent to alpha motor neuron is compatible with the interposition of two synapses. An inhibitory interneuron in the path to the antagonistic muscle converts the signal from excitatory to inhibitory. Other collaterals from the primary afferent axon travel upward and downward through the spinal cord, making ipsilateral intersegmental and supraspinal interconnections through appropriate interneurons. Still other projections cross the cord to influence contralateral muscles. Here, then, is an example of a simple but highly important circuit preprogrammed to produce appropriate excitation and inhibition within a functional muscle group.

As already described, the primary ending is sensitive to both phasic and tonic stretch. When a muscle is stretched, the primary neurons fire a rapid burst of impulses *during* stretching, with frequency directly related to the *velocity* of stretch: the phasic response. When the muscle reaches a length which is maintained, the frequency drops to a lower level appropriate to the new length: the tonic response.

Example of Phasic Response. The knee jerk represents an isolated phasic response evoked by primary afferents. The stimulus, a single sharp blow on the patellar tendon, constitutes phasic stimulation only, because no new length is attained and held. The blow on the tendon puts a small momentary stretch upon the quadriceps muscle group and coincidentally upon its spindles. The stretching of the nuclear regions of the intrafusal fibers, slight though it is, is sufficient to cause distortion of the primary endings and evoke a burst of impulses in the primary afferent neurons. This burst of impulses travels into the spinal cord and is conveyed across synapses to the cell bodies of the alpha motor neurons of the same muscle. Their axons return the discharge to the same muscle, indeed to the vicinity of the muscle where the excited spindles are located. As a result the muscle contracts quickly (i.e., the knee joint momentarily extends), the magnitude of the contraction being determined by the velocity of the stretch. Because the shortening of the muscle relieves the original stretch, the spindle is "unloaded" so the primary neurons stop firing. Since the alpha motor neurons are no longer excited, the muscle relaxes and the limb drops immediately to its original position. The knee jerk is a purely artificial response, useful to the physiologist as an example of an isolated stretch reflex and to the neurologist as an index of the general state of the nervous system. The jerk as such has no functional value in neuromuscular activity.

Example of Tonic Response: Postural Reaction. An example of a purely tonic response may be found in the postural reaction to stretch, i.e., the simple **stretch** or **myotatic reflex**. Slow stretching of muscles as a result of the shifting of the center of gravity causes tonic response in spindle primary afferents and a contraction of the stretched muscles which will correct the displacement. For example, the vertical projection of the center of gravity of the human body usually falls in front of the medial malleoli (ankle joint "centers") by a distance of 5 to 8 cm, when in a normal standing posture. This concentration of weight (force) on the ventral side of the ankle joints places a tonic stretch on the calf muscles, especially the soleus, which involves the myotatic reflex in these muscles. By servo-action, they contract that amount necessary to counter the forward force and "pull" the body backward over its base of support. Such postural adjustments go on continually in the living animal.

Example of Phasic plus Tonic Responses: Adjustment to Load. Another illustration of the activity of the primary afferents, one that is more complex and more functional than the knee jerk, is found in the automatic adjustment of muscle contraction to an added load. This is a normal functional reaction related to both phasic and tonic changes in muscle length and evoked by corresponding phasic and tonic signaling of the primary afferent neurons. The muscle reactions to the two primary afferent neural responses, one to the velocity of spindle stretch and the other to the length *per se*, can be readily distinguished in electromyograms (EMGs) recorded during the automatic adjustment of an outstretched arm when a load is added (Fig. 11.10). Before the addition of the load, the electromyographic record shows a steady discharge in the forearm flexors consistent with the maintained muscle length and the weight of the arm. Addition of the load puts stretch on the muscles, and the EMG first shows a sudden burst of electrical activity in the muscle, reflecting the phasic response in the primary afferents of the spindle and directly related to the velocity of the stretch (abruptness of the load addition). This is followed by a lower steady firing rate consistent with the new length and load and is a result of the tonic response of the spindles. These spindle influences are also reflected in the overt movements of the forearm. As the load is dropped suddenly into the basket, the forearm is depressed by the added weight, and the flexor muscles and their spindles are stretched. The muscles, reacting to the velocity-dependent phasic response of the spindles, contract more strongly than the load requires. As a result the forearm and basket are raised momentarily. Except when the load is light, muscle response more than compensates for the added load, thus giving evidence of the phasic response in the primary afferent neurons. The more abrupt the application of the load the greater the phasic response to the same load, hence the greater the extent of the muscular contraction and therefore the greater the overcompensation. Because the overcompensation partially unloads the spindles, the impulses following the reflex pathways to the muscle are decreased, the muscle's contraction lessens, and the limb begins to drop. This again puts stretch on the muscle which will again respond appropriately to the velocity of the drop. A tendency to oscillate develops but is quickly damped, and the limb shortly comes into equilibrium at a level consistent with the new load. This is then maintained by the tonic response. However, if the load is placed slowly into the basket, the spindles respond tonically only, with no phasic burst but with suffi-

cient increase in activity to support the increased load. There is little or no overcompensation.

Vibration Response. A unique method of applying stretch to muscle spindles is by vibrating the belly of the muscle. Vibration appears to excite the spindles by deforming their equatorial regions. The vibrated muscle contracts slowly and maintains the contraction throughout the vibration, a purely tonic response. The phasic response may be reduced or absent if the muscle is tested shortly thereafter for the jerk response. When applied to the muscle *belly*, low frequency vibration seems to be a specific stimulus for the primary afferents. High frequencies will excite the secondary endings as well, except that when vibration is applied to the *tendon*, regardless of frequency, only primary endings respond. It appears that only the chain intrafusals are involved in the vibration response, but it is too early to make positive statements on this subject. Vibration also is an effective stimulus for the Vater-Pacinian corpuscle and is discussed under that section.

Functions of Secondary Afferent Neurons. Secondary afferent neurons, as already mentioned, are less sensitive than primary afferent endings and produce only the tonic response. They reflect the mean length of the muscle at any instant. Their influences upon their own and other muscles differ markedly from those of the primary afferents. The effect of secondary afferent excitation is to facilitate flexor muscles and inhibit extensor muscles regardless of the type of muscle in which the spindle lies. In other words, if the stimulated spindle lies in a flexor muscle, then its effect is to reinforce both the activation of the muscle itself and the inhibition of antagonistic extensors, supplementing the action of its primary afferents. If, however, the spindle lies in an extensor muscle, it will tend to inhibit its own and other extensor muscles and facilitate the antagonistic flexors. Nevertheless, stretch of an extensor muscle which is sufficient to activate its secondary endings will at the same time so strongly excite its primary endings that excitation of the muscle will

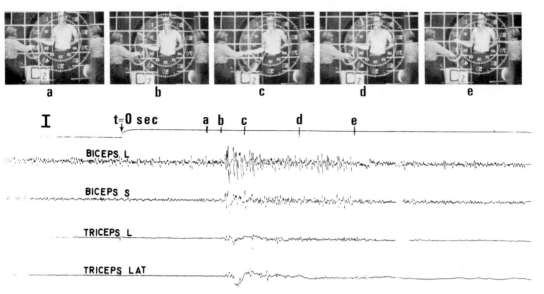

Figure 11.10. Adjustment to load. Electromyograms with syncrhonized cinematography. Subject is supporting basket in the hand, forearm supinated, with the elbow flexed approximately 90° and instructed to maintain this position as load is added and removed. Upper trace synchronizes camera and EMG; rise in baseline indicates start of clock: time = 0.000 second. EMGs recorded from biceps and triceps; L, long; S, short; and LAT, lateral heads. Photograph identification letters indicate situation at corresponding lettered points marked by vertical lines on the EMG. Before b the EMG records tonic activity in muscles to sustain weight of forearm, hand, and basket in prescribed position, reflecting gamma bias under voluntary control. I. b, load is dropped abruptly into basket, load is removed at e.

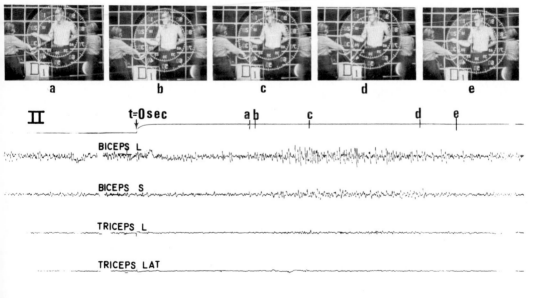

Figure 11.10. II. b–c, load is added gradually to basket; d–e, load is removed. See text.

be maintained in spite of the autoinhibition. Consequently the facilitation of the flexors results in co-contraction, i.e., simultaneous contraction of flexors and extensors. This is especially true of stretch applied to a single joint extensor muscle. Most of the single joint extensors are primarily antigravity muscles, and their spindles are richly supplied with secondary afferents. Rather than being a useless interference with muscle activity, co-contraction acts to provide stabilization of joints for both weight bearing and movement. It is particularly important at proximal joints.

In the case of a muscle which passes over two joints, acting as a flexor of one and an extensor of the other, the spindles tend to act as though in a flexor muscle, the secondary afferents facilitating their own muscle and inhibiting antagonistic single joint extensors.

Functions of Efferent Innervation. The importance of gamma neurons in spindle function cannot be overemphasized. If the spindle had afferent innervation only, all of the responses that it evoked would be of the jerk type with a tendency to produce oscillation. However, impulses over the gamma neurons can, by maintaining contraction of the polar ends, "set" the spindle at any prescribed firing level, thus adjusting its sensitivity independently of absolute muscle length or load. The gamma-innervated spindle, with nuclear regions already under internal stretch and primary afferents already firing in a manner appropriate to the initial length, will be far more sensitive to any further length change which may be imposed by external forces (Fig. 11.8). This setting of spindle sensitivity by the gamma neurons is known as **gamma bias.** At rest, gamma bias holds the spindle just below motor neuron firing level, but, if gamma outflow is raised sufficiently, the muscle will be activated.

The internal stretch of spindles under gamma bias obviates the jerkiness of muscle response which would otherwise result and makes possible smoothness and steadiness of movement or position. Figure 11.11 compares diagrammatically the responses

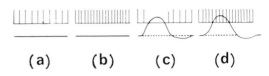

(a)   (b)   (c)   (d)

Figure 11.11. Response of the primary afferent neuron of the spindle to stretch under various conditions. The muscle was weighted with moderate load. Upper trace: frequency response of the primary afferent; lower trace: tension in the muscle. (a), response of the unstimulated muscle to the stretch imposed by the load. (b), stimulation of fusimotor (gamma) neurons induced no change in muscle tension but caused an increase in the firing frequency of the spindle afferent. (c), stimulation of alpha motor neurons caused contraction of the muscle, and the "unloaded" spindle ceased firing. (d), stimulation of alpha and gamma neurons simultaneously resulted in tension development in the muscle but without unloading the spindle. (After Hunt, C. C., and Kuffler, S. W., 1951. Further study of efferent small-nerve fibres to mammalian muscle spindles, multiple spindle innervation and activity during contraction. *J. Physiol. (Lond.) 113:* 283. Cf. Eyzaguirre, C., 1968. Some functional characteristics of muscle spindles. In *The Role of the Gamma System in Movement and Posture,* Revised edition. Fig. 11, p. 22. New York: Association for Aid of Crippled Children.)

of a spindle primary afferent neuron in a weighted muscle with and without gamma stimulation. In (a) the afferent neuron was firing (upper trace) in response to the external stretch of the load. In (b) stimulation of fusimotor neurons caused increased firing of the primary afferent without any change in the muscle tension (lower trace). In (c) stimulation applied to alpha motor neurons resulted in a twitch contraction of the muscle (lower trace), accompanied by temporary cessation of firing in the primary afferent neuron as the contraction unloaded the spindle. In (d) alpha and gamma neurons were stimulated simultaneously. A contraction occurred but the spindle continued to fire throughout the twitch because the internal stretch of the spindle was maintained by the fusimotor discharge.

The shortening and lengthening reactions, long recognized in muscle, are explained by gamma bias. In resting muscle, gamma innervation maintains a tonic firing at low frequency in the primary afferents. If the limb is passively put into a position which shortens the muscle, the primary discharge decreases for a short time, and this is reflected in a reduction of the muscle's contractile capacity. Soon, however, the primary afferent discharge is resumed at the original resting rate and functional capability is improved. This is known as the **shortening reaction**. The change in position has caused a temporary slack in the spindle and hence a reduction in afferent firing. Almost immediately, however, gamma outflow is increased, causing the intrafusal fibers to contract and take up the "slack." In other words, the gamma bias, and hence the spindle sensitivity, is re-established at the new, shorter length. If a limb is passively placed so as to put the muscle in a lengthened state, the firing of the primary afferents increases at first, as a result of passive stretch, producing an increased resistance to stretch, but then drops back to its resting frequency as gamma bias is reset. This is the **lengthening reaction**. Note that in each case the spindle response briefly opposes the change in length.

When a subject voluntarily holds a load in a prescribed position, the spindle gammas are already under cortical stimulation to produce appropriate gamma bias to maintain that position. When a load is added gradually, the static response alone adjusts muscle activity to the increased load. However, if that load is added abruptly, the velocity of stretch produces a phasic response which evokes overcontraction, as described earlier under "Example of Phasic plus Tonic Responses: Adjustment to Load." In addition to the spindle responses to the rate and extent of external stretch, information regarding limb position is fed back into the central nervous system by other sensory receptors (visceral, joint, skin). If the new position deviates from that voluntarily prescribed, cortical outflow alters gamma bias appropriately until the desired position is regained. Gamma activity acts as an important factor in damping out the tendency to oscillate resulting from overcompensation.

Efferent innervation confers on the spindle its capacity for minute and precise control of muscle contraction in accordance with the requirements which unconscious coordination and conscious direction prescribe. Fusimotor neurons receive impulses from corticospinal nerve fibers, from neurons of reticulospinal tracts, and from collaterals of all sensory pathways, including those of the proprioceptors. As a result they discharge tonically, holding intrafusal muscle fibers in a continuous but highly variable state of gamma bias. The greater the state of gamma bias the greater the sensitivity of the spindle afferents to external stretch. Wooldridge (1963) states that the reticular formation of the brainstem acts on the gamma efferents "to adjust the operating parameter" of the muscles by setting the zero point of the spindles so that they will fire minimally when the muscle is at a desired length. As a result the muscle "automatically seeks out just the degree of contraction that will minimize firing frequency of its stretch receptors." Since the two types of gamma neurons exert distinguishable influences upon spindle activity, the two gamma systems would be expected to have separate supraspinal controls.

Role of Spindle Innervation in Voluntary Movement. Granit (1955) has proposed that cortical signals for the programming of voluntary movement may reach muscles over two pyramidal pathways: (1) directly to the alpha motor neurons and thence to the muscle, or (2) indirectly via gamma motor neurons to the spindles, from which they are relayed back to the alpha motor neurons by spindle afferents. Figure 11.12 shows diagrammatically the indirect pathway, which is known as the **gamma loop**. Signals from the pyramidal tract activate the proper gamma neurons. Impulses passing out over the gamma neurons cause intrafusal fibers to contract,

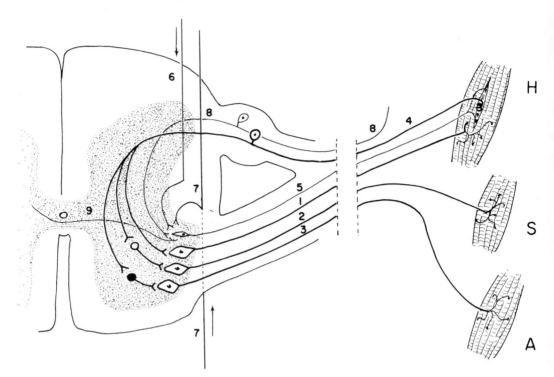

Figure 11.12. Spindle primary afferent circuits and the gamma loop. The spindle primary afferent circuits (as in Fig. 11.9) are shown with the gamma circuits added. As in Figure 11.9, 1, 2, and 3 are alpha motor neurons to the homonym, synergist, and antagonist, respectively. 4, primary afferent from the spindle; 5, gamma (fusimotor) neuron to the spindle; 6 and 7, fibers descending from supraspinal centers; 8, ipsilateral afferent neuron from a joint or skin receptor situated near another muscle; 9, interneuron from a contralateral source. (From Gardner, E. B., 1969. Proprioceptive reflexes and their participation in motor skills. *Quest XII*: Fig. 2, p. 6.)

putting stretch upon the afferent endings. The activated afferents transmit impulses back to the spinal cord and across the synapses to the alpha motor neurons of the muscle. These convey the signal to the muscle fibers, which then respond to a degree commensurate with the amount of stimulation introduced by the corticospinal neurons. Granit points out that, by appropriate cortical influences, the spindle can be used to regulate the amount and rate of muscle shortening. In isometric contraction (i.e., position) the cortex sets the spindle and the muscle follows by servo-action. This position-controlling servomechanism will maintain muscle length regardless of changes in load or resistance.

In isotonic contraction (i.e., movement) the cortex progressively increases the downflow to the gamma neurons, their outflow keeps the spindles progressively activated, and the muscles follow the spindles, thus serving the dictates of volition. Cortical impulses can also hold gamma bias just below firing level in anticipation of the stimuli which will finally trigger activity.

Partridge (1961) gives experimental support to Granit's hypothesis. He concludes that the central nervous system controls voluntary movement by sending simple length-setting signals of proper sign (+ or −) via the gamma system to the spindles of both agonists and antagonists and then allows the stretch reflex to determine the necessary adjustment of muscle force to produce the desired position or movement.

Although transmission over the gamma loop is obviously slower, it is probably the preferred route for voluntary activation and cessation of muscle activity. Rushworth (1968) has shown that the gamma delay, about 50 msec, is not inconsistent with man's fastest repetitive movement (piano trills at 6 to 8/sec), whose latency is about 120 to 150 msec. Voluntary activation via gamma neurons also has the advantage that reciprocal activity of the muscles concerned is assured from the very first contraction.

It is not clear whether impulses from muscle receptors contribute to kinesthetic perception. Muscle afferents do project to the cerebral cortex: they relay directly from the thalamus to the *motor* portion of the sensorimotor area where the effect may be facilitatory or inhibitory. An extensive and diffuse convergence occurs upon these cortical cells from muscles of different joints and from antagonists at the same joint. Since these cortical cells are not activated antidromically by stimulation of pyramidal tract fibers, they cannot be pyramidal cells. It is assumed that they are the cells which *activate* the pyramidal cells. Because the region of convergence rules out the possibility of a role in kinesthetic perception, it is suggested that this afferent projection provides the cortical motor area with information regarding the state or tone of muscles, and this information is used in adjustment of the voluntary descending signal. The spindle afferent projection would thus provide the ascending limb, and the pyramidal tract (the descending limb), of a large servoloop. Over this loop impulses would travel from motor cortex to muscles (probably via both alpha and gamma motor neurons) and back to the cortex from the spindle afferents. Although the information thus relayed would probably be crude owing to the diffuse nature of the convergence, nevertheless it would serve as an important feedback for the regulation of voluntary activity (Oscarsson (1966); Swett and Bourassa (1967)).

Of particular interest to students of motor learning should be the work of Buchwald and Eldred (1961) in which classic conditioning of the gamma efferent system was demonstrated in cats. The investigators found that gamma fibers developed a conditioned response, as shown by accelerated discharge, after 5 to 10 presentations of a conditioned stimulus (a normally ineffectual audible tone) with the unconditioned stimulus (an electric shock to the toe sufficient to induce flexion withdrawal). In a follow-up series of experiments (Buchwald et al. (1964)) further evidence was obtained suggesting that conditioning in the gamma system may play an essential role in the learning of new motor responses.

Speculations and Questions on Spindle Innervation

Spindle afferents provide two routes for feedback to the central nervous system: movement information (phasic length changes in muscles) and position (static length). The two gamma systems can provide separate systems for activation of muscles. If the voluntary objective is movement, the central nervous system may use predominantly the dynamic gamma system, while if the objective is position, the outflow may use the static gamma system. Granit et al. (1955) demonstrated that the two systems are under separate supraspinal controls.

Matthews (1968) points out that alpha excitation alone evokes a load-dependent tension which is length-independent and produces uncontrolled movement if alpha activation exceeds the load. (An example of this is found in the unnecessary speed and extent of movement which occurs when one picks up a very light object which was expected to be heavy.) On the other hand, gamma stimulation results in a cortically determined amount of shortening which is independent of load. Therefore, the central nervous system may control muscle length, both for position and during movement, through the gamma system.

The secondary afferents activated by

the static gammas may provide a simple signal of misalignment between muscle length and gamma bias which is fed back to the central nervous system and which may be used to appraise the success of the voluntary muscular activity. The primary afferent would be useless for such appraisal because its phasic response is not related to length.

Basic control theory emphasizes the need for damping the tendency to oscillate in a servoloop. This is important in the adjustment of muscle to sudden changes in position or load and in pendular movements which tend to overshoot and which if undamped would develop positive feedback. The dynamic response of the spindle probably serves to damp oscillation in the servoloop through gamma control of spindle sensitivity to the phasic response. Both Matthews (1968) and Eldred (1967b) present good discussions of this subject. Damping not only reduces oscillation but minimizes jerkiness and smooths out contractions.

There are a number of questions which still need answers before the correlation of the two types of gamma fibers to the two types of endings on the two types of intrafusal fibers can be finally determined. We must first ascertain that the intrafusals are exclusively and separately innervated, each by a different type of gamma neuron. Second, there must be more evidence regarding differences in contractile properties, if any, in the intrafusals. There are some spindles which have both kinds of intrafusal fibers but no secondary endings, and yet they still evoke the two responses (phasic and tonic) in their primary afferents. This is not difficult to reconcile with the theory previously discussed that the two responses arise as a result of contrasting contractile characteristics in bag and chain fibers. Third, investigation is needed of the possibility that different types of motor endings may produce different types of contraction in the same intrafusal fibers. Fourth, if the primary ending responds to acceleration as well as to velocity of lengthening, this would add further complexity. At present

this does not seem to be the case. Finally, the role of the collaterals from alpha motor neurons, the so-called beta innervations of the spindles, needs elucidation. It seems reasonable that alpha neurons should be concerned in the regulation of spindle activity and that the beta collaterals supply such a pathway.

Instances of Spindle Activity

Spindles serve to coordinate the activity of muscles through the entire course of a movement. As the movement progresses, feedback from spindles changes in precise relationship to the changing lengths of their muscles. Thus, contraction in any muscle is always appropriate to the conditions of the moment.

There are many instances in daily living in which proprioceptive reflexes show up inadvertently, isolated from the context of any complex skill pattern. Some familiar examples of spindle responses follow.

Jerk Responses. Almost everyone has experienced simple jerk responses (knee, ankle, elbow, wrist, etc.) evoked by external stretch applied as a sharp, light blow to a tendon. Although these responses are functional artifacts, they are nevertheless coordinated responses and as such involve reciprocal inhibition and synergic cooperation of the muscles in the functional group concerned. They are thus excellent examples of simple automatic intermuscular regulation, demonstrating that a single proprioceptive reflex not only activates the responding muscles but simultaneously assures coordinated cooperation of muscles without intervention of conscious nervous mechanisms. Muscle coordination in a jerk response appears to be accomplished entirely by feedback from the primary afferents of the stretched muscles. Since the limb is hanging relaxed before its tendon is struck, there is no gamma involvement other than the resting background discharge.

Reciprocal Inhibition. a. When a person standing with the knees slightly hyperextended is kicked lightly in the popliteal space (behind the knee), the stretch reflex thus evoked in the knee flexors excites them, at the same time inhibiting the knee extensors so that the knees buckle. Fortunately, the buckling stretches the extensor muscles which then contract and save the subject from falling. If, however, the buckling of knees is extreme and the stretch of the extensors becomes sufficient to activate the secondary afferents of the extensor spindles, extensor inhibition and flexor facilitation will result and the subject may actually fall.

b. Cramps in the leg muscles are readily relieved by strong, active contraction of antagonistic muscles, whose spindles will evoke reflex inhibition of the affected muscles.

Self-inhibition of Extensors by Their Secondary Afferents. a. Inhibition of extensors by their own secondaries may be a prime factor in causing the novice to crumple to the floor if the knees become extremely flexed on landing from the vault.

b. Therapeutic technique involving passive manipulation of limbs for the purpose of reducing muscle spasm in extensors often includes slow and sustained stretch of the affected muscles; it is possible that the threshold of the spindle secondaries is reached, which reflexly inhibits the extensors, hence the spasm, and restores joint flexibility to acceptable ranges.

Evidence of Anticipatory Settings of Gamma Bias. This may be seen when one reaches to push open a heavy door. Voluntary impulses set the gamma bias to hold spindles just below firing level. The touch of the fingers on the door will trigger the gamma neurons to deliver just the amount of cortically programmed outflow to the muscles which experience has shown to be required in such circumstances, and the muscles will respond with suitable force of contraction. Occasionally someone pulls the door from the opposite side just as one's fingers touch the surface. Since the contracting muscles do not meet the resistance anticipated, the force is converted into forward propulsion. At the least, one's balance is upset; at the worst,

one goes flying through the door and collides with the person on the opposite side.

Pressure as a Spindle Stimulus. Pressure applied to the belly of a muscle has been shown to activate primary afferents. Advantage is unconsciously taken of this fact by older persons, especially when arising from a chair. Firm pressure on the ventral surface of the thighs increases the contractile force of the quadriceps muscles, markedly assisting in extension of the knees. Conversely, when sitting down, similar pressure on the quadriceps will aid in controlling the rate of knee flexion by facilitating the eccentric or lengthening contraction in these muscles.

Positive Supporting Reaction. It should be mentioned that muscle spindles are responsible, at least in part, for the positive supporting reaction. The weight of the body pressing upon the feet spreads the interphalangeal joints, stretching the interosseus muscles. Their spindles transmit signals which reach the extensor muscles of the limbs and induce a tonic contraction response, probably by means of gamma facilitation of extensor spindles. Evidence for the existence of this reflex is found in the situation that occurs when it is absent. For example, if one sits cross-legged on the floor for a long time, afferent nerves from the feet may become anesthetized as a result of circulatory deficiency. Upon standing, the lack of the feedback from the receptors in the feet results in loss of the positive supporting response and the legs flex under the body weight. The person stumbles and will fall unless visual feedback and/or the stretch reflexes evoked in leg extensors by the flexion come to the rescue.

Another instance illustrating the effect of the loss of the positive supporting reaction is supplied by an experiment carried out in the laboratory by the authors of the first edition of this text. A swing seat was suspended from the ceiling with a pulley arrangement which enabled a subject seated in it to be raised and lowered. At some time during the lift the seat was suddenly tilted so that the subject slid off to land on a gymnasium mat on the floor.

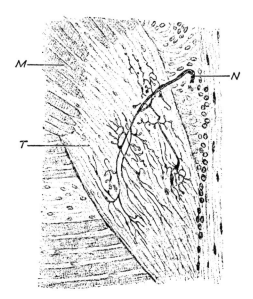

Figure 11.13. Drawing of a Golgi tendon organ. The Ib afferent neuron (N) supplies flower-spray endings lying among tendon fibers (T). Muscle fibers (M) attach to the tendon fibers. (From Truex, R. C., and Carpenter, M. B., 1969. *Human Neuroanatomy*, Ed. 6, Fig. 9–15, p. 187. Baltimore: The Williams & Wilkins Company.)

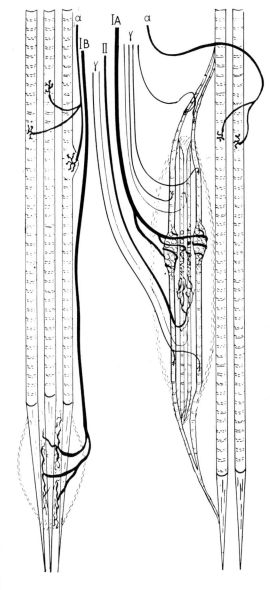

Figure 11.14. Diagrammatic drawing of muscle spindle and tendon organ. The muscle spindle is seen on the right attached to extrafusal muscle fibers and tendon. It consists of small diameter intrafusal muscle fibers which are largely enclosed in a connective tissue capsule. Longitudinally the drawing is not to scale (the length of a spindle may be 50 times its width). Transversely in the drawing the width of the extrafusal muscle fibers represents a diameter of 40 μ; the intrafusal fibers are drawn to the same scale and represent diameters of about 20 μ for the two long fibers with nuclear bags at the equator of the spindle and about 10 μ for the two short fibers with nuclear chains at the equator. The group of nerve fibers shows the relative diameters of these fibers to each other. The largest fiber, IA, supplies the main primary afferent ending lying over the nuclear bags and chains. Fiber II goes to a secondary afferent ending on the nuclear chain fibers adjacent to the primary ending. Six small gamma motor fibers of varying sizes supply motor endings on the intrafusal muscle fibers. The motor endings on the extrafusal muscle fibers are supplied by larger alpha nerve fibers. The remaining IB nerve fiber goes to the encapsulated tendon organ on the left; the branches of the afferent nerve ending lie between the tendons of a group of extrafusal muscle fibers. (Drawing by Sybil Cooper. From Bell, G. H., Davidson, J. N., and Scarborough, H., 1969. *Textbook of Physiology and Biochemistry*, Fig. 40.4, p. 912. Baltimore: The Williams & Wilkins Company.)

During control "spills" subjects had no difficulty regaining the erect stand. In a second set of lifts and spills the subject, blindfolded, was alternately raised and lowered several times and to varying distances until he lost an accurate sense of distance from the floor. When dropped, he regained equilibrium readily but with some slight delay as compared with the controls. Finally the subject's feet were immersed in ice water for a period (20 minutes) sufficient to produce local anesthesia in the feet. Immediately after immersion the subject was raised, lowered, raised, etc., and spilled as before. In this case, however, there was no extension of the lower limbs upon landing and he crumpled to the mat. It is assumed that chilling interfered with the sensory feedback essential for the positive supporting reaction and also for the extensor thrust reflex.

Golgi Tendon Organs

A second type of proprioceptor intimately incorporated into the gross muscle structure is the Golgi tendon organ. Unlike the spindle, its effect upon its own muscle is inhibitory. It consists of an encapsulation of tendon fibers (Fig. 11.13) located at the musculotendinous junction and hence lying in series with the contractile muscle fibers. Figure 11.14 provides a comparison of the locations of spindles and tendon organs in relation to the extrafusal contractile muscle fibers. While the muscle spindles are situated *in parallel* with the contractile fibers, the tendon organs lie *in series* with them. A tendon organ usually involves several to many muscle fiber tendons.

Golgi tendon organs may be excited by strong passive stretch but are much less sensitive than the muscle spindles. They are, however, highly sensitive to the stretch imposed upon them by the *contraction* of the muscle in which they lie. Whereas contraction of its own muscle tends to relieve the stretch on a spindle and hence to result in a decrease or cessation of its discharge, the Golgi tendon organ, because of its location in series with

the contracting muscle fibers, is increasingly stimulated. Its discharge accelerates as contractile tension mounts (Fig. 11.15). Houk and Henneman (1967) found that some of the Golgi tendon organs in the cat soleus had absolute thresholds of less than 0.1 g and were caused to fire during contraction even in a shortened muscle where no tension was developed. In other words, these receptors were sensitive to internal forces developed in the muscle. Apparently the Golgi tendon organs continuously feed back to the spinal cord a filtered sample of the active forces operating in the muscle, information which is vital to coordinated response.

The tendon receptors are supplied by large group I afferents similar in size and conduction velocity to the primary spindle afferents (Fig. 11.14). Tendon afferents may be identified in spinal nerves, however, by two response characteristics which distinguish them from the spindle afferents: (1) they are not unloaded by contraction of the muscle, and (2) they are unaffected by gamma stimulation. Spindle afferents have been designated as Ia and tendon afferents as Ib neurons.‡ While a single Ib neuron may supply more than one Golgi tendon organ, more usually each is restricted to a single receptor, the average being 1.2 tendon organs per Ib afferent.

Afferents of tendon organs enter the cord and make di- or polysynaptic connections with alpha, and probably gamma, efferent neurons to their own muscle and to synergists and antagonists. Their effects are opposite to those of spindle primary afferents: Golgi tendon organs inhibit their own muscles (**autogenic inhibition**) and synergists and facilitate their antagonists. It has been suggested that Golgi afferents

---

‡ The symbols 1A and 1B are also used for designation of the afferents to spindles and tendon organs. The authors consider the Ia and Ib designation as more applicable: "I" indicates the size and conduction characteristics of the neurons (Lloyd classification of sensory neurons), and "a" and "b" serve simply to distinguish large afferents from two specific receptors, obviating any erroneous association with the A and B groups of the Erlanger-Gasser classification.

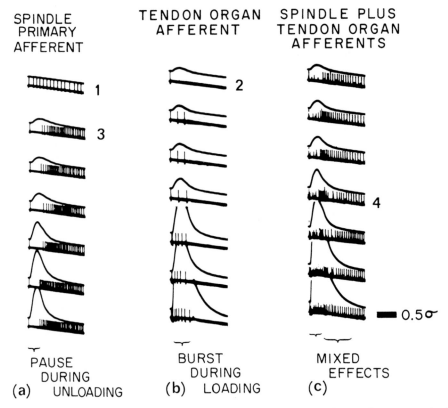

Figure 11.15. Afferent discharge responses from a spindle and a Golgi tendon organ during a graded twitch contraction. Muscle shortening is recorded by the tension trace. The application of the stimulus to the muscle is reflected in the single spike at the left of each record. (a), during contraction, the spindle primary afferent discharge ceases as the spindles are unloaded, resuming again during relaxation. 1, resting discharge; 3, note the phasic burst which is characteristic of the primary afferent during resumption of spindle firing and which may be followed by a secondary pause, 4. (b), as muscle contraction exerts a pull on the tendon organs, the tendon afferent shows a burst of activity which subsides and disappears during muscle relaxation. This is in direct contrast to the response of the spindle. The higher threshold of the tendon organ which is seen by comparing spindle and tendon organ traces, is also evidenced by the lower firing frequency of the latter. The tendon organ's high degree of adaptation is indicated by its failure to persist in firing throughout the twitch, even when tension was increased. (c), a simultaneous record from spindle and tendon afferents reflects a mixture of the responses from the "in parallel" and "in series" receptors. (From Eldred, E., 1967. Peripheral receptors: Their excitation and relation to reflex patterns. *Am. J. Phys. Med.* 46: Fig. 15, p. 84.)

may exert their inhibitory effects by means of presynaptic inhibition of the primary spindle afferents at their synapses on the anterior horn cells. This idea is not completely accepted.

If, as seems likely, all contractions activate tendon organs, their inhibitory effect must be offset during voluntary movement, either by voluntary exertion in excess of the opposing resistance or by some built-in reflex mechanism. Some evidence exists that during rapid voluntary movements in man the interneurons in the autogenic inhibitory circuits from the tendon organs to their own muscles are inhibited, thus canceling the inhibitory effect. Hufschmidt (1966) found that a stimulus sufficient to excite both Ia and Ib afferents produced only facilitation, indicating that the inhibitory action of the Ib neurons had been removed. Muscles have been shown to be capable of greater contractile force than their own structural makeup can withstand. Therefore it may be that the inhibitory reflex initiated by the tendon afferents is a safety measure to protect the muscle from dangerously high tensions. Loofbourrow (1960) points out that it is logical that the action of Golgi tendon organs inhibits not only the alpha motor neurons but the gamma system as well.

Three examples of Golgi tendon organ reflexes may be cited.

1. The immediate relaxation of muscles when volition ceases is probably due to the fact that the full inhibitory effect of the Golgi tendon organs is exerted as soon as their inhibition, which is associated with the voluntary movement, is no longer operative.

2. In Indian arm wrestling the loss of the contest usually occurs abruptly, when mounting tendon organ inhibition finally overcomes the voluntary effort to maintain contraction.

3. The "breaking point" in muscle strength testing is probably due to tendon organs and, if this is so, it suggests that maximal strength is dependent upon the individual's ability to voluntarily oppose the inhibition of his own tendon organs.

Vater-Pacinian Corpuscles

The Pacinian corpuscles located in muscle have been identified by Gray and Matthews (1951) and Hunt (1961). Pacinian corpuscles are widely distributed throughout the body and are usually associated with joint receptors (where they will be discussed more fully). The Pacinian bodies, located in the muscle respond to vibration and pressure (Powers, 1976). In response to steady pressure, they respond with a rapidly adapting discharge; if they are stimulated in vibratory fashion, they

apparently respond by following the stimulus frequency (as evidenced by the action potentials recorded from large myelinated afferent fibers). It would appear that the Pacinian corpuscles supply information to the central nervous system concerning the contractile state of muscle and can, therefore, contribute to the reflex control of movement (Powers, 1976).

Involvement in Motor Skills

Spindle and Golgi Tendon Organ Reflexes

Spindle and tendon organ reflexes can be recognized contributing their effects to facilitation, reinforcement, or inhibition of muscle contraction in almost any movement. The advantages which accrue from spindle activity by putting a muscle on stretch before its activation are obvious. There are many skills in which preparatory movements are made in a direction opposite to the anticipated functional movement. Common examples are backswings and crouches. These put stretch upon the muscles about to be used. Contractile capacity of the muscle tissue is increased by the well known length-tension phenomenon but, more importantly, the external stretch increases spindle activity, reinforcing the resting background of gamma discharge. As a result, impulse outflow over afferent neurons to the motor pools of the functional or prime mover muscles is increased. The speed and extent of the preparatory movement determine, through the phasic and tonic response frequencies of spindle primary afferents, the amount of involuntary discharge to the muscle. When the voluntary command is given, impulses descending by way of supraspinal pathways are added to this background; impulse frequency rises and a stronger contraction results. (Of course, other physical and mechanical advantages are involved as well.)

In sports which involve a single alternating movement, the nature of the backswing should have certain characteristics consistent with the main objective of the forward swing, whether to achieve force or to obtain precision. When the objective is force, spindle function suggests (1) that the backswing should be fast to gain the advantage of the phasic increase in spindle discharge; (2) that there should be minimal hesitation between backswing and forward drive movements to capitalize on the maximal frequency attained as a result of the velocity of the stretch; and (3) that the swing should be as long as practical to increase the tonic response of the spindles. In anticipation of the forceful hit, an increased outflow of voluntary impulses may be expected to occur over fusimotor neurons, enhancing spindle sensitivity to both phasic and tonic stretch. Consequently, as the muscle is rapidly stretched in the backswing, spindle frequency will rise sharply. The faster the backswing the higher the frequency of impulses transmitted to the muscles and the stronger the force of the forward driving movement. This principle finds application in the long shots of golf, assuming, of course, that a good swing has already become standardized by practice.

If the primary objective of the forward movement is something other than force, such as accuracy in golf approach shots (also tennis placements, billiards, etc.), spindle physiology would recommend that the backswing should be relatively slower and shorter, with velocity and length both carefully tailored to produce only that amount of force consistent with the distance to the green and the club being used.

In target throwing or shooting, a short hesitation at the end of a preliminary movement of optimal length should be advisable to allow the phasic frequency of spindle afferents to subside to the tonic level. Coincidentally, it would permit time for discriminative last minute adjustments to be made for accuracy, distance, and direction. It should be mentioned, however, that readjustment can be faulty and too much cerebration may hinder rather than help.

A centipede was happy, quite,
    Until a frog in fun said
"Pray which leg comes after which?"

Which raised her mind to such a pitch
She lay distracted in a ditch
    Considering how to run!

In tennis and other racket games the player intuitively times his backswing with the ball's approach so that the ballistic movements of back and forward swings become a smooth continuum. He may thus take advantage of the phasic stretch effect to achieve maximal force in his drive or smash. When he desires to diminish the force of his return, as when playing at the net, he may hold his racket in a static "ready" position and allow the simple stretch reflex to drain most but not all of the force from the oncoming ball and drop it just over the net.

In instances in which force is the main factor but a preparatory position must be held to await an outside event, the advantage of the stretch response is still available. In the case of the baseball batter awaiting a pitch, he can seek the advantages of the tonic spindle response to supplement his own muscle power by taking as long a backswing as possible within the limits of mechanical efficiency. As a basketball player awaits a tie ball toss-up, he should take as deep a crouch as is consistent with the strength of his antigravity muscles and the mechanical factors and force vectors involved.

Golgi tendon organs may be used to contribute to increased strength of contraction in their antagonists during alternating movements. For example, the greater the contraction of the backswinging extensor muscles the greater will be the activation of their tendon organs, whose effect upon the forward-swinging flexor muscles is one of facilitation. The tendon organ facilitation added to that of the stretch effects discussed above results in an increased strength of flexor contraction as the movement reverses.

For beginners, follow-through in a swing often requires specific voluntary effort to offset the inhibitory influence which the Golgi tendon organs impose upon their own shoulder flexor muscles, especially as the forward swing proceeds.

Furthermore, this autogenic flexor inhibition is being simultaneously reinforced by inhibition from the spindles in the stretched extensor muscles. This is probably why many students must be taught to extend the swing voluntarily beyond the impact point, until with practice follow-through finally becomes part of the learned motor pattern.

The detrimental effects of applying force too early in a striking movement may be related to the fact that the tendon organs become activated at a point where their effects interfere with the continuation of contraction. When the ball is finally struck, inhibition has already reduced the forcefulness of the swing.

In sedentary individuals in whom muscle weakness and imbalance are of common occurrence, spindle and tendon organ reflexes may be used for corrective purposes. By invoking maximal contraction in the lengthened or weak muscles, reciprocal inhibition may be imposed upon their shortened and too strong antagonists. Maintaining the contraction in the weakened muscles for several minutes will cause gamma bias to be reset, increasing the sensitivity of their spindles and improving resting muscle tone (Stockmeyer (1967)). Diminished proprioceptive feedback is characteristic of sedentary living and results in a reduction of the sensitivity of spindle secondary afferents in antigravity muscles. Without the stabilizing effect of co-contraction, the joints are unprotected and the jars of walking and exercise may produce discomfort and pain. Corrective procedures consist of exercising the extensors against resistance to increase their spindle sensitivity and to improve joint stability.

Vater-Pacinian Corpuscle Reflexes

"The laying on of hands," performed by the therapist or medical doctor for a specific diagnostic or rehabilitative purpose, should be done so that the intended goal is accomplished. If facilitation of muscle is intended, then, pressure on affected muscle or vibratory techniques sufficient to reach the threshold of the Pacinian bodies in muscle, could be used. If, however, one is testing for relaxed muscles, then every attempt should be made to place hands so that reflex facilitation from the Pacinian corpuscles does not interfere with the subject's effort or intent to relax muscles.

Summary of Muscle Proprioceptors

Eldred (1965) summarizes the importance of the muscle reflexes by saying that many of the finer attributes of sensitivity, restraint, and coordination which distinguish the activity of muscles must be derived from the feedback of data on the moment-to-moment status of muscle, much of which is provided by the three afferent transducers: spindle primary, spindle secondary, and Golgi tendon afferents. She further states, "ultimate understanding of the composition of movement must take into account the interplay by triple sensory influence, dual motor-sensory modulation, and the reaction of the alpha motor neuron final common pathway."

JOINT AND CUTANEOUS RECEPTORS

Joint Receptors

Another group of reflexes which play upon both alpha and gamma neurons to modify muscle activity are initiated by receptors in the joints and in the skin. Most joint receptors resemble either Ruffini spray receptors or small Pacinian corpuscles, although several size and shape variations are identifiable. Receptors consist of nerve endings, straight or branching, that are usually enclosed in a connective tissue capsule (Fig. 11.16). They are found in or near ligaments, joint capsules and adjacent connective tissues, and muscle fascia. Some which appear to be acceleration detectors are highly sensitive to movement, adapting quickly when movement ceases. Paciniform corpuscles in the

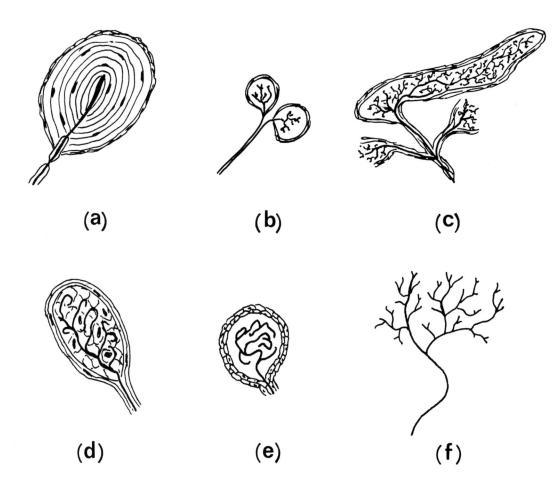

Figure 11.16. Receptors of joints and skin. Many of these receptors play a dual role as both exteroceptors and proprioceptors. (a), Pacinian corpuscle. A lamellated capsule enclosing an unbranched ending of a group I sensory neuron. These are fast-adapting receptors found near joints and tendons and in the deeper parts of the dermis; they are sensitive to quick movement, vibration, and pressure, depending in part on location. Smaller paciniform corpuscles are also present around joints. (b) Golgi-Mazzoni corpuscle. These are nerve sprays commonly situated near joint capsules, each with its own afferent neuron. They are pressure sensitive. (c), Ruffini end organs. These are slowly adapting flower-spray receptors in joint capsules and in the middle layers of the dermis. They signal joint position and may be sensitive to warmth. They are served by group II neurons. (d), Meissner's corpuscle. An encapsulated nerve ending located just under the epidermis; sensitive to touch. Touch stimuli affect skin areas of variable size and may involve many end organs. The pattern of input thus evoked provides the central nervous system with coded information for discriminative and integrative processing. (e), Krause's end bulb. Situated in the outer regions of the dermis, this may be sensitive to cooling. (f), free nerve endings. Widely distributed through the outer areas of the dermis and around joints, these probably contribute to pain sensation and to flexion reflexes.

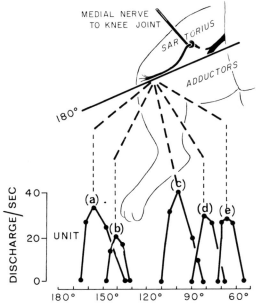

Figure 11.17. Range fractionation in the signaling of joint position. Discharge curves are shown for five joint receptor units isolated from the medial nerve to the knee joint of a cat. Each unit fired only over a restricted range of joint angles with a maximum at one fairly sharply defined point. Discharge was constant at a stable position of the joint but became elevated during movement within the response range for that unit. (From Eldred, E., 1967. Peripheral receptors: Their excitation and relation to reflex patterns. *Am. J. Phys. Med.* 46: Fig. 16, p. 85.)

neighborhood of tendons and joints also respond to deformation by firm pressure.

Static joint position produces firing in slowly adapting receptors which are insensitive to movement but which discharge in proportion to joint angle. Among these are flower-spray organs, known as Ruffini endings, found widely dispersed throughout joint capsules and able to maintain firing rates for as long as 90 minutes without decline. Their sensitivity is such that a change of 2° in joint angle is sufficient to alter rate of discharge. In any given position of the joint, some sensors are under intensive stimulation, some are less stimulated, and others are unstimulated (Fig. 11.17). Impulse frequencies and discharge patterns vary accordingly. Sensitivities of joint receptors are such that there is no position in which all joint receptors are silent. The integrative significance of such specificity in sensory coding is obvious. Receptors in joints have structures similar to those found in other tissues. In order to avoid drawing erroneous analogies, Freeman and Wyke (1967) proposed a grouping of joint receptors according to their functional properties. Table 11.1 summarizes the four joint receptor types.

Joint receptors, especially those of the interphalangeal joints of the feet, contribute to the positive supporting reflexes, along with the spindles of the interosseus

Table 11.1
Joint Receptors

Type	Description	Resemblance	Location	Adaptability	Stimulus
I	Ovoid, corpuscle with thin connective tissue capsule	Ruffini	Fibrous joint capsules, extrinsic ligaments	Fast and slow	Direction of movement; speed of movement
II	Single terminal within a thick laminated capsule	Pacinian	Fibrous joint capsules	Fast	Vibration or phasic pressure; accelerated movement
III	Large and fibrous	Golgi type	Intrinsic ligaments of joint capsules	Slow	Position of joints
IV	Fine, branching unmyelineated fibers	"Free" nerve endings	Fibrous joint capsules, joint ligaments, synovial capsules, fat pads	Slow	Pain

muscles. Apparently the complex combination of stimuli entering the cord facilitates appropriate stretch reflexes of extensor muscles to convert the limb into a firm but compliant pillar. If the phalanges are squeezed together or flexed instead of being abducted, all joints of the limb tend to flex.

When any one of the three joints of a limb is flexed, the other two joints also flex. When one is extended, the others also extend. Compression of a joint, as by weight bearing, reflexly evokes contraction of extensor muscles, while traction, as in hand suspension, excites flexors. These responses are mediated by joint receptors.

Since the afferent neurons of joint receptors interconnect in the spinal cord with interneurons of broad distribution, their influence is more diffuse than that of the muscle proprioceptors. They are especially effective in modifying activity in contralateral muscles. Their effects vary with the existing state of the joint. In other words, if the joint is in flexion, a given stimulus will produce a different pattern of response than if the joint happens to be in extension.

Interlimb joint receptors evoke responses of mutual facilitation. Sway to one side unbalances the force couples at the proximal joints, decreases the angle on the side of the sway, and increases the angle contralaterally. These changes evoke corrective action in the ab- and adductor muscles of the hip and the inverters and everters of the foot to neutralize the force couples. If, however, the sway exceeds the corrective capacity of the limb responses, equilibrium reflexes are elicited and the animal steps or hops sideward to realign center of gravity and base of support.

Signals from receptors in the joints between individual vertebrae and between the vertebral column, pelvis, and shoulder are coordinated to bring about cooperative muscle action to support the body. In combination with signals from limb joint receptors, they contribute importantly to the forces necessary to support the body weight and distribute it properly among the limbs.

The forelimbs act powerfully on the hind limbs (e.g., clenching the fists to jump) through both spinal and supraspinal circuits and both ipsilaterally and contralaterally. In contrast, the hind limbs are less effective in their action on the forelimbs; they appear to act only through supraspinal routes and to induce mainly ipsilateral responses.

Manipulation of ipsilateral joints facilitates labyrinthine responses and contributes increased stability and strength to the corresponding limb during standing and walking. Simultaneous work executed by a contralateral limb facilitates and improves performance.

Impulses arising in joint receptors are projected by multisynaptic pathways to the sensorimotor cortex, and the information thus provided appears to be the major factor in kinesthetic sensation.

Cutaneous Receptors

Most sense organs of the skin consist of a lamellated connective tissue capsule surrounding a soft cellular core in which the ending of the nerve axon is embedded. They vary somewhat in the complexity of this general design. Those responding to touch, light touch, pressure, or pain function both as somesthetic exteroceptors and as proprioceptors. In the former capacity they contribute to sensation by transmitting information to the thalamus and sensory cortex. In the latter capacity they initiate basic reflexes and contribute to the body righting reflexes. Their signals combine with those from receptors of joints and muscles to coordinate body movements. Cutaneous reflex pathways include both spinal and supraspinal loops and, at least in some instances and to some extent, responses require an intact cerebral cortex.

Cutaneous receptors initiate many of the fundamental inborn reflexes, such as the following.

Extensor Thrust Reflex. A strong but transient extension of the limb follows touch stimulation of the sole of the foot (or palm of the hand). This reflex may play an important role in running. It is a spinal reflex and can be elicited in the hind limb of a dog with a completely transected spinal cord.

Withdrawal or Flexion Reflex. Noxious stimulation to any part of the skin evokes withdrawal of the part from the stimulating agent, usually involving flexion.

Magnet Reaction. If a stimulus is moved across the sole of the foot, reflex movements are made so that the foot follows the stimulus, maintaining contact with it.

Grasp Reflex. Stimulus to the palm of the hand is followed by grasping of the stimulating object.

Placing Reactions. When a foot is placed upon an uneven receiving surface, corrective movements occur to position the foot so that it is ready to support the body weight (Fig. 11.18). These responses are considered "reflex" because of their stereotyped nature, but there is some evidence that they are learned rather than inborn.

Stimulation of the Skin. Pinching or squeezing the skin of the back produces generalized inhibition of limb extension. Stimulation of the skin of the abdomen facilitates extension of the limbs.

The location of a receptor is important in determining what muscles are activated or inhibited, and the most important factor determining direction of reflex movement is the location of skin stimulation (Sherrington's "**local sign**"). The significance of receptor locus is evident in the fact that the response tends to do something about the stimulus which aroused it. The dog's scratch reflex is an example. Hagbarth (1952) demonstrated in the cat that stimulation of the skin over an extensor muscle produced facilitation of extensors with inhibition of ipsilateral flexors, while stimulation of most other skin areas produced the opposite effect.

Impulses arising in pressure receptors as a result of contact with a surface play an important part in the orientation of the body in space, evoking the body-on-head and the body-on-body components of the

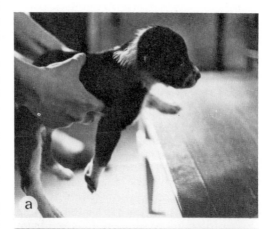

Figure 11.18. Placing reactions in a puppy and a kitten. (a), the puppy reaches for the table with his left front foot. (b), the kitten extends both front legs to place them on the table surface. (c), the puppy extends all four legs ready for weight bearing.

righting reflexes (discussed in the next section). Touch receptors also facilitate muscular adjustments for the maintenance of equilibrium, especially in subjects who are in poor physical condition. In the O'Connell-Gardner laboratory some individuals, if permitted to touch a piece of vertically hanging paper, could maintain a one-foot stance significantly longer than without touch stimulus. In fact, some blindfolded subjects stood longer when permitted the touch stimulus than they did with eyes uncovered but lacking touch. Slightly tipsy persons use touch, without taking any real support from the contact, to reinforce their depressed equilibrium reflexes. Perhaps the touch stimuli act to damp oscillation in postural correction and to avoid positive feedback.

Both skin and joint receptors exert a broad influence, acting in the interregulation of muscles at joints other than the joint acted upon by the prime mover muscles and including contralateral joints. Their afferents affect the outflow to muscles via both alpha and gamma routes, the gamma neurons having lower thresholds than the alpha neurons. Holmqvist (1961), however, found the contralateral actions by joint neurons to be stronger than those of either skin or muscle afferents. Since facilitation of both flexion and extension was common, she suggested that there must exist alternative pathways by which one afferent fiber can act upon a contralateral motor pool. Other studies support the complexity of reflex interconnections and the fact that a muscle's response is dependent upon its state at the time of stimulation, i.e., upon the sum total of the afferent influences converging upon it as a result of position and movement in any and all parts of the body.

Involvement in Motor Skills

Joint and Cutaneous Reflexes

Proprioceptive influences from joint receptors are so integral a part of movement that it is not easy to isolate specific examples of their exclusive operation. Facilitative feedback by irradiation may be used to increase the strength of contraction, as in clenching the fists to jump high or gritting the teeth while striking or lifting forcefully. Strong contraction in another part of the body has been shown to increase impulse outflow over subconscious channels to the motor synapses of prime mover muscles. Hellebrandt and Waterland (1962), who studied the phenomenon of irradiation extensively, said, "irradiation is probably a normal concomitant of pushing volitional effort into unfamiliar zones of activity and probably the excessive innervation thus aroused is just as necessary for the development of motor skill as overloading is for the development of strength."

The use of starting positions to direct impulse flow is common. Most starting positions place joints in the position opposite to that of the upcoming movement. This not only puts the muscles on stretch but also favors irradiation of impulses to the proper muscles.

Since feedback from joint receptors is important in the coordination of arms with legs, instruction may be given to foster limb reciprocity in take-off for apparatus vaulting, diving, trampoline, etc. The limbs on one side of the body may be used to influence contralateral limbs. Irradiation from joints is useful in augmenting the flow of impulses to weak muscles in the poorly skilled.

Feedback from joints may contribute consciously as well as unconsciously to the guidance and timing of movements in patterned sequences. Gymnasts rely strongly on kinesthetic perception for assuring good form. The diver times his maneuvers by the "feel" of body position in the air. Mirrors permit the dancer to see herself in action. By visual feedback she adjusts limb and trunk movements to her satisfaction. Joint receptors feed back information on the desired angulation and movement, and this kinesthetic image is eventually impressed on her memory. As a result, she can later repeat the same positions and movements without the visual stimuli. Perfecting form in any skill depends strongly on feedback from joint receptors.

Cutaneous stimuli are often used to evoke specific parts of a movement pattern. Stimuli are supplied by contact with the environment (floor, trampoline, diving board, water, horse, box, rings, etc.); contact with a sports implement (racket, golf club, ski, ball, bat, etc.); and contact between parts of the body. Such stimuli inform the nervous system of the instant of impact (in tumbling, apparatus, vaulting, diving, dancing, trampolining, swimming, racket games, golf, baseball, football, etc.); of the changing contact with surface or implement (in skiing, skating, diving, golf, swimming, etc.); and the chance or deliberate contact between the body parts in all activities. These feedbacks are employed consciously or unconsciously (depending upon the learning stage) to time the abrupt changes in a movement sequence and to achieve the smooth transitions in force and direction which are basic components of a skilled performance.

Knowledge of certain facts regarding skin stimulation should be kept in mind when assisting a gymnast to perform. Strong stimulation of the skin over a muscle is excitatory to the muscle, but stimulation over a nonmuscular area excites extensors only. (Pressure on the belly of a muscle also activates its spindles.) Therefore the spotter should be careful to place his hands so that he does not elicit undesirable reflex response but, if possible, reinforces the desirable muscle action. When extension of the arms is essential (parallel bars, horse, box, etc.) avoid pressure upon flexor muscles and, if feasible, place the hands upon extensors or upon bony areas so as to reinforce extension. When flexion is required, the opposite hand placement is recommended.

LABYRINTHINE AND NECK RECEPTORS

The fact that the head exerts important influences upon movements of the trunk and limbs can hardly be overlooked by anyone who deals professionally with muscular activity. These orienting and reinforcing effects are mediated by two groups of proprioceptors: the labyrinthine and the neck receptors.

Labyrinthine Receptors

Labyrinthine receptors are specialized proprioceptors located in the inner ear (Fig. 11.19). They consist of complex arrangements of cells with projecting hairs whose orientation is displaced either by the pull of stony concretions or the movement of endolymph. They include the semicircular canals, which are stimulated by angular acceleration, and the utricles (perhaps also the saccules), which are stimulated by linear acceleration and by changes in the orientation of the head in relation to gravity. The semicircular canals serve a major function in the maintenance of equilibrium during movement, while the utricles appear to be concerned primarily with postural reflexes and in the differential distribution of muscle tone.

Utricles

In the early life of the child, reflexes arising from the utricular receptors are present in primitive and highly predictable form known as the **tonic labyrinthine reflexes**, or TLR. Stimulation induced by body position or head inclination produces stereotyped effects (Fig. 11.20). The supine position or corresponding orientation of the head with gravity facilitates extension in all limbs and inhibits flexion, while the prone position or corresponding head orientation produces the opposite response. Side lying or an equivalent head position induces extensor facilitation of the under limbs and flexor facilitation of the top limbs with reciprocal inhibition of antagonists. As growth and development proceed, the TLR are supplanted by more complex reflexes, the **labyrinthine righting reflexes**, whose purpose seems to be to orient the head correctly with gravity. This is accomplished mainly by contraction of

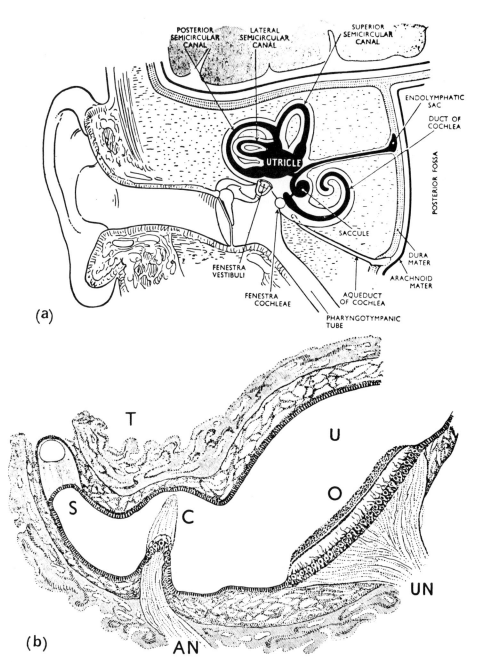

Figure 11.19. Labyrinthine receptors. (a), the bony and membranous labyrinths. The tubes shown as solid black contain endolymph. (Drawing by J. D. B. MacDougall. From Bell, G. H., Davidson, J. N., and Scarborough, H., 1969. *Textbook of Physiology and Biochemistry*. Fig. 45.15, p. 1022. Baltimore: The Williams & Wilkins Company.) (b), diagrammatic cross-section through a semicircular canal (S) and utricle (U). The sensitive areas are the otolith organ at O and the crista ampullaris (C) in the ampulla of the semicircular canal. Both contain bottle-shaped and cylindrical hair cells separated by supporting cells. The sensory nerve cells attached to the hair cells send impulses into myelinated nerve fibers. The hairs pass into canals in a jelly-like material in which, in the case of the otolith organs, there are particles of calcium carbonate forming the otolith membrane. The crista ampullaris is formed into a crest surmounted by a mass of jelly-like material, the cupola, which probably reaches across the ampulla to the opposite wall. T, temporal bone; AN, ampullary nerve; UN, utricular nerve. (From Bell, G. H., Davidson, J. N., and Scarborough, H., 1969. *Textbook of Physiology and Biochemistry*, Fig. 45.16, p. 1023. Baltimore: The Williams & Wilkins Company.)

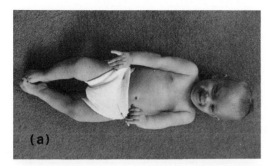

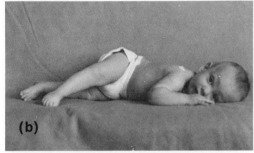

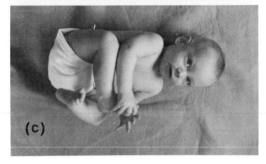

Figure 11.20. Tonic labyrinthine reflexes. The "primitive reflexes" often persist beyond the first 2 or 3 months of life, as demonstrated by this child of 7 months, and can even be elicited in the adult. The tonic neck (Fig. 11.23) and tonic labyrinthine reflexes are inseparably interrelated; they reinforce each other in their effects on the upper extremities and oppose each other in their effects on the lower extremities. However, the tonic labyrinthine reflex has been acknowledged as the more dominant reflex for lower extremity patterning, which may explain the knee flexion in (a) and the right knee extension in (b). (a), supine lying: all limbs extended. (b), side lying: under limbs (left upper and lower extremities) extended, upper limbs (right upper and lower extremities) flexed. (c), prone lying: all limbs flexed; the child was encouraged to play with his feet; a much younger child, when placed face down, would adopt a similar position with the elbows alongside the trunk and the knees drawn up under the hips.

the neck muscles, although secondary influences are also exerted on trunk and limb muscles to supplement the raising of the head. These utricular reflexes cooperate with other reflexes stemming from stimulation of skin, neck, and visual receptors to constitute, as a group, the **righting reflexes**, whose normal operation is essential in the child's achievement of vertical posture. The righting reflexes (Fig. 11.21) include the following components.

Labyrinthine Righting Reflexes. Stimulation of the labyrinthine receptors evokes contractions of neck muscles to orient the head in relationship to gravitational force (Fig. 11.22, (a) and (b)).

Body-on-Head Reflexes. Asymmetrical stimulation of skin receptors resulting from differential contact with the supporting surface (side or back lying) results in activity of trunk and limb muscles which raises the head into an upright position. If skin stimulation is equalized by placing a board on the upper side of the animal, no righting occurs.

Neck Righting Reflexes. Impulses arising in joint receptors in the neck produce contraction of limb and body muscles which align the body with the head (Fig. 11.22, (b) and (c) and Fig. 11.24, (a) and (b)).

Body-on-Body Righting Reflexes. Asymmetrical stimulation of skin receptors causes contraction of trunk muscles which raise the body toward the upright position.

Visual Righting Reflexes. Visual feedback is used to orient the head and body correctly with the environment and is especially important when other sensory input is deficient or excluded.

Righting reflexes are later dominated by the **equilibrium reflexes**.

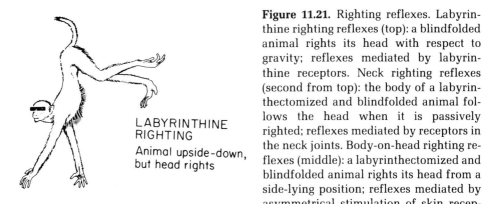

LABYRINTHINE
RIGHTING

Animal upside-down,
but head rights

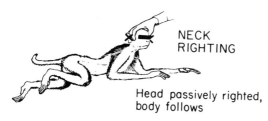

NECK
RIGHTING

Head passively righted,
body follows

BODY RIGHTING
ON HEAD

Assymetric contact
to body,
head rights

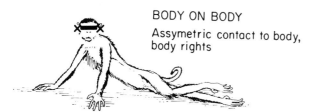

BODY ON BODY

Assymetric contact to body,
body rights

OPTICAL RIGHTING

Animal upside-down,
rights by vision

Figure 11.21. Righting reflexes. Labyrinthine righting reflexes (top): a blindfolded animal rights its head with respect to gravity; reflexes mediated by labyrinthine receptors. Neck righting reflexes (second from top): the body of a labyrinthectomized and blindfolded animal follows the head when it is passively righted; reflexes mediated by receptors in the neck joints. Body-on-head righting reflexes (middle): a labyrinthectomized and blindfolded animal rights its head from a side-lying position; reflexes mediated by asymmetrical stimulation of skin receptors of the two sides of the body. Body-on-body righting reflexes (second from bottom): as a result of asymmetrical skin stimulation, a labyrinthectomized and blindfolded animal rights its body from a side-lying position. Optical (visual) righting reflexes (bottom): a labyrinthectomized animal rights its head, using visual feedback. (From Twitchell, T. E., 1965. Attitudinal reflexes. In *The Child with Central Nervous System Deficit*, Fig. 3, p. 81. Washington, D. C.: U. S. Government Printing Office.)

 BLINDFOLDED

 LABYRINTHECTOMIZED

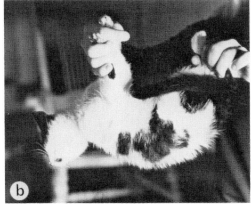

Figure 11.22. Labyrinthine righting reflexes in a kitten. (a), the kitten is held horizontally, belly up. (b), the labyrinthine righting reflexes reorient the head into proper relationship with gravity. (c), as the kitten is dropped, the neck righting reflexes correct the relation of body to head and it lands squarely on its four feet.

Semicircular Canals

The equilibrium reflexes are evoked by stimulation of the semicircular canals. Appropriate muscular responses maintain or regain body balance either by redistributing body segments to keep the center of gravity over the base of support or by shifting the base of support to keep it under the center of gravity.

Movements of endolymph during angular or rotatory motions of the head stimulate the ampullae of the canals which are situated in line with the movement. Impulses traverse pathways to antigravity muscles, particularly the extensors of the limbs and neck, to evoke alterations of tone in these muscles and to produce movements which oppose the angular or rotatory acceleration of the head.

In the following account, we shall ignore the coincidental utricular responses of the neck muscles for the sake of simplicity. The reflex responses to a diagonally forward tilt are induced by the stimulation of ampullae of the ipsilateral anterior and contralateral posterior vertical canals. These responses consist of increased extensor tone on the same side, accompanied by decreased extensor tone on the opposite side. Thus in a four-legged animal tipped toward its right front leg, that leg extends strongly while the left hind leg partially flexes.

Lateral tilting of the supporting surface excites ampullae of both ipsilateral vertical canals, anterior and posterior, and produces extension of the ipsilateral limbs with concurrent reduction of extension contralaterally so that the body retains its original position in space. A straight forward or backward tilt excites the ampullae of both corresponding anterior vertical or posterior vertical canals, respectively, and produces suitable changes in extensor tonus.

Whirling about a vertical axis involves both horizontal canals and induces changes in all four limbs, which act to oppose the acceleration of the head and to keep the head stationary in space. The acceleration evokes extension of the limbs on the side of the direction of rotation; i.e., rotation to the left is accompanied by extension of the limbs on the left side. If the rotation is suddenly stopped (deceleration), inertia reverses the flow of the endolymph, producing reactions directed opposite to those during the rotary acceleration. Hence the post-rotatory movements seen in past-pointing and in staggering occur in the direction of the preceding rotation, indicating that the deceleration has evoked extension in the limbs of the side opposite to the rotatory direction (the right side in our example).

The acceleratory reflexes provide trunk and limb movement which will have a velocity-damping effect upon the rotational movements of the head. The objective seems to be to maintain a constant orientation of the head in space. However, because movement of the head upon the neck is restricted by anatomical limitations, a nystagmic jerk to return the head to its central position occurs when the limit is reached. Jerks are noticeable in the dancer's pirouette. Their absence in the spinning of a figure skater is a result of his training to hold the head erectly immobile voluntarily. Doing so restricts the labyrinthine responses to those of the horizontal canals, avoiding any responses which might otherwise be induced by labyrinthine reflexes arising from accelerations in forward and lateral planes. Furthermore, by obviating the changes in neck angulation which would result from stimulation of the utricles, the erect posture of the head provides a stable point of reference for the body in maintaining vertical equilibrium.

Semicircular canals are stimulated by angular acceleration about any axis and in any direction. Excitation is not limited to a particular pair of canals, but presumably all may be concerned in any angular movement, with the extent of involvement determined by the resolved components of the planes involved in the acceleration. Therefore, it is difficult to correlate precisely individual canals with specific muscles during the complex movements of locomotion and motor skills. To complicate

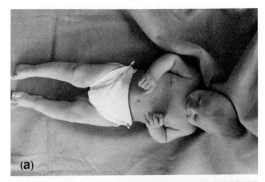

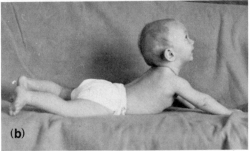

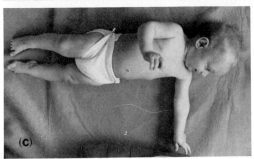

Figure 11.23. Tonic neck reflexes, demonstrated by a child of 7 months of age. The tonic neck and tonic labyrinthine reflexes (Fig. 11.20) are inseparably interrelated; they reinforce each other in their effects on the upper extremities and oppose each other in their effects on the lower extremities. However, the tonic labyrinthine reflex has been acknowledged as the more dominant reflex for lower extremity patterning, which may explain the lack of flexion in the right lower extremity of (c). (a), head ventriflexed: upper extremities flexed, lower extremities extended. (b), head dorsiflexed: upper extremities extended, lower extremities flexed. (c), head rotated to the left: left upper and lower extremities abducted and extended, right upper and lower extremities adducted and flexed.

the problem further, acceleration of the head usually simultaneously affects the utricles, thus evoking the attitudinal or postural reflexes at the same time.

Neck Reflexes

The neck reflexes arise as a result of stimulation of joint receptors in the cervical spine (especially around the atlantooccipital and atlantoaxial joints) when the head is inclined forward, backward, or sideward, or rotated to either side. These reflexes, like the TLR, are also present at birth in a stereotyped form, the **tonic neck reflexes** or TNR, which persist postnatally in compulsive form for a short period. The following are typical tonic neck reflex responses (Fig. 11.23): ventriflexion of the neck evokes flexor facilitation of the front limbs with concomitant inhibition of antagonists (Fig. 11.24(a)), while dorsiflexion of the neck produces opposite responses (Fig. 11.24(b)); rotation of the head facilitates extension and abduction of the limbs on the face side and flexion and adduction of the contralateral limbs (a posture reminiscent of the fencer's *en garde* position), with reciprocal inhibition of antagonistic muscles. The tonic neck reflexes become less apparent as motor development proceeds and are no longer compulsive by the 6th to 8th week. In the older child they finally assume their mature role as **neck righting reflexes** which, by evoking suitable activity in body and limb muscles, assure that the body follows the head.

The TNR and TLR are inseparably interrelated, continually modifying each other, sometimes reinforcing, sometimes diminishing the effect of the other. For example, in ventriflexion, front limb flexion is facilitated by both (reinforcement), whereas the hind limb effects oppose each other: the TLR favor flexion but the TNR favor extension.

The TNR and TLR are not completely eradicated in the adult. Their circuits are still intact; only their pattern of synaptic facilitation and inhibition has been altered as they have become dominated by later developing and more useful patterns.

Figure 11.24. Tonic neck reflexes in a kitten and a puppy. (a), ventriflexion induces flexion of the front limbs and extension of the hind limbs (kitten). (b), dorsiflexion induces extension of the front limbs and flexion of the hind limbs (puppy).

Twitchell (1965) says, "wholly new and distinct reactions are not added at successively higher levels of the nervous system but more primitive reactions become modified and elaborated as the stimulus for their response becomes more discriminating." Brunnstrom (1964) mentions the work of a group of Japanese scientists whose electromyographic studies in 1951 showed the presence of TNR in normal subjects. Their presence in normal adults was also demonstrated by Hellebrandt and co-workers in 1962. Her work with Waterland (1962) indicated that the primitive reflexes operate normally during stressful activity to reinforce muscular contractions and to extend endurance.

It seems a reasonable speculation that the simple labyrinthine and neck reflexes are concerned at all times with the integration of limb movements into a total body pattern. Their presence in the primitive form is detectable when other more complex responses are not masking, inhibiting, or precluding their automatic modification of limb muscle activity. It is difficult to separate the TLR and TNR since any movement of the head unavoidably activates the receptors of both the labyrinth and the neck. However, since it has been shown that neck reflexes act more effectively upon the upper limbs and the labyrinthine reflexes on the lower limbs, we may reasonably attribute influences on the arms to neck reflexes and those on the legs to labyrinthine reflexes. A few examples follow.

1. There is a stunt in which the performer stands in a doorway, pressing the backs of his hands forcefully against the door jambs by shoulder abduction for about a minute, and then steps forward out of the doorway. In subjects who are able to relax, the arms spontaneously rise to at least 90° abduction. If the head is rotated to one side, a clearly defined TNR pattern occurs in responsive subjects.

The abduction may be explained physiologically as follows: neural transmission to the abductors persists in reverberating circuits, probably at the spinal level, and the muscle contraction, now unresisted, is converted from isometric to isotonic. (If the arms are dropped to the sides as the subject steps from the door frame, there is a brief pause before abduction begins.)

2. When placing a water-filled ice cube tray into the freezer compartment at the top of a refrigerator, TNR may cause one to spill the water; i.e., if the head is turned away from the hand holding the ice tray to open the freezer door, inadvertent flexion tends to spill water down the arm, while if the head is turned toward the tray-holding hand, extension will spill water on the floor.

3. Anyone who has taught diving to beginners has seen the labyrinthine reflexes thwart a diver's volitional attempt to enter the water head first. In a forward dive the labyrinthine righting reflexes can raise the head as balance is lost, causing a "bellyflop," or, in response to equilibrium reflexes, the diver may jump in feet first. In a backward dive, labyrinthine righting reflexes can cause ventriflexion of the head so that the diver lands on his back or trunk flexion so that he sits into the water.

4. The presence of head righting movements is obvious in maintaining and regaining balance. These movements occur in combination with the equilibrium movements to bring about proper alignment of the head, center of gravity, and base of support.

Involvement in Motor Skills

Neck and Labyrinthine Reflexes

The fact that the head exerts important influences upon the trunk and limbs can hardly be overlooked by anyone who deals professionally with muscular activity. With the possible exception of the spindle reflexes, neck reflexes are probably the most important single reflex mechanism used in sports skills, and they provide proprioceptive influences commonly employed by both performer and teacher.

Positions and movements of the head are used to reinforce contractions of arm muscles by invoking tonic neck and labyrinthine reflexes. A ventriflexed head favors bilateral elbow flexion; a dorsiflexed head favors extension. The symmetrical bilateral influences of the neck reflexes are obvious in weight lifting. One sees ventriflexion reinforcing the arm flexion of the lift and dorsiflexion reinforcing the shoulder flexion and elbow extension of putting up the barbell. Ventriflexion combined with contralateral rotation of the head favors unilateral pulling (flexion) movements, while dorsiflexion combined with ipsilateral rotation favors unilateral pushing (extension) movements.

Hellebrandt et al. (1962) showed that a lowered position of the head amplifies the effects of the tonic neck reflexes. In the handstand, therefore, dorsiflexion of the head probably contributes to the maintenance of arm extension as well as to balance.

Use of head and shoulder maneuvers is especially noticeable in rebound skills to accomplish turns about the long body axis. Reflex facilitation probably reinforces the mechanical advantages involved.

Neck reflexes facilitate essential trunk and limb movements in somersaults. The diver and tumbler use strong movements of the head to initiate and maintain a spin. Ventriflexion reflexly facilitates trunk and limb flexion for the forward somersault; dorsiflexion facilitates extension for the backward somersault. Resumption of the normal head-neck angle automatically assures that the body in general will be aligned with the head for a correct finish.

In balance work, fixation of the head by focusing the eyes on a point stabilizes gravitational effects upon the labyrinths so as to maintain an unchanging inflow from these receptors. The neck reflexes then make any necessary adjustments of the body to keep it aligned with the correctly postured head.

ROLE OF REFLEXES IN SKILLED MOVEMENT

A motor skill is a group of simple, natural movements combined in a new or unusual manner to achieve a predetermined objective. Skilled movement requires both mobility and stability of body parts. Each movement consists of a coordinated combination of several to many joint movements, and each joint movement further consists of a coordinated combination of muscle actions: contraction of prime movers, relaxation of antagonists, and supporting contraction of synergists and stabilizers. These are mediated by afferent information received and processed in the central nervous system and converted into appropriate signals which finally converge upon the motor neuron pools, evoking the proper activity in each

muscle (Fig. 11.25). All of these muscles must be interregulated as to intensity, speed, and duration and as to their changes in activity from the beginning to the end of the movement. This demands precise integrative function, too precise to be left to man's conscious efforts. Hence that regulation has been delegated to automatic mechanisms, among which the proprioceptive reflexes occupy a place of importance.

The proprioceptive reflexes mediated at spinal and lower brain levels, however, must be integrated into larger, coordinated patterns at centers in the brainstem, cerebellum, and cerebral cortex. It has been demonstrated that the reticular formation exerts powerful influences, both excitatory and inhibitory, upon the spindle reflexes, the pons-midbrain areas being excitatory and the medulla areas being inhibitory to the fusimotor neurons. The gamma biasing of the spindles appears to be accomplished largely via the descending reticular system. Of special importance is the role of the cerebellum in maintaining the proper balance of muscle activity for coordinated movement. And finally, all reflex responses are subject to control by the sensorimotor areas of the cerebral cortex. While a thorough discussion of supraspinal regulation of movement is beyond the scope of this book, the reader should keep in mind the fact that the proprioceptive reflexes, important as they are, require interregulation from superior sources. The very nature of skilled movement makes this self-evident.

Motor skills are learned. They are the result of operant (instrumental) conditioning whereby inborn and proprioceptive reflexes are conditioned to be triggered by other than their natural (unconditioned) stimuli, to occur in new combinations or sequences, or not to occur at all under the usual unconditioned circumstances (to be inhibited). Conditioning involves a rechanneling of impulse flow within the central nervous system by the establishment of new patterns of synaptic resistance. Eventually the new (conditioned) response substitutes for the natural (unconditioned) response.

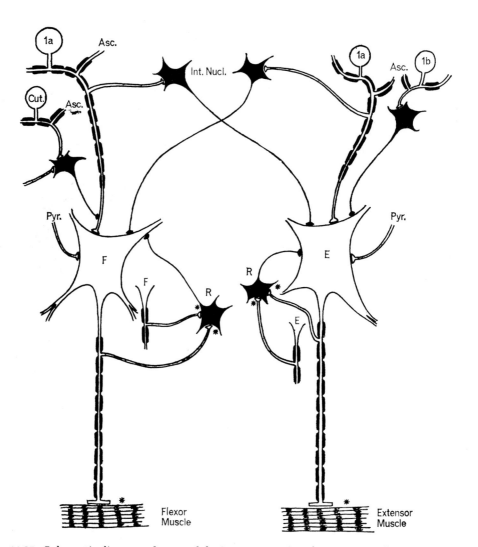

Figure 11.25. Schematic diagram of some of the interconnections between two flexor motoneurons (F) of flexor muscle and extensor motoneurons (E) of antagonist extensor muscle. Inhibitory neurons and their synaptic terminals are shown in black. The recurrent collaterals make cholinergic (*) connections with Renshaw cells (R). 1a, cell bodies of afferent fibers from annulospiral endings on muscle spindles within the corresponding muscles; 1b, cell body of an afferent fiber from a Golgi organ in the tendon of the extensor muscle; Cut., cell body of a cutaneous afferent which is connected to an inhibitory neuron having a presynaptic inhibitory terminal on one of the 1a excitatory endings; Int. Nucl., inhibitory neurons in the intermediate nucleus; Pyr., excitatory terminals of the descending pyramidal tract; Asc., branches of the afferents joining the ascending sensory fiber tracts. (From Bell, G. H., Davidson, J. N., and Scarborough, H., 1969. *Textbook of Physiology and Biochemistry*, Fig. 39.41, p. 901. Baltimore: The Williams & Wilkins Company.)

Consider the catching of a ball as an example. The visual stimulus of an object flying through the air toward one's head naturally evokes an *avoidance* response; one ducks, closes his eyes, and raises his hands protectively to ward off possible impact. With conditioning, however, the approach of a flying ball elicits a *catching* reaction. One no longer ducks, that movement having been inhibited. The eyes are kept open and the hands are purposefully raised to entrap the ball. The new, acquired skill, rather than the old, unconditioned avoidance response, then occurs automatically whenever a ball approaches.

The ultimate aim of motor learning is automaticity, the performance of a skill with a minimum of conscious involvement. To that end, the proprioceptive feedback from each movement or position becomes the cue (conditioned stimulus) for the next; and the movement proceeds smoothly with a minimum of conscious direction. The cortex sits back and monitors the situation ready to impose modification as need arises, unconcerned with directing individual movements unless correction is required, and, relieved of integrational details, it is free to concern itself with strategy.

MOTOR LEARNING
IMPLICATIONS

Each new motor skill requires a rechanneling of impulse flow between receptors and effectors to produce new spatial and temporal relationships among the component movements. A different pattern of movement sequences must be established, these sequences being precisely triggered by feedback mechanisms which have been conditioned to evoke the new responses. During the early stages of learning a difficult skill, the learner must be consciously concerned with the individual parts of the total pattern. Because conscious direction often interferes with the smooth operation of subcortical mecha-

nisms, the attempted performance usually includes a number of movements which serve no useful purpose and which may even interfere with the achievement of the objective. As learning progresses there is a gradual reduction in the amount of cortical involvement as one after another of the pattern components becomes conditioned, until finally a single thought may be sufficient to initiate the whole train of motor events.

Once fully established, the skill becomes a feedback-regulated pattern which is somehow stored in the central nervous system to be called forth in its entirety by a simple but specific (within some permissable range of variation) input signal. Once initiated, the sequence of muscular actions proceeds almost entirely under the control of subcortical mechanisms. The major portion of movement control has been shifted by the learning process from cortical to peripheral mechanisms. Instances of the operation of proprioceptive responses can be readily identified in motor skills; their advantageous use may be inferred from specific portions of skilled movement patterns. Their involvement is implicit in many instructive cues. Some examples are discussed below.

Conscious Inhibition of Reflexes

The essential role of the proprioceptive reflexes in skilled movement is unquestionable, but in motor learning there is a negative side to be considered. Many skills involve movements in opposition to the natural inherited reflexes, especially those directed by the labyrinths. Body inversions, somersaults, cartwheels, etc., are outstanding examples. It seems obvious that a large part of the difficulty in learning such stunts relates to the need *consciously* to impose inhibition upon these responses. It is often necessary at first to emphasize a position or movement antagonistic to the natural movement and to elicit active muscular contraction against the movement. Paillard (1960) states, "learning requires before anything else the disrupting of some preexisting functional units, . . . then the selective choice of useful motor combinations and finally assembling them into a new working unit." Instructions such as "keep your head down," "hold your chin on your chest," "lock your knees," etc., are frequently used in teaching of diving.

Basmajian (1978) further confirms the concept of inhibiting the unnecessary movements and the muscles producing them. In his work with biofeedback, he uses the technique of electromyography to demonstrate for the patient and the therapist, the electrical activity present in muscles not recognized as contributors to the intended movement. Patients can learn to inhibit the unnecessary muscles when they are provided with a visual or auditory display of the action potential which is unwanted. As the action potential is reduced, the tension in muscle also reduces and offers significant reduction in and relief of pain.

The need for deliberate inhibition is common. In the golf swing inhibitions are reflected in the need for conscious effort to keep the left elbow straight, to keep the left hand on top of the club, to pivot without swaying, to keep the head down throughout the swing. In all of these, voluntary effort is required to inhibit "what comes naturally." As learning progresses, the need for conscious contractile antagonism diminishes as synaptic inhibition of the reflex circuits becomes an integral part of the conditioned response. Instances of involvement of conscious inhibition are abundant in the learning of most sports.

NEUROKINESIOLOGICAL ANALYSES

Although the physical education student may have had some undergraduate training in neuromuscular physiology, all too frequently he has not been made aware of its practical applications, nor has he consciously applied such information. He has scant understanding of the neuromus-

cular physiological reasons for taking a long backswing to achieve a good hit, or a big wind-up to make a good throw; he only knows that these work. He neither knows nor cares that he is making use of the stretch reflex to augment the contractile force of the muscles which produce the desired forward motion.

The student of kinesiology should be aware of the neural mechanisms by which motor performance is integrated and controlled. He should be familiar with the various reflexes which may help or hinder a performance so that he may make use of them. Practice in trying to detect the operation of reflexes in simple responses such as postural response and in complex motor patterns may include direct observation, study of motion pictures and electromyograms (EMGs), and combinations of these methods.

Analyses of Simple Responses

Adjustment to Load

Films and electromyograms taken during the adjustment to load (discussed earlier in this chapter) provide an objective basis for analysis of the spindle reflexes pertinent to that response (Fig. 11.10). The subject was asked to hold a basket in his hand, palm up, so that a 4-lb dumbbell might easily be added and removed, and instructed to maintain the elbow at an angle of approximately 90° throughout. The muscular responses were recorded by bipolar surface electrodes placed on the long and short heads of the biceps and on the lateral and long heads of the triceps. Motion pictures, synchronized with the EMG, were taken simultaneously to record the gross movements of the limb. The load, introduced at b on the EMG record, was added abruptly in I and gradually in II. For abrupt loading (I), the starting position is shown in photo Ia. The activity in the muscle which was recorded in the EMG before b (loading) was that set by gamma bias to maintain the prescribed position under the dictation of the cortex, plus the spindle response to the weight of forearm and basket. When the load was dropped into the basket (photo Ib), there was momentarily a slight increase in elbow extension (Ic), followed by a return of the forearm to the original position (Id). Upon addition of the load (with a slight lag partially attributable to the inertia of the pens of the inkwriting polygraph) the EMG exhibited a sudden burst of activity in both heads of the biceps, showing that the sudden stretch of the muscle had evoked a phasic response in the primary afferents of its spindles. The muscle responded by an appropriate contraction. (Contraction of the muscle will momentarily unload the spindles, causing contraction to lessen and the arm to drop. The tendency to oscillate which results is, fortunately, damped out by the gamma bias. The magnitude of the oscillatory activity with this subject and this load is not sufficient to show in these gross records.) Muscle activity then settled down to a lower level consistent with the new load and reflecting the tonic spindle response. This was maintained until the load was removed at e, at which time muscle activity returned to its initial level. During removal, the small unexpected and transient increase in activity was probably due to momentary pressure on the load exerted by the assistant in grasping the dumbbell.

In II, the same load was placed gradually into the basket, beginning at b (photo IIb) and completed by c (photo IIc). Activity in the biceps gradually increased to a new tonic level but without any phasic burst. Upon removal of the load (d to e and photos IId and IIe), the initial tonic level of activity was resumed.

Postural Reflexes: The Scale

In the gymnast's scale or the dancer's arabesque the performer stands poised with his weight balanced over one foot (Fig. 6.3). Because the pose does not permit adjustment of the base of support, equilibrium must be maintained by proper counterbalancing of muscle tone at the hip, trunk (intervertebral), and shoulder joints. In order to assume and maintain this pose, a number of reflexes are functioning.

These are controlled by the visual, labyrinthine, and neck righting reflexes and by interplay of joint reflexes to balance the force couples at the joints. Postural corrections are made continuously, mediated by the myotatic (stretch) reflex. The labyrinthine and visual righting reflexes maintain the head's orientation with gravity; as the pose is taken and held, extension of the neck (dorsiflexion of the head) is obvious, and this in turn aids in maintaining the arched back (spinal extension). Eye fixation helps to stabilize the head so that any change in the orientation of the body in relation to the head will be corrected by neck righting reflexes.

The performer in Figures 5.19 and 6.3 is balancing 110.95 lb on the head of the right femur, 131.0 lb on the right knee, and 138.0 lb on the head of the right talus (see Chapter 5). Compression of these joints evokes a strong extensor response so that the joints of the lower extremity can support the weight. At the same time the pressure exerted by the total body weight (140 lb) on the right foot stimulates pressure receptors in the skin of the sole and the spindles of the intrinsic muscles of that foot, thus eliciting contraction of the antigravity muscles of the extremity (the positive supporting reaction).

Analyses of More Complex Skills

The muscle response patterns of well-learned motor skills are largely integrated and controlled at the subcortical level, and consequently they involve the integrated action of many reflexes and usually the inhibition of some. Although the movements are more complex, neurokinesiological analysis of these skills is accomplished by the same procedure. There will be, however, an increase in the requirement for inhibition of natural reflexes in most instances. The following are some sample discussions taken from Figures 6.5, 11.26, 11.27, and 11.28.

Dive for Height

Figure 6.5 is a series of tracings made from a film of a dive for height. From these tracings it is possible to make postulations concerning reflex activity which assists the performer to accomplish the dive. Frames 1 and 5 show the subject in the air and landing on both feet for the take-off. This indicates that there is a strong stimulus to the pressure receptors in the skin of the soles of the feet, which should trigger an *extensor thrust* reflex as an initial part of the take-off. Also, in these same frames, the fact that the hips and knees are flexed and the ankles dorsiflexed indicates that all the antigravity muscles of the lower extremities are put on stretch, thus activating the spindles in these muscles. The resulting *stretch reflex* contributes to the forcefulness of the voluntary contraction of these same muscles projecting the body forward and upward in frames 8 through 11.

During take-off and early stages of the dive (frames 8 to 14), there is little change in the head and its cervical angulation with the trunk but shortly thereafter, by frame 18, the head is dorsiflexed, probably as a result of an effective *labyrinthine head righting reflex* resulting from stimulus to the utricle receptors, and also to some extent as a result of visual feedback. The angulation of the head (neck extension) in turn stimulates the joint receptors in the upper cervical region, which triggers a *neck righting reflex*, facilitating extension of the trunk and lower limbs to ensure a *body-following-head* response. By frame 39, as the performer approaches landing, the head is ventriflexed in preparation for a forward roll. This ventriflexion may evoke a *tonic labyrinthine reflex* which in turn facilitates the flexion of the trunk and upper extremities necessary for the performance of the roll and again stimulates the joint receptors in the cervical region which, by the *neck righting reflex*, ensure that the body will follow the head.

Contact of the hands with the floor (frame 41) initiates an extensor thrust reflex through the pressure receptors in the skin of the palms of the hands, a reflex which in this case is probably limited to an eccentric contraction of the shoulder flexors§ and elbow extensors as these muscles aid in absorbing the force of landing.

Standing Long Jump

The presence of the labyrinthine head righting reflex is evident in the standing long jump (Fig. 11.26) in frames 1 through 24 as the performer prepares for take-off; as the angle of trunk inclination increases, the head becomes more and more dorsiflexed and cervical extension is evident. During the early part of this same period, frames 1 through 17, hips and knees are flexing and ankles are dorsiflexing, evoking stretch reflexes in the antigravity muscles which facilitate their contraction in take-off. The coincident movement of the upper extremities as they are swung from hyperextension to flexion (frames 1 to 28) contributes facilitation to the lower limbs by irradiation of neural impulses arising from the resulting stimulation of the shoulder joint receptors.

At frame 29, as the jumper's toes leave the floor, the labyrinthine head righting reflex has been inhibited and the head is ventriflexed. This inhibition of the head righting (labyrinthine) reflex is undesirable during the early stages of the flight as it could elicit a body-following-head response which can cut down the height of the trajectory and so decrease the distance of the jump. The inhibition is not total during the jump itself as the head can be seen to dorsiflex slightly in both frames 34 and 44.

Badminton Smash

Figure 11.27 depicts a few of the positions relative to time adopted by the badminton player during the force producing and follow-through phases of the forehand smash. The first plate, (a), is taken from a "side-view" position while the concomitant frames of the "back-view" are shown

---

§ Since the arms are almost fully flexed at the shoulders as the performer reaches for the floor, they are moving into extension as the hands accept the body weight.

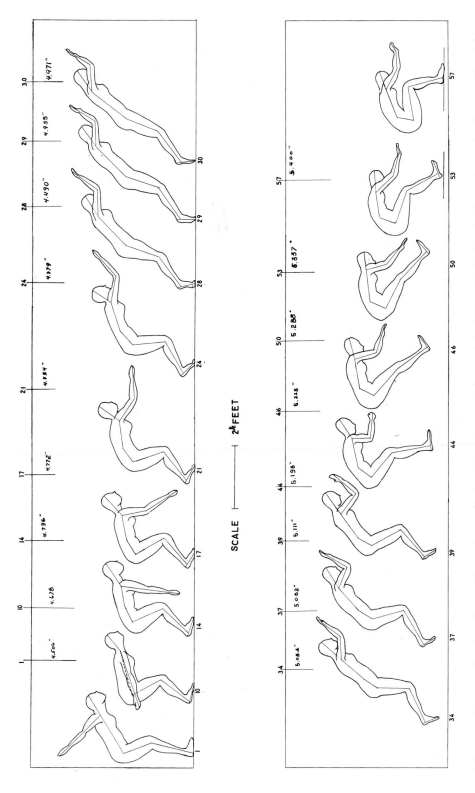

SCALE ├───┤ 2½ FEET

Figure 11.26. Standing long jump. (From Boston University, Sargent College of the Allied Health Professions Biomechanics and Kinesiology Laboratory.)

Figure 11.27. (a), badminton forehand smash, "side-view": contact with the shuttlecock at frame no. 20. (From McMaster University Kinesiology/Biomechanics Laboratory, School of Physical Education and Athletics.)

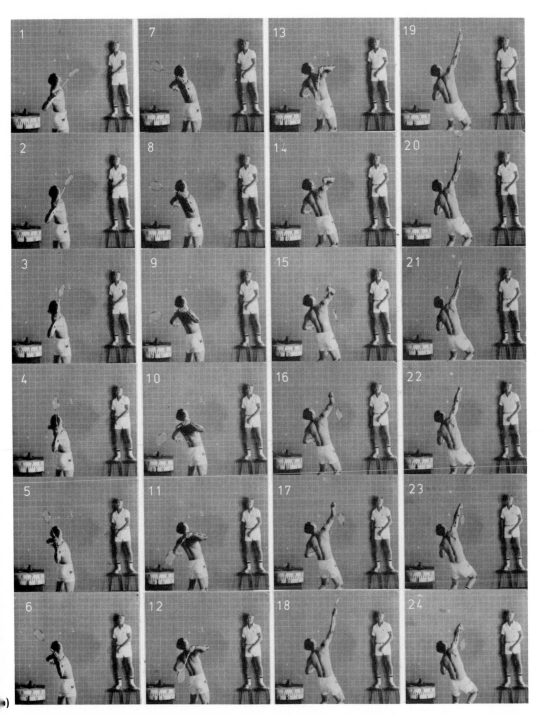

Figure 11.27. (b), badminton forehand smash, concomitant frames of the "back-view." (From McMaster University Kinesiology/Biomechanics Laboratory, School of Physical Education and Athletics.)

in the second plate (b). Time between frames is approximately 100 msec. [||] Since one of the objectives of the smash is to hit with power, it seems logical that the player makes use of all elements and reflexes which augment the force of the muscles involved. It can be seen from frames 1 through 12 that the medial rotators and adductors of the shoulder joint and the pronators of the radioulnar joints have been put on stretch. Further, these same frames show that the arm (and racket) are whipped back quickly into this position (i.e., in a time span of 0.03 second). By stretching the muscle groups that are about to become involved in the forceful part of the stroke and stretching them quickly, the player gains force from the length-tension relationship of muscle and the phasic and tonic components of the neuromuscular spindle.

Gowitzke and Waddell (1977, 1978) have shown that a major force component for the smash emanates from the medial rotation of the arm at the shoulder joint and pronation of the forearm at the radioulnar joints. These two actions start at approximately frame 13 (0.02 second before contact), continue through contact with the shuttlecock (frame 20), and have not stopped 0.01 second after contact (frame 24). At the same time, elbow extension occurs, changing to flexion just before contact. This combination of movements (medial rotation at the shoulder joint) suggests that the second diagonal arm extension pattern, discussed in Chapter 1, is involved. The badminton player uses elements of muscle mechanics and neuromuscular reflexes to advantage when force is the objective.

Practice Exercises

The student's interest in and understanding of subcortical neural function in

---

[||] Two high-speed cameras were operated simultaneously at a film transport speed of 400 frames/sec and a shutter speed of 1/1200 second; every fourth frame was selected to display in Figure 11.27.

complex sequential skills can be fostered by studying films or film sequences, or tracings made from films as in Figures 6.5, 11.26, 11.27, and 11.28. After such perusal the student should be able to identify, explain, and discuss the reflexes which are or could be operating. This discussion can be somewhat simplified if the student follows an outline format such as the following.

1. a. The name of the reflex.
   b. Identification of the frame or frames' (by number or time) in which evidence of the reflex is found.
2. The receptors involved and where they are located.
3. The evidence which suggests the presence or absence of the reflex.
4. The expected effect of the reflex if it occurs.
5. The actual result as seen in the subsequent frame or frames.
6. Discussion.

Types of Reflex Activity to be Identified In Motor Skills

The following list suggests a number of reflexes which the student may identify as helping or hindering the performance in his neurokinesiological analysis of skilled movement patterns. The student should follow the outline for the presentation of each reflex. The discussion (item 6 in the outline above) is especially important since it is here that the student displays his understanding of the reflex and its participation in the skill under study.

Spindle reflexes
   Stretch reflex
      Primary afferent responses: phasic and/or tonic
      Reciprocal activity in antagonists and synergists
      Secondary afferent responses: supplementary facilitation of flexors; inhibition leading to co-contraction and joint stabilization in simple extensors
      Gamma bias: setting the voluntary limits of the performance

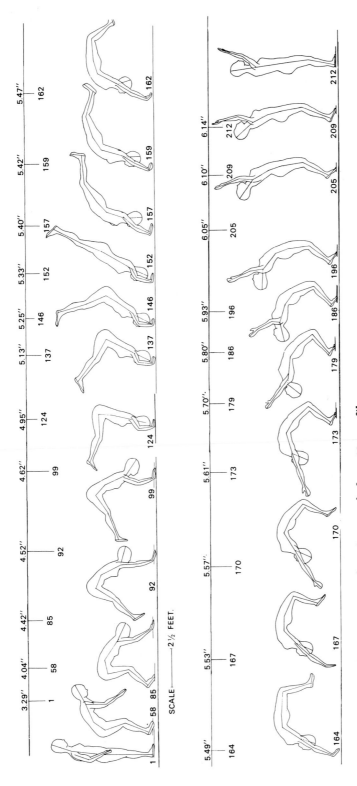

Figure 11.28. Headspring. Series of tracings made from 16 mm film.

Positive supporting reaction; weight bearing
Golgi tendon organ reflex
Feedback of muscle tension level
Relaxation at the end of volitional movement (termination of active contraction)
Limitation of use of force by autogenic inhibition
Joint reflexes
Kinesthetic feedback
Limb-trunk angles
Compression
Traction
Positive supporting reaction
Interlimb facilitations: forelimbs on hind; contralateral; ipsilateral
Balancing of force couples
Cutaneous responses
Extensor thrust
Grasp reflex
Placing responses
Local sign
Body-on-head righting
Body-on-body righting
Labyrinthine reflexes
Labyrinthine righting reflex
Equilibrium reflexes
Shift of base of support
Shift of center of gravity
Inhibition of righting reflex
Neck reflexes
Tonic neck reflexes
Neck righting reflexes
Visual righting reflexes

The use of such an outline for neurokinesiological analysis is illustrated below with examples drawn from tracings of the headspring (Fig. 11.28).

Headspring
1. a. Reflex: stretch reflex
   b. Frame 127
2. The receptors are the nerve endings in the muscle spindles located in the hip extensors (gluteus maximus and hamstrings).
3. Evidence: the hips are flexed more than 90°.
4. Expected effect: there should be a strong contraction of these muscles.
5. Actual results: frames 146 to 152 show

a sharp extension of the hips; the reflex is effective.
6. Discussion (this should be an essay discussion which displays the student's comprehension of the reflex.) (N.B. Other stretch reflexes can be similarly identified in frame 170, rectus abdominis; frame 173, quadriceps femoris; and frame 205, the anterior tibial muscles.)

APPLYING KNOWLEDGE OF PROPRIOCEPTION

The ultimate aim of motor learning is the performance of a skill with a minimum of conscious concern for the details of the movement. Smooth and harmonious coordination of the body segments participating in the pattern must be assured by the relatively automatic action of muscles under the self-regulating control of proprioceptive feedback. The proprioceptive responses must become conditioned so that each movement supplies the stimulus for the next part of the pattern, leaving the cortex free to direct strategy rather than movement.

Most, if not all, teaching cues make direct or indirect use of proprioceptive reflexes, but this is usually more by accident than by design. However, proprioceptive reflexes are better understood than most other components of the motor learning process and they are readily accessible to manipulation. Physical therapists have successfully used techniques of proprioceptive facilitation and inhibition in the rehabilitation of patients with neuromuscular disabilities since World War II. The inclusion of neurokinesiological analysis in the study of motor patterns could contribute to the student's repertory of teaching and coaching techniques. Teaching cues and preliminary exercises based on understanding of neuromuscular mechanisms and reflex control as well as on sound principles of biomechanics and applied anatomy have been deliberately used in the training of the undergraduate student (Gowitzke (1968)). The student (prospective teacher, coach, therapist) uses this type of approach as a catalyst for evolving teaching, coaching, or therapeutic methodology. As a result, new methods evolve which have the potential for improving motor performance, shortening the learning time and the extent of trial and error learning, and enhancing the skilled performer's accomplishments possibly beyond presently accepted levels.

References

Adal, M. N., and Barker, D., 1965. Intramuscular branching of fusimotor fibers. *J. Physiol. (London)* 177: 288.

Adrian, E. D., 1959. Sensory mechanisms, introduction. In *Handbook of Physiology*, Section 1: Neurophysiology, Vol. I, edited by J. Field and H. W. Magoun. Washington, D.C.: American Physiological Society.

Alexander, R. M., 1968. *Animal Mechanics*. Seattle: University of Washington Press.

American Academy of Orthopaedic Surgeons, 1965. *Joint Motion—Method of Measuring and Recording*. Edinburgh: E. and S. Livingstone, Ltd.

Andrews, B. J., and Jarrett, M. O., 1976. On-line kinematic data acquisition. Glasgow, Scotland: Internal report, University of Strathclyde.

Appelberg, B., and Emonet-Denand, F., 1967. Motor units of the first superficial lumbrical muscle of the cat. *J. Neurophysiol.* 30: 154.

Asmussen, E., 1962. Muscular performance. In *Muscle as a Tissue*, edited by K. Rodahl and S. M. Horvath. New York: McGraw-Hill Book Company, p. 161.

Atkins, K. R., 1966. *Physics*. New York: John Wiley & Sons, Inc.

Baon, L., Kenwright, J., and Hirsch, C., 1972. Gait analysis after prosthetic replacement of the hip joint. In *Digest of 3rd International Conference on Medical Physics Including Medical Engineering*. Goteborg, Sweden: Chalmers University of Technology, 10–3.

Bárány, M., 1967. ATPase activity of myosin correlated with speed of muscle shortening. In *The Contractile Process: Proceedings of a Symposium Sponsored by the New York Heart Association*. Boston: Little, Brown and Company, p. 197.

Barker, D., 1966. The motor innervation of the mammalian muscle spindle. In *Muscle Afferents and Motor Control*, edited by R. Granit. New York: John Wiley & Sons, Inc.

Barter, J. T., 1957. *Estimation of the Mass of Body Segments*, WADC Technical Report 57–260. Wright-Patterson Air Force Base, Ohio: Aero Medical Laboratory, Wright Air Development Center.

Basmajian, J. V., 1974. *Muscles Alive*. Baltimore: The Williams & Wilkins Company.

Basmajian, J. V., 1978. Personal communication.

Basmajian, J. V., and Stecko, G. A., 1962. A new bipolar indwelling electrode for electromyography. *J. Appl. Physiol.* 17: 849.

Bell, G. H., Davidson, J. N., and Scarborough, H., 1969. *Textbook of Physiology and Biochemistry*. Baltimore: The Williams & Wilkins Company.

Bendall, J. R., 1970. *Muscles, Molecules and Movement*. London: Heinemann Education Books, Limited.

Bennett, M. V. L., 1968. Similarities between chemical and electrical mediated transmission. In *Physiological and Biochemical Aspects of Nervous Integration*, edited by F. D. Carlson. Englewood Cliffs, N.J.: Prentice-Hall, Inc.

Bessou, P., Emonet-Denand, F., and Laporte, Y., 1963. Occurrence of intrafusal muscle fibre innervation by branches of slow alpha motor fibres in the cat. *Nature* 198: 594.

Bessou, P., and LaPorte, Y., 1966. Observations on static fusimotor fibres. In *Muscle Afferents and Motor Control*, edited by R. Granit. New York: John Wiley & Sons, Inc.

Birdwhistell, R. L., 1970. Essays on body motion communication. In *Kinesics and Context*, edited by R. L. Birdwhistell. Philadelphia: University of Pennsylvania Press.

Blievernicht, D. L., 1967. Multi-dimensional timing device for cinematography. *Res. Q.* 38: 146.

Bobath, B., 1965. *Abnormal Postural Reflex Activity Caused by Brain Lesions*. London: William Heinemann Medical Books Limited.

Bodian, D., 1967. Neurons, circuits and neuroglia. In *The Neurosciences*, edited by G. C. Quarton, T. Melnechuk, and F. O. Schmidt. New York: The Rockefeller University Press.

Bowsher, D., 1967. *Introduction to the Anatomy and Physiology of the Nervous System*. Oxford: Blackwell Scientific Publishers.

Boyd, I. A., 1962. The structure and innervation of the nuclear-bag muscle fibre system and the nuclear-chain muscle fibre system in mammalian muscle spindles. *Philos. Trans. R. Soc. Lond., B* 245: 81.

Boyd, I. A., 1968. The morphology of muscle spindles and tendon organs. In *The Role of the Gamma System in Movement and Posture*, Revised edition, edited by C. A. Swinyard. New York: Association for Aid of Crippled Children, p. 2.

Boyd, I. A., and Davey, M. R., 1968. *The Composition of Peripheral Nerves*, London: E. & S. Livingstone Ltd.

Boyd, I. A., Eyzaguirre, C., Matthews, P. B. C., and Rushworth, G., 1968. *The Role of the Gamma System in Movement and Posture, Revised edition, edited by C. A. Swinyard*. New York: Association for the Aid of Crippled Children.

Boyd, I. A., Gladden, M. F., McWilliam, P. N., and Ward, J., 1975. 'Static' and 'dynamic' nuclear bag fibers in isolated cat muscle spindles. *J. Physiol. (Lond.)* 250: 11.

Bresler, B., and Frankel, J. P., 1950. The forces and moments in the leg during level walking. *Trans. Am. Soc. Mech. Eng.* 72: 27.

Broer, M. R., 1973. *Efficiency of Human Movement.* Philadelphia: W.B. Saunders Company.

Brooks, V. B., 1959. Contrast and stability in the nervous system. *Trans. N.Y. Acad. Sci. Ser. II* 21: 387.

Brown, M. C., and Matthews, P. B. C., 1966. On the subdivision of the efferent fibres to muscle spindles into static and dynamic fusimotor fibres. In *Control and Innervation of Skeletal Muscle,* edited by B. L. Andrew. Dundee, Scotland: D.C. Thomson and Company, Ltd.

Bruegger, W., and Milner, M., 1978. Computer-aided tracking of body motions using a CCD-image sensor. *Med. Biol. Eng. Comput.* 16: 207.

Brunnstrom, S., 1962. *Clinical Kinesiology.* Philadelphia: F.A. Davis Company.

Brunnstrom, S., 1964. Historical approach. In *Approaches to Treatment of Patients with Neuromuscular Dysfunction,* edited by S. Sattely, Dubuque, Iowa: William C. Brown Book Company.

Buchthal, F., Rosenfalck, P., and Erminio, F., 1960. Motor unit territory and fiber density in myopathies. *Neurology* 10: 398.

Buchwald, J. S., and Eldred, E., 1961. Conditioned response in the gamma efferent system. *J. Nerv. Ment. Dis.* 132: 146.

Buchwald, J. S., Standish, M., Eldred, E., and Hallas, E. S., 1964. Contribution of muscle spindle circuits to learning as suggested by training under Flaxedil. *Electroenceph. Clin. Neurophysiol.* 16: 582.

Buller, A. J., 1965. Mammalian slow and fast skeletal muscle. In *Studies in Physiology,* edited by D. R. Curtis and A. K. McIntyre. New York: Springer-Verlag Inc.

Buller, A. J., Eccles, J. C., and Eccles, R. M., 1960. Interactions between motoneurones and muscles in respect to the characteristic speed of their responses. *J. Physiol. (Lond.)* 150: 417.

Bullock, M. I., and Harley, I. A., 1972. The measurement of three-dimensional body movements by the use of photogrammetry. *Ergonomics* 15: 309.

Bullock, T. H., 1959. The neuron doctrine and electrophysiology. *Science* 129: 17.

Bunge, M. B., Bunge, R. P., and Peterson E. R., 1967. The onset of synapse formation in spinal cord cultures as studied by electron microscopy. *Brain Res.* 6: 728.

Burke, R. E., 1967. Motor unit types of the cat's triceps surae muscle. *J. Physiol. (Lond.)* 193: 141.

Burke, R. E., Rudomin, P., and Kajac, F. E., 1970. Catch property in single mammalian motor units. *Science* 168: 122.

Carlsöö, S., 1972. *How Man Moves.* London: William Heinemann, Limited.

Cheng, I.-S., 1974. Computer-television analysis of biped locomotion, Ph.D. Thesis. Columbus: Ohio State University.

Close, R., 1965. Effects of cross union of motor nerves to fast and slow skeletal muscles. *Nature* 206: 4986.

Close, R., 1967. Properties of motor units in fast and slow skeletal muscles of the rat. *J. Physiol. (Lond.)* 193: 45.

Close, R., and Hoh, J., 1968. The after-effects of repetitive stimulation on the isometric twitch contraction of rat fast skeletal muscle. *J. Physiol. (Lond.)* 197: 461.

Coers, C., and Woolf, A. L., 1959. *Innervation of Muscle.* Springfield, Ill.: Charles C Thomas, Publisher.

Cook, T. M., and Cozzens, B. A., 1976. The effects of heel height and ankle-foot-orthosis configuration on weight-line location: A demonstration of principles. *Orthot. Prosthet.* 30: No. 4, 43.

Cook, T. M., Cozzens, B. A., and Kenosian H., 1977. Real-time force line visualizations: Bioengineering applications. In *Advances in Bioengineering,* edited by E. S. Grood. New York: American Society of Mechanical Engineering.

Cooper, J. M., and Glassow, R. B., 1968. *Kinesiology.* St. Louis: The C.V. Mosby Company.

Costantin, L. L., Franzini-Armstrong, C., and Podolsky, R. J., 1965. Localization of calcium-accumulating structures in striated muscle fibers. *Science* 147: 158.

Cotton, F. S., 1932. Studies in center of gravity changes. *Aust. J. Exp. Biol.* 10: 16–34, 225–247.

Counsilman, J. E., 1968. *The Science of Swimming.* Englewood Cliffs, N.J.: Prentice-Hall, Inc.

Couteaux, R., 1960. Motor end-plate structure. In *Structure and Function of Muscle,* Vol. I, edited by G. H. Bourne. New York: Academic Press, p. 337.

Cunningham, D. M., and Brown, G. W., 1952. Two devices for measuring the forces acting on the human body during walking. *Proc. Soc. Exp. Stress Anal.* 2: 75.

Curtis, D. R., and Eccles, J. C., 1960. Synaptic action during and after repetitive stimulation. *J. Physiol. (Lond.)* 150: 374.

Damon, A., Stoudts, H. W., and McFarland, R. A., 1966. *The Human Body in Equipment Designs.* Cambridge, Mass.: Harvard University Press.

Davis, H., 1959. Excitation of auditory receptors. In *Handbook of Physiology,* Section 1: Neurophysiology, Vol. I, edited by J. Field and H. W. Magoun. Washington, D.C.: American Physiological Society.

Davis, H., 1961. Some principles of sensory receptor action. *Physiol. Rev.* 41: 391.

Dawson, P., 1935. *The Physiology of Physical Education.* Baltimore: The Williams & Wilkins Company.

Dempster, W. T., 1955. *Space Requirements for the Seated Operator,* WADC Technical Report 55159. Wright-Patterson Air Force Base, Ohio: Wright Air Development Center.

Dempster, W. T., 1961. Free body diagrams as an approach to the mechanics of posture and motion. In *Biomechanical Studies of the Musculoskeletal System,* edited by F. G. Evans, Springfield, Ill.: Charles C Thomas, Publisher.

Dowling, J. E., and Boycott, B. B., 1965. Neural connections of the retina. *Cold Spring Harbor Symp. Quant. Biol.* 30: 393.

Doyle, A. N., and Mayer, R. F., 1969. Studies of the

motor units in the cat. *Bull. Sch. Med. Maryland* 54: 11.

Dreizen, P., and Gershman, L. C., 1970. Molecular basis of muscular contraction: Myosin. *Trans. N.Y. Acad. Sci. Ser. II 32*: 170.

Duggar, B. L., 1966. The center of gravity and moment of inertia of the human body. In *The Human Body in Equipment Design*, by A. Damon, H. W. Stoudts, and R. A. McFarland. Cambridge, Mass.: Harvard University Press.

Dyson, G. H. G., 1977. *Mechanics of Athletics*. London: Hodder and Stoughton.

Eccles, J. C., 1957. *The Physiology of Nerve Cells*. Baltimore: The Johns Hopkins Press.

Eccles, J. C., 1960. Neuron physiology, introduction. In *Handbook of Physiology*, Section 1: Neurophysiology, Vol. I, edited by J. Field and H. W. Magoun. Washington, D.C.: American Physiological Society.

Eccles, J. C., 1964. Ionic mechanism of postsynaptic inhibition. *Science 145*: 1140.

Eccles, J. C., 1965. The synapse. *Sci. Am. 212*: No. 1, 56.

Eccles, J. C., 1967. Postsynaptic inhibition in the central nervous system. In *The Neurosciences*, edited by G. C. Quarton, T. Melnechuk, and F. O. Schmidt. New York: The Rockefeller University Press.

Eccles, J. C., Magni, F., and Willis, W. D., 1962. Depolarization of central terminals of Group I afferent fibers from muscle. *J. Physiol. (Lond.) 160*: 62.

Elder, G., 1978. Personal communication.

Eldred, E., 1960. Posture and locomotion. In *Handbook of Physiology*, Section 1: Neurophysiology, Vol. II, edited by J. Field and H. W. Magoun. Washington, D.C.: American Physiological Society.

Eldred, E., 1965. The dual sensory role of muscle spindles. In *The Child with Central Nervous System Deficit*. Washington, D.C., U.S. Government Printing Office.

Eldred, E., 1967a. Peripheral receptors: Their excitation and relation to reflex patterns. *Am. J. Phys. Med. 46*: 69.

Eldred, E., 1967b. Functional implications of dynamic and static components of the spindle response to stretch. *Am. J. Phys. Med. 46*: 129.

Elftman, H., 1939. The function of muscles in locomotion. *Am. J. Physiol. 125*: 357.

Erlanger, J., and Gasser, H. S., 1937. *Electrical Signs of Nervous Activity*. Philadelphia: University of Pennsylvania Press.

Esch, D., and Lepley, M., 1974. *Evaluation of Joint Motion: Methods of Measurement and Recording*. Minneapolis: University of Minnesota Press.

Eshkol, N., and Wachman, A., 1958. *Movement Notation*. London: Weidefeld and Nicholson.

Everett, N. B., 1971. *Functional Neuroanatomy*. Philadelphia: Lea and Febiger.

Eyzaguirre, C., 1966. Properties of intrafusal muscle fibres of amphibians and mammals. In *Control and Innervation of Skeletal Muscle*, edited by B. L. Andrew. Dundee, Scotland: D. C. Thomson and Company, Ltd.

Eyzaguirre, C., 1968. Some functional characteristics of muscle spindles. In *The Role of the Gamma System in Movement and Posture*, Revised edition, edited by C. A. Swinyard. New York: Association for Aid of Crippled Children, p. 18.

Eyzaguirre, C., 1969. *Physiology of the Nervous System*. Chicago: Year Book Medical Publishers, Inc.

Fernandez-Moran, H., 1967. Membrane ultrastructure in nerve cells. In *The Neurosciences*, edited by G. C. Quarton, T. Melnechuk, and F. O. Schmidt. New York: The Rockefeller University Press.

Freeman, M. A. F., and Wyke, B., 1967. The innervation of the knee joint, an anatomical and histological study in the cat. *J. Anat. 101*: 505.

Gaddum, J., 1965. The neurological basis of learning. *Perspect. Biol. Med. 8*: 436.

Ganong, W. F., 1965. *Review of Medical Physiology*. Los Altos: Lange Medical Publications.

Gardner, E., 1963. *Fundamentals of Neurology*. Philadelphia: W. B. Saunders Company.

Gardner, E. B., 1965a. The neuromuscular base of human movement: Feedback mechanisms. *J. Health Phys. Educ. Rec. 36* (Oct.): 61.

Gardner, E. B., 1965b. Physiological aspects of motor learning. In Proceedings, Fall Conference of Eastern Association for Physical Education of College Women.

Gardner, E. B., 1967. The neurophysiological basis of motor learning—A review. *J. Am. Phys. Ther. Assoc. 47*: 1115.

Gardner, E. B., 1969. Proprioceptive reflexes and their participation in motor skills. *Quest XII*: 1.

Garrett, R. W., Widule, C. J., and Garrett, G. E., 1968. Computer aided analysis. *Kinesiology Rev.*, edited by A. L. O'Connell, p. 1.

Garrett, G. E., Widule, C. G., Reed, W. S., and Garrett, R. E., 1969. Human movement via computer graphics. Paper presented at the convention of the American Association for Health, Physical Education, and Recreation, Boston, 1969.

Gelfan, S., 1955. Functional activity of muscle. In *Textbook of Physiology*, Ed. 17, edited by J. F. Fulton. Philadelphia: W. B. Saunders Company, p. 123.

Gernandt, B. E., 1959. Vestibular mechanisms. In *Handbook of Physiology*, Section 1: Neurophysiology, Vol. I, edited by J. Field and H. W. Magoun. Washington, D. C.: American Physiological Society.

Gernandt, B. E., and Shimamura, M., 1961. Mechanisms of interlimb reflexes in the cat. *J. Neurophysiol 24*: 665.

Gibbs, C. L., and Ricchiuti, N. V., 1965. Activation heat in muscle: Method of determination. *Science 147*: 162.

Glassow, R. B. Personal communication, 1964.

Goldberg, J. M., and Lavine, R. A., 1968. Nervous system: Afferent mechanisms. *Ann. Rev. Physiol. 30*: 319.

Gollnick, P. D., and King, D. W., 1969. Energy release in the muscle cell. *Med. Sci. Sports 1*: 23.

Gordon, A. M., Huxley, A. F., and Julian, F. J., 1966.

Variation in isometric tension with sarcomere length in vertebrate muscle fibers. *J. Physiol. (Lond.)* 184: 170.

Gowitzke, B. A., 1968. Kinesiological and neurophysiological principles applied to gymnastics. In *Kinesiology Rev.*, edited by A. L. O'Connell, p. 22.

Gowitzke, B. A., 1969. An analysis of levers contributing to the glide kip. Unpublished paper prepared at the University of Wisconsin.

Gowitzke, B. A., 1979. Biomechanical principles applied to badminton stroke production. In *Science in Racquet Sports*, International Congress of Sports Sciences, Edmonton, Alberta. Del Mar, Calif.: Academic Publishers.

Gowitzke, B. A., and Waddell, D. B., 1977. The contributions of biomechanics in solving problems in badminton stroke production. Presented at the First International Coaching Conference, Malmo, Sweden.

Gowitzke, B. A., and Waddell, D. B., 1979. Technique of badminton stroke production. In *Science in Racquet Sports*, International Congress of Sports Sciences, Edmonton, Alberta. Del Mar, Calif.: Academic Publishers.

Granit, R., 1955. *Receptors and Sensory Perception.* New Haven, Conn.: Yale University Press.

Granit, R., Holmgren, B., and Merton, P. A., 1955. Two routes of excitation of muscle and their subservience to the cerebellum. *J. Physiol. (Lond.)* 130: 213.

Grant, J. C. B., and Smith, C. G., 1953. In *Morris' Human Anatomy*, edited by J. P. Shaeffer. New York: The Blakiston Company.

Gray, J. A. B., 1959. Initiation of impulses in receptors. In *Handbook of Physiology*, Section 1: Neurophysiology, Vol. I, edited by J. Field and H. W. Magoun. Washington, D.C.: American Physiological Society.

Gray, J. A. B., and Matthews, P. B. C., 1951. Response of Pacinian corpuscles in the cat's toe. *J. Physiol. (Lond.)* 113: 475.

Grieve, D. W., 1968. Gait patterns and the speed of walking. *Biomech. Eng.* 3: 119.

Grundfest, H., 1959. Synaptic and ephaptic transmission. In *Handbook of Physiology*, Section 1: Neurophysiology, Vol. I, edited by J. Field and H. W. Magoun. Washington, D.C.: American Physiological Society.

Güth, V., Abbink, F., and Heinrichs, W., 1973. Eine methode fur chronozyklographischen bewegungsaufzeichnung mit einem prozessrechner. *Int. Z. Angew Physiol.* 151.

Guyton, A. C., 1972. *Structure and Function of the Nervous System.* Philadelphia: W.B. Saunders Company.

Hagbarth, K. E., 1952. Excitatory and inhibitory skin areas for flexor and extensor motoneurones. *Acta Physiol. Scand.* 26: Suppl. 94, 5.

Hagbarth, K. E., and Eklund, G., 1966. Motor effects of vibratory stimuli in man. In *Muscle Afferents and Motor Control*, edited by R. Granit, New York: John Wiley & Sons, Inc.

Halverson, L., 1966. Development of motor patterns in young children, Quest 6: 44.

Hanson, J., and Lowy, J., 1960. Structure and function of the contractile apparatus in muscles of invertebrate animals. In *Structure and Function of Muscle*, Vol. I, edited by G. H. Bourne. New York: Academic Press, p. 312.

Harvey, R. J., and Matthews, P. B. C., 1961. The response of de-efferented muscle spindle endings in the cat's soleus to slow extension of the muscle. *J. Physiol. (Lond.)* 157: 370.

Hellebrandt, F., 1956. Physiological effects of simultaneous static and dynamic exercise. *Am. J. Phys. Med.* 35: 106.

Hellebrandt, F. A., Genevieve, G., and Tepper, R. H., 1938. The relation of the center of gravity to the base of support in stance. *Am. J. Physiol.* 119: 331.

Hellebrandt, F. A., Kubin, D., Longfield, W. M., and Kelso, L. E. A., 1937. The base of support in stance. *Phys. Ther. Rev.* 17: 231.

Hellebrandt, F., Schade, M., and Carns, M., 1962. Methods of evoking TNR in normal individuals. *Am. J. Phys. Med.* 41: 90.

Hellebrandt, F., and Waterland, J., 1962. Motor patterning in stress. *Am. J. Phys. Med.* 41: 56.

Henneman, E., and Olson, C. B., 1965. Relations between structure and function in the design of skeletal muscle. *J. Neurophysiol.* 28: 581.

Henneman, E., Somjen, G., and Carpenter, D. O., 1965. Functional significance of cell size in spinal motoneurons. *J. Neurophysiol.* 28: 560.

Hicks, J. H., 1953. The mechanics of the foot. I. The joints. *J. Anat.* 87: 345.

Hill, A. V., 1965. *Trails and Trials in Physiology.* Baltimore: The Williams & Wilkins Company.

Holmqvist, B., 1961. Crossed spinal reflex actions evoked by volleys in somatic afferents. *Acta Physiol. Scand.* 52: Suppl. 1, 81.

Houk, J., and Henneman, E., 1967. Response of Golgi tendon organs to active contraction of the soleus muscle of the cat. *J. Neurophysiol.* 30: 466.

Houtz, S. J., and Fischer, F. J., 1959. An analysis of muscle action and joint excursion during exercise on a stationary bicycle. *J. Bone Joint Surg.* 41-A: 123.

Hoyle, G., 1970. How is muscle turned on and off? *Sci. Am.* 222: No. 4, 84.

Hubbard, J. I., and Willis, W. D., 1963. The effect of use on the transmitter release mechanism at the mammalian neuromuscular junction. In *Effect of Use and Disuse on Neuromuscular Functions*, edited by E. Gutman and P. Hnik. Amsterdam: Elsevier Publishing Company.

Hufschmidt, H. J., 1966. Demonstration of autogenic inhibition and its significance in human voluntary movement. In *Muscle Afferents and Motor Control*, edited by R. Granit. New York: John Wiley & Sons, Inc.

Hunt, C. C., 1961. On the nature of vibration receptors in the hind limb of the cat. *J. Physiol. (Lond.)* 155: 175.

Hunt, C. C., and Kuffler, S. W., 1951. Further study of efferent small-nerve fibres to mammalian muscle spindles, multiple spindle innervation and activity during contraction. *J. Physiol. (Lond.)* 113: 283.

Huxley, A. F., and Niedergerke, R., 1954. Structural changes in muscle during contraction; interference microscopy of living muscle fibers. *Nature (Lond.)* 173: 971.

Huxley, A. F., and Taylor, R. F., 1958. Local activation of striated muscle fibers. *J. Physiol. (Lond.)* 144: 426.

Huxley, H. E., 1958. The contraction of muscle. *Sci. Am.* 199: No. 5, 66.

Huxley, H. E., 1965. The mechanism of muscular contraction. *Sci. Am.* 213: No. 6, 18.

Huxley, H. E., 1969. The mechanism of muscular contraction. *Science* 164: 1356.

Huxley, H. E., and Hanson, J., 1954. Changes in cross-striations of muscle during contraction and stretch and their structural interpretation. *Nature (Lond.)* 173: 973

Huxley, H. E., and Hanson, J., 1960. The molecular basis of contraction. In *Structure and Function of Muscle*, Vol. I, edited by G. H. Bourne. New York: Academic Press, p. 183.

Infante, A. A., Klaupiks, D., and Davies, R. E., 1964. ATP changes in muscle doing negative work. *Science* 144: 1577.

Inman, V. T., 1947. Functional aspects of the abductor muscles of the hip. *J. Bone Joint Surg.* 29: 607.

Jansen, J. K. S., 1966. On fusimotor reflex activity. In *Muscle Afferents and Motor Control*, edited by R. Granit. New York: John Wiley & Sons, Inc.

Jansen, J. K. S., and Matthews, P. B. C., 1962. The central control of the dynamic responses of muscle spindle receptors. *J. Physiol. (Lond.)* 161: 357.

Jones, M. H., Steinfeldt, S., and Delwiche, R., 1966. A simple timing device combined with surface electromyography in the study of phasic muscle activity. *Develop. Med. Child. Neurol.* 8: 269.

Joseph, J., 1960. *Man's Posture, Electromyographic Studies.* Springfield, Ill.: Charles C Thomas, Publisher.

Kandel, E., and Wechtel, H., 1968a. Neural aggregates in aplysia. In *Physiological and Biochemical Aspects of Nervous Integration*, edited by F. D. Carlson. Englewood Cliffs, N.J.: Prentice-Hall, Inc.

Kandel, E., and Wechtel, H., 1968b. The functional organization of neural aggregates in aplysia. In *Physiological and Biochemical Aspects of Nervous Integration*, edited by F. D. Carlson. Englewood Cliffs, N.J.: Prentice-Hall, Inc.

Karpovich, P. V., 1950. Mechanics of rising on the toes. Abstract of paper presented at the National Convention of Health, Physical Education, and Recreation, Dallas, April 1950.

Karpovich, P. V., and Manfredi, T. G., 1971. The mechanism of rising on the toes. *Res. Q.* 42: 395.

Kasvand, T., and Milner, M., 1972. Pattern recognition applied to measurement of human limb positions during movement. *J. Cybernetics* 2: No. 1, 66.

Kasvand, T., Milner, M., Quanbury, A. O., and Winter, D. A., 1976. Computers and the Kinesiology of gait. *Comput. Biol. Med.* 6: 111–120.

Katz, B., 1966. *Nerve, Muscle and Synapse.* New York: McGraw-Hill Book Company.

Keele, C. A., and Neil, E., 1965. *Samson Wright's Applied Physiology*, Edition XI. London: Oxford Press.

Kenwright, J., Baon, L., and Hirsch, C., 1972. Gait biomechanics after arthrodesis in the lower limb. In *Digest of 3rd International Conference on Medical Physics Including Medical Engineering.* Goteborg, Sweden: Chalmers University of Technology, 10-2.

Knott, M., and Voss, D. E., 1968. *Proprioceptive Neuromuscular Facilitation.* New York: Harper and Row.

Korr, I. M., Wilkinson, P. N., and Chornock, F. W., 1967. Axonal delivery of neuroplasmic components to muscle cells. *Science* 155: 342.

Lamaster, M. A., and Mortimer, E., undated. A device to measure body rotation in film analysis. Paper presented at research section of the American Association for Health, Physical Education, and Recreation.

Lamoreux, L. W., 1971. Kinematic measurements in the study of human walking. *Bull. Prosth. Res.* 10-15: 3.

Langman, J., 1969. *Medical Embryology: Human Development, Normal and Abnormal.* Baltimore: The Williams & Wilkins Company.

Lemkuhl, D., 1966. Local factors in muscle performance. *J. Am. Phys. Ther. Assoc.* 46: 473.

Lester, H. A., 1977. The response to acetylcholine. *Sci. Am.* 236: No. 2, 106.

LeVeau, B., 1977. *Biomechanics of Human Motion.* Philadelphia: W.B. Saunders Company.

Lloyd, D. P. C., 1943. Neuron patterns controlling transmission of ipsilateral hind limb reflexes in the cat. *J. Neurophysiol.* 6: 293.

Lloyd, D. P. C., 1960. Spinal mechanisms involved in somatic activities. In *Handbook of Physiology*, Section 1: Neurophysiology, Vol. II, edited by J. Field and H. W. Magoun. Washington, D.C.: American Physiological Society.

Loofbourrow, G. N., 1960. In *Science and Medicine of Exercise and Sports*, edited by W. R. Johnson. New York: Harper and Brothers.

Lorand, L., and Molnar, J., 1962. Biochemical control of relaxation in muscle systems. In *Muscle as a Tissue*, edited by K. Rodahl and S. M. Horvath. New York: McGraw-Hill Book Company, p. 97.

Lowenstein, W. R., 1960. Biological transducers. *Sci. Am.* 203: No. 2, 98.

MacConaill, M. A., and Basmajian, J. V., 1969. *Muscles and Movements—A Basis for Human Kinesiology.* Baltimore: The Williams & Wilkins Company.

MacDougall, J. D., Elder, G. C. B., Sale, D. G., Moroz, J. R., and Sutton, J. R., 1977. Abstract: Skeletal muscle hypertrophy and atrophy with respect to fiber type in humans. *Can. J. Appl. Sports Sci.* 2: No. 4, 229.

MacDougall, J. D., Ward, G. R., Sale, D. G., and Sutton, J. R., 1977. Biochemical adaptation of human skeletal muscle to heavy resistance training and immobilization. *J. Appl. Physiol.* 43: 700.

Magoun, H. W., 1958. *The Waking Brain.* Springfield, Ill.: Charles C Thomas, Publisher.

Marchiafava, P. L., 1968. Activities of the central nervous system: Motor. *Ann. Rev. Physiol.* 30: 359.

Matthews, P. B. C., 1962. The differentiation of two types of fusimotor fibre by their effects on the dynamic response of muscle spindle primary endings. *Q. J. Exp. Physiol.* 47: 324.

Matthews, P. B. C., 1968. Central regulation of the activity of skeletal muscle. In *The Role of the Gamma System in Movement and Posture,* Revised edition, edited by C. A. Swinyard. New York: Association for Aid of Crippled Children, p. 29.

McComas, A. J., 1977. *Neuromuscular Function and Disorders.* London: Butterworths.

McLennon, H., 1963. *Synaptic Transmission.* Philadelphia: W.B. Saunders Company.

McPhedran, A. M., Wuerker, R. B., and Henneman, E., 1965. Properties of motor units in a homogeneous red muscle (soleus) of the cat. *J. Neurophysiol.* 28: 71.

Megirian, D., 1962. Bilateral facilitatory and inhibitory skin areas of the cat. *J. Neurophysiol.* 25: 127.

Melzack, R., and Wall, P. D., 1965. Pain mechanisms: A new theory. *Science* 150: 971.

Miller, W. H., Ratliff, F., and Hartline, H. K., 1961. How cells receive stimuli. *Sci. Am.* 205: No. 2, 222.

Milner, M., Basmajian, J. V., and Quanbury, A. O., 1971. Multifactorial analysis of walking by electromyography and computer. *Am. J. Phys. Med.* 50: 235.

Milner, M., Dall, D., McConnell, V. A., Brennan, P. K., and Hershler, C., 1973a. Angle diagrams in the assessment of locomotor function: (1) Normal subjects for various speeds: (2) Some preliminary work on patients requiring total hip reconstruction (Charnley low-friction arthroplasty). *S. Afr. Med. J.* 47: 951.

Milner, M., Dall, D., Ruff, A. L., and Brennan, P. K., 1974. Pre- and post-operative angle diagrams in cases of total hip reconstruction. *Digest 5th Can. Med. Biol. Eng. Conf.* 17.3: a, b.

Milner, M., and Quanbury, A. O., 1970. Some facets of control in human walking. *Nature* 227: 734.

Milner, M., Wilberforce, C. B. A., and Brennan, P. K., 1973b. Stroboscopic polaroid photography in clinical studies of human locomotion. *S. Afr. Med. J.* 47: 948.

Moore, J. C., 1969. *Neuroanatomy Simplified.* Dubuque, Iowa: Kendall/Hunt Publishing Company.

Morehouse, L. E., and Cooper, J. M., 1950. *Kinesiology.* St. Louis: C. V. Mosby.

Morton, D. J., 1952. *Human Locomotion and Body Form.* Baltimore: The Williams & Wilkins Company.

Mosso, A., 1884. Application de la balance a l'etude de la circulation du sang chez l'homme. *Arch. Ital. Biol.* 5: 130.

Mountcastle, V. B., 1967. The problem of sensing and the neural coding of sensory events. In *The Neu-* rosciences, edited by G. C. Quarton, T. Melnechuk, and F. O. Schmidt. New York: The Rockefeller University Press.

Mountcastle, V. B., and Powell, R. P. S., 1959. Central nervous mechanisms subserving position sense and kinesthesis. *Johns Hopkins Hosp. Bull.* 105: 173.

Nathanson, J. A., and Greengard, P., 1977. "Second messengers" in the brain. *Sci. Am.* 237: No. 2, 108.

Noback, C. R., and Demarest, R. J., 1972. *The Nervous System: Introduction and Review.* New York: McGraw-Hill Book Company.

Nystrom, B., 1968. Effect of direct tetanization on twitch tension in developing cat leg muscles. *Acta Physiol. Scand.* 74: 319.

O'Connell, A. L., 1958. Electromyographic study of certain leg muscles during movements of the free foot and during standing. *Am. J. Phys. Med.* 37: 289.

O'Connell, A. L., 1966. Ingredients of coordinate movement. *Am. J. Phys. Med.* 46: 334.

O'Connell, A. L., 1969. *Laboratory Manual for Kinesiology.* Boston: Boston University Bookstore.

Oscarsson, O., 1966. The projection of Group I muscle afferents to the cat cerebral cortex. In *Muscle Afferents and Motor Control,* edited by R. Granit. New York: John Wiley & Sons, Inc.

Oscarsson, O., Rosen, I., and Sulg, I., 1966. Organizations of neurones in the cat cerebral cortex that are influenced from Group I muscle afferents. *J. Physiol. (Lond.)* 183: 189.

Page, E. W., and Infante, J., 1973. A new instrumentation technique for gait analysis. *Digest 26th Ann. Conf. Eng. Med. Biol.* 14.4: 126.

Paillard, J., 1960. Patterning of skilled movement. In *Handbook of Physiology,* Section 1: Neurophysiology, Vol. III, edited by J. Field and H. W. Magoun. Washington, D.C.: American Physiological Society.

Palay, S. B., 1967. Principles of cellular organization in the nervous system. In *The Neurosciences,* edited by G. C. Quarton, T. Melnechuk, and F. O. Schmidt, New York: The Rockefeller University Press.

Parsons, C., and Porter, K. R., 1966. Muscle relaxation: Evidence for and intrafibrillar restoring force in vertebrate striated muscle. *Science* 153: 3734.

Partridge, L. D., 1961. Motor control and the myotatic reflex. *Am. J. Phys. Med.* 40: 96.

Paul, J. P., 1965. Bioengineering studies of the forces transmitted by joints. In *Biomechanics and Related Bioengineering Topics,* Vol. II, edited by R. M. Kenedi. London: Pergamon Press.

Paul, J. P., 1966. The biomechanics of the hip joint and its clinical relevance. *Proc. R. Soc. Med.* 59: 943.

Paul, J. P., 1967. Forces transmitted by joints in the human body. *Proc. Inst. Mech. Eng.* 181: part 3, 1.

Peachey, L. D., 1965. The sarcoplasmic reticulum and transverse tubules of frog's sartorius. *J. Cell Biol.* 25: 209.

Peachey, L. D., and Porter, K. R., 1959. Intracellular impulse conduction in muscle cells. *Science* 129: 721.

Penfield, W., and Rasmussen, T., 1968. *The Cerebral Cortex of Man*. New York: The Macmillan Company.

Perry, S. V., 1960. Introduction to the contractile process in striated muscle. In *The Contractile Process: Proceedings of a Symposium Sponsored by the New York Heart Association*. Boston: Little, Brown and Company, p. 63.

Plagenhoef, S. C., 1962. An analysis of the kinematics and kinetics of selected symmetrical body actions, Doctoral dissertation. University of Michigan.

Pletta, D. H., and Frederick, D., 1964. *Engineering Mechanics, Statics and Dynamics*. New York: The Ronald Press.

Porter, K. R., and Franzini-Armstrong, C., 1965. The sarcoplasmic reticulum. *Sci. Am.* 212: No. 3, 72.

Powers, W. R., 1976. Nervous system control of muscular activity. In *Neuromuscular Mechanisms for Therapeutic and Conditioning Exercise*, edited by H. G. Knuttgen. Baltimore: University Park Press.

Prives, M. G., 1960. Influence of labor and sport upon skeletal structure in man. *Anat. Rec.* 136: 261.

Ralston, H. J., 1957. Recent advances in neuromuscular physiology, *Am. J. Phys. Med.* 26: 94.

Rasch, P. J., and Burke, R. K., 1967. *Kinesiology and Applied Anatomy*. Philadelphia: Lea Febiger.

Resnick, R., and Halliday, D., 1960. *Physics—For Students of Science and Engineering*, Part I. New York: John Wiley & Sons, Inc.

Reuleaux, F., 1875. *Theoretische Kinematik: Grundig einer Theorie des Maschinenwesens*. Braunschweig: I. F. Vieweg und Sohn. (Also translated by Kennedy, A. B. W., 1876: *The Kinematic Theory of Machinery: Outline of a Theory of Machines*. London: Macmillan Publishing Company.)

Reuschlein, P. L., 1962. An analysis of the speed of rotation for the forward somersault on the trampoline. Unpublished paper prepared at the University of Wisconsin.

Roberts, T. D. M., 1966. The nature of the controlled variable in the muscle servo. In *Control and Innervation of Skeletal Muscle*, edited by B. L. Andrew. Dundee, Scotland: D. C. Thomson & Company, Ltd.

Roberts, T. D. M., 1967. *The Neurophysiology of Postural Mechanisms*. New York: Plenum Press.

Rodahl, K., and Horvath, S. M., 1962. *Muscle as a Tissue*. New York: McGraw-Hill Book Company.

Roebuck, J. A., Jr., 1966. Kinesiology in engineering. Paper presented at the Kinesiology Council, Convention of the American Association for Health, Physical Education, and Recreation, March 1966.

Roebuck, J. A., Jr., 1968. A system of notation for space suit mobility evaluations. *Hum. Factors 10*.

Rogers, E. M., 1960. *Physics for the Inquiring Mind*. Princeton, N.J.: Princeton University Press.

Rooney, J. R., 1969. *Biomechanics of Lameness in Horses*. Baltimore: The Williams & Wilkins Company.

Rose, J. E., and Mountcastle, V. B., 1959. Touch and kinesthesis. In *Handbook of Physiology*, Section 1, Neurophysiology, Vol. I, edited by J. Field and H. W. Magoun. Washington, D.C.: American Physiological Society.

Rushworth, G., 1968. The modification of gamma system activity in man. In *The Role of the Gamma System in Movement and Posture*, Revised edition, edited by C. A. Swinyard. New York: Association for Aid of Crippled Children, p. 48.

Rydell, N., 1972. Normal gait—A biomechanical study on walkways. In *Digest of 3rd International Conference on Medical Physics Including Medical Engineering*. Goteborg, Sweden: Chalmers University of Technology, 10-1.

Sage, G. H., 1977. *Introduction to Motor Behavior: A Neuropsychological Approach*. Reading, Mass.: Addison-Wesley Publishing Company.

Schaeffer, J. P. (Editor), 1953. *Morris' Human Anatomy*. New York: The Blakiston Company.

Scott, M. G., 1963. *Analysis of Human Motion*. New York: Appleton-Century-Crofts.

Sechenov, I., 1935. *Selected Works*. Moscow: State Publishing House (1863 quote).

Selcom, A. B., 1976. Trade Literature on the Selspot System.

Shambes, G. M., 1968. The influence of the muscle spindle on posture and movement. *J. Am. Phys. Ther. Assoc.* 48: 1094.

Sherrington, C. S., 1906. *The Integrative Action of the Nervous System*. New Haven, Conn. Yale University Press.

Simpson, J. A., 1966. Control of muscle in health and disease. In *Control and Innervation of Skeletal Muscle*, edited by B. L. Andrew. Dundee, Scotland: D.C. Thomson & Company, Ltd.

Singh, M., and Karpovich, P. V., 1966. Isotonic and isometric forces of forearm flexors and extensors. *J. Appl. Physiol.* 21: 1435.

Smith, R. S., 1966. Properties of intrafusal muscle fibres. In *Muscle Afferents and Motor Control*, edited by R. Granit. New York: John Wiley & Sons, Inc.

Spears, B., 1966. *Fundamentals of Synchronized Swimming*. Minneapolis: Burgess Publishing Company.

Stein, J. M., and Padykula, H. A., 1962. Histochemical classification of individual muscle fibers of the rat. *Am. J. Anat.* 110: 103.

Steindler, A., 1973. *Kinesiology of the Human Body*. Springfield, Ill.: Charles C Thomas, Publisher.

Stockmeyer, S. A., 1967. An interpretation of the approach of Rood to the treatment of neuromuscular disfunction. *Am. J. Phys. Med.* 46: 900 (NUSTEP Proceedings).

Stockmeyer, S. A., 1970. Personal communication.

Swearingen, J. J., 1949. *Determination of the Most Comfortable Knee Angle for Pilots*. Report No. 1, Biotechnology Project No. 3-48. Oklahoma City: Civil Aeronautics Medical Research Laboratory.

Swett, J. E., and Bourassa, C. M., 1967. Short latency activity of pyramidal tract cells by Group I afferent volleys in the cat. *J. Physiol. (Lond.)* 189: 101.

Szent Gyorgyi, A., 1953. *Chemical Physiology of Body and Heart Muscle*. New York: Academic Press.

Szent Gyorgyi, A. G., 1960. Proteins of the myofibril. In *Structure and Function of Muscle*, Vol. II, edited by G. H. Bourne. New York: Academic Press, p. 1.

Tasaki, I., 1959. Conduction of the nerve impulse. In *Handbook of Physiology*, Section 1, Neurophysiology, Vol. I, edited by J. Field and H. W. Magoun. Washington, D.C.: American Physiological Society.

Terry, R. J., and Trotter, M., 1953. In *Morris' Human Anatomy*, edited by J. P. Schaeffer. New York: The Blakiston Company.

Thompson, R. F., 1967. *Foundations of Physiological Psychology*. New York: Harper and Row.

Timoshenko, S., and Young, D. H., 1956. *Engineering Mechanics*, Edition 4. New York: McGraw-Hill Book Company.

Tribe, M. A., and Eraut, M. R., 1977. *Nerves and Muscle* (Basic Biology Course, Book 10). Cambridge: Cambridge University Press.

Tricker, R. A. R., and Tricker, B. J. K., 1967. *The Science of Movement*. New York: American Elsevier Publishing Company, Inc.

Truex, R. C., and Carpenter, M. B., 1969. *Human Anatomy*, Edition 6. Baltimore: The Williams & Wilkins Company.

Twitchell, T. E., 1965. Attitudinal reflexes. In *The Child with Central Nervous System Deficit*. Washington, D.C.: U.S. Government Printing Office.

Waddell, D. B., and Gowitzke, B. A., 1977. Analysis of overhead badminton power strokes using high speed bi-plane cinematography. Presented at the First International Coaching Conference, Malmo, Sweden.

Walker, S. M., and Schrodt, G. R., 1967. Contraction of skeletal muscle, *Am. J. Phys. Med.* 46: 151.

Walls, E. W., 1960. The microanatomy of muscle. In *Structure and Function of Muscle*, Vol. I, edited by G. H. Bourne. New York: Academic Press, p. 21.

Waterland, J. C., and Shambes, G. M., 1970. Biplane center of gravity procedure. *Percept. Motor Skills* 30: 511.

Weiss, P., 1963. Self-renewal and proximo-distal convection in nerve fibers. In *The Effect of Use and Disuse on Neuromuscular Functions*, edited by E. Gutman and P. Hnik. Amsterdam: Elsevier Publishing Company.

Wells, J. B., 1965. Comparison of mechanical properties between slow and fast mammalian muscles. *J. Physiol. (Lond.)* 178: 252.

Wells, K. F., 1966. *Kinesiology*. Philadelphia: W.B. Saunders Company.

Wells, K. F., and Luttgens, K., 1976. *Kinesiology: Scientific Basis of Human Motion*. Philadelphia: W.B. Saunders Company.

Wenger, M. A., Jones, F. N., and Jones, M. H., 1956. *Physiological Psychology*. New York: Holt, Rinehart and Winston, Inc.

Williams, M., and Lissner, H. R., 1962. *Biomechanics of Human Motion*. Philadelphia: W.B. Saunders Company.

Willows, A. O. D., and Hoyle, G., 1969. Neural network triggering a fixed action pattern. *Science* 166: 1549.

Wilson, V. J., 1966a. Inhibition in the central nervous system. *Sci. Am.* 214: No. 5, 102.

Wilson, V. J., 1966b. Regulation and function of Renshaw cell discharge. In *Muscle Afferents and Motor Control*, edited by R. Granit. New York: John Wiley & Sons, Inc.

Winter, D. A., Greenlaw, R. K., and Hobson, D. A., 1972a. A microswitch shoe for use in locomotion studies. *J. Biomech.* 5: 553.

Winter, D. A., Hobson, D. A., and Greenlaw, R. K., 1972b. TV-computer analysis of kinematics of gait. *Comp. Biomech. Res.* 5: 498.

Woltring, H. D., 1974. New possibilities for human motion studies by real-time light spot position measurement. *Biotelemetry* 1: 132.

Wooldridge, D. E., 1963. *The Machinery of the Brain*. New York: McGraw-Hill Book Company.

Wuerker, R. B., McPhedran, A. M., and Henneman, E., 1965. Properties of motor units in a heterogeneous pale muscle (m. gastrocnemius of the cat). *J. Neurophysiol.* 28: 85.

Appendix 1
Description of Movement Patterns

The authors' description of the movements depicted on page 38, Figure 1.44 follows:

MOVEMENT DESCRIPTION-

	Figure 1.44 (a)	**Figure 1.44 (b)**
RIGHT LOWER EXTREM-ITY		
I-P* joints	Anatomical position	Extension
M-P* joints	Anatomical position	Extension
Ankle	Anatomical position	Plantar flexion
Knee	Slight hyperextension	Slight flexion
Hip	Slight abduction	Slight abduction
	Slight lateral rotation	Slight lateral rotation
LEFT LOWER EXTREMITY		
I-P joints	Flexion	Flexion
M-P joints	Flexion	Flexion
Ankle	Plantar flexion	Plantar flexion
Knee	Slight hyperextension	Slight hyperextension
Hip	Flexion	Hyperflexion
	Slight medial rotation	
VERTEBRAL COLUMN		
	Lateral flexion to the left	Slight flexion of lumbar spine
	Hyperextension of lumbar spine	
CERVICAL SPINE		
	Slight rotation to the left	Slight dorsiflexion
LEFT SCAPULA		
	Protraction	
	Rotation upward	Rotation upward
	Depression	
LEFT UPPER EXTREMITY		
Shoulder	Flexion	Abduction
	Medial rotation	Medial rotation
Elbow	Extension	Slight hyperextension
Radioulnar	Pronation	Marked pronation
Wrist	Flexion	Flexion
M-P* joints	Flexion (holding onto foot)	Extension
I-P joints	Flexion (holding onto foot)	Extension

* Note: I-P = Interphalangeal, M-P = Metatarsal-phalangeal, and metacarpal-phalangeal respectively.

Appendix 2

International System of Units

The International System of Units or SI System (from the French Systeme Internationale D'Unites) derives from the metric system of physical units and was designated in 1960 at the 11th General Conference on weights and measures. Six basic units were adopted and defined; and other units are derived from them. They are the meter, kilogram, second, ampere, degree kelvin and candela, for length, mass, time, electric current, temperature and luminous intensity, respectively.

Table A2.1 reflects some elemental and some derived SI units and symbols.

Table A2.2 renders a number of conversion factors relating typical United States units and those in the SI system.

Table A2.1
Some Elemental and Derived SI Units and Symbols

Quantity	SI Units		
	Unit	Formula	Symbol
Length	meter		m
Mass	kilogram		kg
Time	second		s
Plane angle	radian		rad
Solid angle	steradian		sr
Acceleration	meter/sec^2	m/s^2	
Area	square meter	m^2	
Density	kilogram/cubic meter	kg/m^3	
Energy	joule	$N \cdot m$	J
Force	newton	$kg \cdot m/s^2$	N
Frequency	hertz	s^{-1}	Hz
Kinematic viscosity	square meter/second	m^2/s	
Momentum	kilogram-meter/second	kg-m/s	
Power	watt	J/s	W
Pressure	newton/square meter	N/m^2	
Stress	newton/square meter	N/m^2	
Velocity	meter/second	m/s	
Viscosity	Newton-second/square meter	$N \cdot s/m^2$	
Volume	cubic meter	m^3	

Table A2.2
Conversion Factors for Typical U.S. and SI systems

inches × 25.4	=	millimeters
feet × 0.3048	=	meters
yards × 0.9144	=	meters
miles × 1.60934	=	kilometers
square inches × 6.4516	=	square centimeters
cubic inches × 16.3871	=	cubic centimeters
gallons* × 0.00378541	=	cubic meters
pounds (avdp) × 0.453592	=	kilograms
pounds (avdp) × 4.4498	=	newtons
horsepower × 0.7457	=	kilowatts

* The Imperial gallon is larger than the U.S. gallon and the relevant conversion factor is 0.004546.

Appendix 3

Trigonometric Functions

Given the right triangle ABC
Let:

angle BAC be known as θ (the Greek letter theta)
side AC be known as side A or the side adjacent to θ
side BC be known as side O or the side opposite θ
side AB be known as H or the hypotenuse of the right triangle

There are certain constant relationships for these three sides, regardless of the size of the triangle, which are dependent solely on the size of the angle θ. These relationships are expressed as:

$$\text{sine } \theta = \frac{\text{side opposite}}{\text{hypotenuse}} \text{ or } \sin \theta = \frac{O}{H}$$

$$\text{cosine } \theta = \frac{\text{side adjacent}}{\text{hypotenuse}} \text{ or } \cos \theta = \frac{A}{H}$$

$$\text{tangent } \theta = \frac{\text{side opposite}}{\text{side adjacent}} \text{ or } \tan \theta = \frac{O}{A}$$

$$\text{cotangent } \theta = \frac{\text{side adjacent}}{\text{side opposite}} \text{ or } \cot \theta = \frac{A}{O}$$

Following from these relationships:

$$\sin \theta = \cos (90 - \theta)$$

$$\tan \theta = \sin \theta / \cos \theta$$

It is also possible to show that $\sin^2 \theta + \cos^2 \theta = 1$

(Values for these functions for any angle are easily obtained using mathematical tables or hand-held calculators with these functions available.)

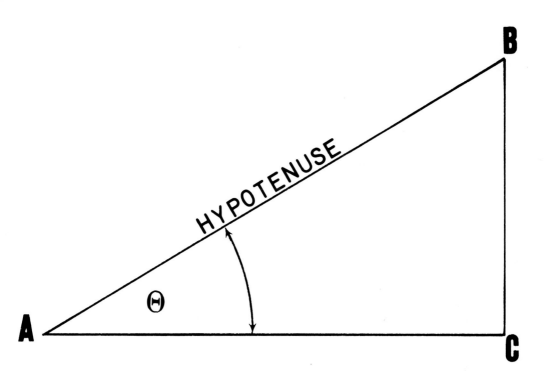

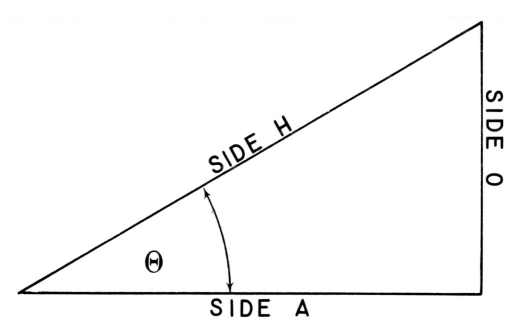

Figure A3.1. Right triangle, ABC θ, angle BAC; AC, the side adjacent (side A) to θ; BC, the side opposite (side O) to θ; AB, the hypotenuse (side H) of the triangle.

Appendix 4
Body Segment Data

Table A4.1*

A. Average Segment Characteristics of Three Cadavers Dissected by Braune and Fischer (1889)

Part	Weight	Percentage of Total Body Weight	Center of Gravity Location with Respect to Joint Axes for Each Limb, Expressed as Percentage of Limb Length	
			From above	From below
	kg			
Upper arm (each)	2.127	3.3	47.0	53.0
Forearm (each)	1.335	2.1	42.1	57.9
Forearm and hand	1.872†	2.9	47.2‡	52.8‡
Hand (each)	0.533	0.85		
Upper leg (each)	6.793	10.75	43.9	56.1
Lower leg (each)	3.025	4.8	41.95	58.05
Foot (each)	1.067	1.7	43.4§	56.6§
Lower leg and foot	4.127†	6.5	51.9¶	48.1¶
Head, neck, and trunk, minus the limbs	33.990	53.0		
Head	4.440	6.95	Average locations not given. See individual cadaver data	
Torso	29.550	46.3		
Entire body	63.85	100.0		

† Combined weights of several segments were slightly larger than the sum of the individual components because of loss of material during sawing.

‡ Percentage of length from cubital axis to lower edge of flexed fingers.

§ Percentage of length from front to rear of foot.

¶ Percentage of length from knee axis to sole of foot.

Table A4.1 *continued*

B. Average Segment Characteristics of Two Cadavers Described by Braune and Fischer (1892)

Part	Weight	Percentage of Total Body Weight	Center of Gravity Location with Respect to Joint Axes for Each Limb, Expressed as Percentage of Limb Length	
			From above	From below
	kg			
Upper arm	1.51	2.92	45.9	54.1
Forearm and hand	1.30	2.54	46.0	54.0
Upper leg	5.78	11.23	43.4	56.7
Lower leg	2.32	4.53	42.4	57.6
Foot	0.95	1.88	41.7	58.3
Lower leg and foot	3.28	6.42	53.4	46.6
Head and trunk	26.12	51.31	58.1	41.9
Entire body	51.28	100.00		

* From Damon, A., Stoudts, H. W., and McFarland, R. A., 1966. *The Human Body in Equipment Design,* Table 69, p. 171. Cambridge, Mass.: Harvard University Press.

Table A4.2
Average Segment Characteristics of Eight Cadavers*

Part		Weight	Percentage of Total Weight
		kg	
Upper arm	R	1.614	2.77
	L	1.536	2.63
Forearm	R	0.954	1.64
	L	0.914	1.57
Hand	R	0.388	0.67
	L	0.383	0.66
Forearm and hand	R	1.343	2.30
	L	1.297	2.22
Entire upper extremity	R	2.973	5.09
	L	2.875	4.93
Upper leg	R	5.756	9.86
	L	5.812	9.95
Lower leg	R	2.741	4.69
	L	2.732	4.68
Foot	R	0.832	1.42
	L	0.872	1.49
Lower leg and foot	R	3.592	6.16
	L	3.625	6.21
Entire lower extremity	R	9.481	16.25
	L	9.408	16.13
Trunk minus limbs		33.626	57.61
Both shoulders		6.174	10.58
Head and neck		4.610	7.90
Thorax		6.763	11.58
Abdomen plus pelvis		16.395	28.09
Sum of segment weights		58.363	100.0

* From Dempster, W. T., 1955 *Space Requirements for the Seated Operator.* WADC Technical Report 55159. Wright-Patterson Air Force Base, Ohio: Wright Air Development Center.

Table A4.3
Location of Centers of Gravity of Body Segments*

Segment or Part and Reference Landmarks	Distance from Center of Gravity to Reference Dimension Stated as Percentage
Hand (position of rest), wrist axis to knuckle III	50.6% to proximal (wrist) axis
Forearm, elbow axis to wrist axis	43.0% to proximal (elbow) axis
Upper arm, glenohumeral axis to elbow axis	43.6% to proximal (glenohumeral) axis
Forearm plus hand, elbow axis to ulnar styloid	67.7% to proximal (elbow) axis
Whole upper limb, glenohumeral axis to ulnar styloid	51.2% to proximal (glenohumeral) axis
Foot, heel to toe II	A ratio of 42.9 to 57.1 along heel to toe distance establishes a point above which the center of gravity lies; the latter lies on a line between ankle axis and ball of foot
Lower leg, knee axis to ankle axis	43.3% to proximal (knee) axis
Thigh, hip axis to knee axis	43.3% to proximal (hip) axis
Leg plus foot, knee axis to medial malleolus	43.3% to proximal (knee) axis
Whole lower limb, hip axis to medial malleolus	43.4% to proximal (knee) axis
Head and trunk minus limbs, vertex to transverse line through hip axes	60.4% to vertex (top of head)
Head and neck, vertex to seventh cervical centrum	43.3% to vertex
Shoulders/thorax† centrum L_1 to centrum T_1	53.6% to first thoracic centrum
Abdominopelvic mass (lower trunk), centrum first lumbar to hip axes	59.9% to centrum first lumbar

* From Dempster, W. T., 1955. *Space Requirements for the Seated Operator*. WADC Technical Report 55159. Wright-Patterson Air Force Base, Ohio: Wright Air Development Center. See Figure A.1, Appendix A.

† Adapted from the above.

Table A4.4
Specific Gravity of Body Segments*

Body Segment		Cadaver Number								Mean
		14815	15059	15062	15095	15097	15168	15250	15251	
Trunk minus limbs		1.04	1.05	1.05	1.04	1.03	1.02	1.02	1.00	1.03
Trunk minus shoulders		1.05	1.06	1.00	1.05	1.06	1.03	1.00	1.00	1.03
Shoulders		1.00	1.04	1.02	1.05	1.03	1.04	1.05	1.05	1.04
Head and neck			1.12	1.13	1.12	1.10	1.10	1.10	1.11	1.11
Thorax			0.81	0.94	0.92	1.01	0.91	0.95	0.90	0.92
Abdominopelvic			1.00	1.00	1.00	1.02	1.03	1.03	1.00	1.01
Entire lower extremity	R	1.06	1.07	1.07	1.07	1.04	1.06	1.04	1.06	1.06
	L	1.06	1.07	1.05	1.07	1.05	1.06	1.05	1.06	1.06
Thigh	R	1.04	1.06	1.04	1.05	1.05	1.05	1.04	1.05	1.05
	L	1.04	1.06	1.03	1.05	1.04	1.05	1.05	1.05	1.05
Leg and Foot	R	1.05	1.11	1.08	1.11	1.07	1.09	1.06	1.07	1.08
	L	1.09	1.11	1.08	1.12	1.07	1.10	1.08	1.08	1.09
Leg	R	1.10	1.12	1.08	1.11	1.08	1.09	1.07	1.09	1.09
	L	1.11	1.12	1.08	1.11	1.08	1.10	1.09	1.09	1.09
Foot	R	1.02	1.17	1.08	1.17	1.12	1.11	1.01	1.07	1.09
	L	1.02	1.17	1.07	1.15	1.14	1.13	1.05	1.09	1.10
Entire upper extremity	R	1.10	1.25	1.07	1.10	1.07	1.10	1.10	1.09	1.11
	L	1.11	1.09	1.08	1.11	1.08	1.10	1.10	1.09	1.10
Arm	R	1.09	1.07	1.06	1.01	1.07	1.09	1.10	1.09	1.07
	L	1.09	1.07	1.06	1.01	1.07	1.08	1.10	1.08	1.07
Forearm and hand	R	1.11	1.18	1.10	1.01	1.12	1.13	1.11	1.13	1.11
	L	1.09	1.14	1.10	1.16	1.11	1.12	1.12	1.13	1.12
Forearm	R	1.14	1.12	1.11	1.20	1.08	1.12	1.14	1.10	1.13
	L	1.14	1.15	1.11	1.18	1.09	1.14	1.14	1.05	1.12
Hand	R	1.14	1.41	1.05	1.33	1.09	1.15	1.13	1.08	1.17
	L	1.14	1.27	1.05	1.28	1.12	1.09	1.10	1.09	1.14

 * Calculated from mass data on cadaver parts and volumetric data derived from immersion in water. (From Dempster, W. T., 1955. *Space Requirements for the Seated Operator*. WADC Technical Report 55159. Wright-Patterson Air Force Base, Ohio: Wright Air Development Center.)

Table A4.5
Segmental Fractions of Body Weight According to Somatotype*

Body Type	Rotund	Muscular	Thin	Median (Air Force)
Head and neck	0.07900	0.07900	0.07900	0.07900
Shoulders-thorax	0.17103	0.18097	0.17361	0.16947
Abdomen-pelvis	0.21507	0.22753	0.21985	0.21309
Thighs	0.29776	0.27376	0.27476	0.28624
Legs	0.09832	0.09374	0.10510	0.10114
Feet	0.02288	0.02680	0.03192	0.02730
Arms	0.07132	0.07010	0.06384	0.07490
Forearms	0.03494	0.03724	0.03856	0.03642
Hands	0.00968	0.01086	0.01336	0.01244

* If a subject seems to match any of these types, the appropriate data may be used to calculate the location of his center of gravity by the segmental method. (Calculated and adapted from data presented by Dempster, W. T., 1955. *Space Requirements for the Seated Operator*. WADC Technical Report 55159. Wright-Patterson Air Force Base, Ohio: Wright Air Development Center.)

Table A4.6
Regression Equations for Computing the Mass (in kg) of Body Segments*

Body Segment	Regression Equation†	Standard Deviation of the Residuals
Head, neck, and trunk	$= 0.47 \times \text{TBW} + 5.4$	±2.9
Total upper extremities	$= 0.13 \times \text{TBW} + 1.4$	±1.0
Both upper arms	$= 0.08 \times \text{TBW} + 1.3$	±0.5
Forearms plus hands‡	$= 0.06 \times \text{TBW} + 0.6$	±0.5
Both forearms‡	$= 0.04 \times \text{TBW} + 0.2$	±0.5
Both hands	$= 0.01 \times \text{TBW} + 0.3$	±0.2
Total lower extremities	$= 0.31 \times \text{TBW} + 1.2$	±2.2
Both upper legs	$= 0.18 \times \text{TBW} + 1.5$	±1.6
Both lower legs plus feet	$= 0.13 \times \text{TBW} + 0.2$	±0.9
Both lower legs	$= 0.11 \times \text{TBW} + 0.9$	±0.7
Both feet	$= 0.02 \times \text{TBW} + 0.7$	±0.3

* From Barter, J. T., 1957. *Estimation of the Mass of Body Segments*. WADC Technical Report 57-260. Wright-Patterson Air Force Base, Ohio: Aero Medical Laboratory, Wright Air Development Center.

† TBW, total body weight.

‡ N = 11; all others, N = 12.

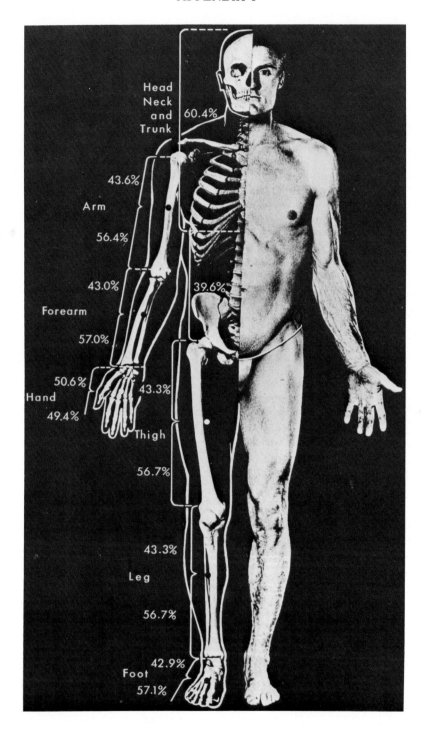

Figure A4.1. Link boundaries (at the joint centers) and percentage of distance of the centers of gravity from link boundaries. (From Williams, M., and Lissner, H.R., 1962. *Biomechanics of Human Motion*, Fig. A-1, p. 132. Philadelphia: W.B. Saunders Company.) See Table A4.3.

Index

A band (*see* Striation bands)
Abduction, 13
Acceleration, 40, 46
  angular, 48
  due to gravity, 56
  from film, 158
  measurement, direct, 185
  radial, 48
  tangential, 49
Accelerometer, 185
Accommodation, 89
Acetylcholine (ACh), 230–231, 238–239
Acetylcholinesterase (ACh-ase), 231
Acromioclavicular, 35
Actin
  filaments (*see also* Contraction, nature of), 193–196, 199
    association with tropomyosin in relaxed muscle, 208–209
    in contraction, 207–208
    in overlap, 207
  subunits of (F-actin, G-actin), 195–196
Action potential (*see* Potential, membrane, and Potential, action)
Activation, of muscle, 199
Active state, in activation of muscle, 199
Active transport mechanism, 217
Actomyosin, 194, 199, 206, 208–209
Adaptation, 89
Adduction, 13, 22
  horizontal, 17
Adenosine diphosphate (ADP)
  condensation of, to produce ATP, 201
  in muscle metabolism, 199–204
  in oxygen debt, 201
  in regulation of muscle metabolism, 205
Adenosine monophosphate (AMP), 201
Adenosine triphosphate (ATP)
  as energy source, 198–204
  high energy bonds in, 198
  hydrolysis of, by ATPase, 199
  in contraction, 198–199, 207
  in excitation, 207
  in muscle fibers, 193
  in muscle metabolism, 199–204
  in oxygen debt, 201
  in regulation of muscle metabolism, 205
  in relaxation, 208–209

  production of, 199–204
  rephosphorylation of by CP, 200–201
ADP (*see* Adenosine diphosphate)
All-or-none, 217, 261
Alpha motor neuron (*see also* Neuron, motor, alpha), phasic, tonic (*see also* Motor unit, fast, slow), 268–269
AMP (*see* Adenosine monophosphate)
Analogy, 51
  between linear and rotational motion, 80
Analysis
  motion, 157
  of motor skills (*see* Neurokinesiological analyses)
  static, 157
Anatomical position, 5
Angle-angle diagram, 178
  amputee, 181
  postoperative, 179
  preoperative, 179
Angular acceleration, determination, 162
Angular momentum, 78
Angular velocity, determination, 162
Ankle axes, 10
Applied mechanics, 54
Arm, rotation, 7
  inward, 7
  outward, 7
Artifacts, 93
Atlantooccipital movement, 23, 30
ATP (*see* Adenosine triphosphate)
ATPase (*see also* Adenosine triphosphate), 99, 102
  in activation, 199
  in contraction, 206, 209
  in hydrolysis of ATP, 199
  in relaxation, 208–209
Auditory receptors (*see also* Cochlea, Labyrinth), 259, 262, 263
Available energy, 204
Axial skeleton, axes, 11
Axis
  ankle, 10
  clavicle, 7
  elbow, 8
  finger, 9
  forearm, 8
  foot, 9, 10
  hip, 9

  interphalangeal, 9
  joint, 7, 9, 10
  knee, 10
  metatarsophalangeal, 9
  shoulder, 7
  tibia, 10
  wrist, 8
Axolemma, 192, 214, 217
Axon(s) (*see also* Neuron)
  axolemma, 214
  axon hillock, 215
  axoplasmic transport, 215
  collateral branches, 215
  cytoplasm, 214–216
  diameter, 214, 215
  function, 214
  initial segment of, 215, 229, 230
  myelin sheath, 214, 215
    segmentation, 215
  nodes of Ranvier, 215
  structure, 214, 215
  telodendria, 215, 216
  terminal(s), 192
  unmyelineated, 215

Basal ganglia (*see also* Cephalon, diencephalon), 231, 245
  amygdaloid nucleus, 245
  caudate nucleus, 245
  claustrum, 245
  globus pallidus, 245
  putamen, 245
  striate body, 245
Biomechanics, 117
Body markers, 175
Body motion, capture, 173
Bridges (*see* Cross-bridges)
Buoyancy, 82
  center of, 82

Calculus, 51
Cations, 217
Center of buoyancy, 82
Center of gravity
  body segments, 120
  locating, 133
  segmental method, 137
  shift, 121
Central control hypothesis, 278
Central nervous system
  coordinative centers, 272
  coronal and cross-sectional anatomy, 248–250
Central regulation of sensory input, 262
  regulation by efferent neurons, 262
  regulation by presynaptic inhibition (*see also* Inhibition feedback, presyn-